# Advances in Intelligent Systems and Computing

Volume 1425

The series "Advances in Intelligent Systems and Computing" contains publications on theory, applications, and design methods of Intelligent Systems and Intelligent Computing. Virtually all disciplines such as engineering, natural sciences, computer and information science, ICT, economics, business, e-commerce, environment, healthcare, life science are covered. The list of topics spans all the areas of modern intelligent systems and computing such as: computational intelligence, soft computing including neural networks, fuzzy systems, evolutionary computing and the fusion of these paradigms, social intelligence, ambient intelligence, computational neuroscience, artificial life, virtual worlds and society, cognitive science and systems, Perception and Vision, DNA and immune based systems, self-organizing and adaptive systems, e-Learning and teaching, human-centered and human-centric computing, recommender systems, intelligent control, robotics and mechatronics including human-machine teaming, knowledge-based paradigms, learning paradigms, machine ethics, intelligent data analysis, knowledge management, intelligent agents, intelligent decision making and support, intelligent network security, trust management, interactive entertainment, Web intelligence and multimedia.

The publications within "Advances in Intelligent Systems and Computing" are primarily proceedings of important conferences, symposia and congresses. They cover significant recent developments in the field, both of a foundational and applicable character. An important characteristic feature of the series is the short publication time and world-wide distribution. This permits a rapid and broad dissemination of research results.

Indexed by DBLP, INSPEC, WTI Frankfurt eG, zbMATH, Japanese Science and Technology Agency (JST).

All books published in the series are submitted for consideration in Web of Science.

For proposals from Asia please contact Aninda Bose (aninda.bose@springer.com).

More information about this series at https://link.springer.com/bookseries/11156

Sukanta Das · Genaro J. Martinez
Editors

# Proceedings of First Asian Symposium on Cellular Automata Technology

## ASCAT 2022

Springer

*Editors*
Sukanta Das
Department of Information Technology
Indian Institute of Engineering Science
and Technology, Shibpur
Howrah, West Bengal, India

Genaro J. Martinez
National Polytechnic Institute
Mexico City, Mexico

ISSN 2194-5357 ISSN 2194-5365 (electronic)
Advances in Intelligent Systems and Computing
ISBN 978-981-19-0541-4 ISBN 978-981-19-0542-1 (eBook)
https://doi.org/10.1007/978-981-19-0542-1



This Springer imprint is published by the registered company Springer Nature Singapore Pte Ltd.
The registered company address is: 152 Beach Road, #21-01/04 Gateway East, Singapore 189721, Singapore

# Committees

## Chief Patron

Parthasarathi Chakrabarti, Director, Indian Institute of Engineering Science and Technology, Shibpur, India

## Patron

Dipanwita Roy Chowdhury, Indian Institute of Technology, Kharagpur, India

## General Co-Chairs

Biplab K. Sikdar, Indian Institute of Engineering Science and Technology, Shibpur
Jarkko Kari, University of Turku, Finland

## Program Co-Chairs

Sukanta Das, Indian Institute of Engineering Science and Technology, Shibpur
Genaro Juarez Martinez, National Polytechnic Institute, Mexico City, Mexico

## Program Committee

Dipanwita Roy Chowdhury, Indian Institute of Technology, Kharagpur, India

Biplab K. Sikdar, Indian Institute of Engineering Science and Technology, Shibpur
Kenichi Morita, Hiroshima University, Japan
Jarkko Kari, University of Turku, Finland
Mihir K. Chakraborty, School of Cognitive Science, Jadavpur University, Kolkata, India
B. R. Shankar, National Institute of Technology, Karnataka, Surathkal, India
N. P. Gopalan, National Institute of Technology, Tiruchirappalli, India
R. Ramanujam, The Institute of Mathematical Sciences, India
Fasel Qadir, University of Kashmir, India
Andrew Adamatzky, University of the West of England, UK
Ebrahim Patel, London Interdisciplinary School, UK
Eric Goles, Universidad Adolfo Ibáñez, Chile
Salvatore Di Gregorio, University of Calabria, Italy
Andreas Malcher, Universität Gießen, Germany
Pabitra Pal Chaudhuri, Indian Statistical Institute, Kolkata, India
K. Ramachandra Rao, Indian Institute of Technology, Delhi, India
Debashish Chowdhury, Indian Institute of Technology, Kanpur, India
Teijiro Isokawa, University of Hyogo, Japan
Chih-Hung Chang, National Chiao Tung University, Taiwan
Sudhakar Sahoo, Institute of Mathematics and Applications, Bhubaneswar, India
Mamata Dalui, National Institute of Technology, Durgapur, India
Nazma Naskar, Kalinga Institute of Industrial Technology University, India
Biswanath Sethi, Indira Gandhi Institute of Technology, Odisha, India
Raju Hazari, National Institute of Technology, Calicut, India
Jimmy Jose, National Institute of Technology, Calicut, India
Kamalika Bhattacharjee, National Institute of Technology, Tiruchirappalli, India
Hiroshi Umeo, University of Osaka Electro-Communication, Japan
Hiep Xuan Huynh, College of Information and Communication Technology, Can Tho University, Vietnam
Samira El Yacoubi, University of Perpignan, France
Nazim Fatès, Inria-Loria, France
Anna Lawniczak, Guelph University, Canada
Stefano Nichele, Oslo Metropolitan University, Norway
Niloy Ganguly, Indian Institute of Technology, Kharagpur, India
Pradipta Maji, Indian Statistical Institute, Kolkata, India
Genaro Juarez Martinez, National Polytechnic Institute, Mexico City, Mexico
Sukanta Das, Indian Institute of Engineering Science and Technology, Shibpur

## Organizing Committee

Biplab K. Sikdar, Indian Institute of Engineering Science and Technology, Shibpur
Abhik Mukherjee, Indian Institute of Engineering Science and Technology, Shibpur

Amit Roy Chowdhury, Indian Institute of Engineering Science and Technology, Shibpur
Chandan Giri, Indian Institute of Engineering Science and Technology, Shibpur
Surajit Roy, Indian Institute of Engineering Science and Technology, Shibpur
Ashish Layek, Indian Institute of Engineering Science and Technology, Shibpur
Kamalika Bhattacharjee, National Institute of Technology, Tiruchirappalli, India
Souvik Roy, Indian Institute of Engineering Science and Technology, Shibpur
Sumit Adak, Indian Institute of Engineering Science and Technology, Shibpur
Sukanta Das, Indian Institute of Engineering Science and Technology, Shibpur

## Registration Chairs

Supreeti Kamilya, Birla Institute of Technology, Mesra, Ranchi, India
Sukanya Mukherjee, Institute of Engineering and Management, Kolkata, India

# Foreword

The European countries (the UK, France, Germany, Italy, Finland, etc.), Japan in Asia, and Brazil in South America, and the USA and Canada in North America established a strong base of theory and applications of Cellular Automata Research (CAR) since 1970s. In the early 1980s, the CAR seed was planted at Indian Institute of Technology (IIT) Kharagpur, India. A group of research scholars came forward to ensure the development of a fully grown tree out of this seed. Along with the development of theory, the group concentrated on different applications of CA Technology—one of the major focuses was testable VLSI design. The seed grown out of this tree was later planted at Bengal Engineering College (currently named as IIEST—Indian Institute of Engineering Science and Technology). The group of research scholars at IIEST carried forward the research, developing theory and wide varieties of applications. In addition to these two institutes (IIT Kharagpur and IIEST), faculties and research scholars of other institutions in India (ISI Kolkata, IIT—Delhi, Kanpur, Madras, NIT—Trichy, Kashmir University) contributed in the field of CAR.

In this background, it is a pleasure for me to write this *Foreword* for the first Asian Symposium on Cellular Automata Technology (ASCAT), 2022. In the last few years, faculty and research scholars of IIEST and IIT Kharagpur have been planning to organize a CA symposium in India.

Looking back at Cellular Automata Research of the last five decades in different regions of the world, we come across publications of a large number of research papers and books on Cellular Automata. This is a natural process in academia due to the universal appeal of CA as a modeling tool, demonstrated by wide varieties of its evolution patterns. In 'Nature', we observe evolving patterns out of the interaction of local physical entities around us. Further, the growth of human civilization in the last few centuries can be viewed as temporal evolution of relevant patterns in all the fields of human society—be it legal system, administrative/political structure, civil society in a country, or international relations among different countries in the world. Researchers of diverse disciplines are naturally attracted to develop models based on Cellular Automata Technology (CAT). However, development of an appropriate model for any physical phenomenon is an upheaval task due to higher complexity of evolving patterns of CA, specifically with neighborhood (larger than three) and change of seed in CA cells prior to evolution. In view of such complexity, we hardly find any fruitful industrial/commercial application of CAT. Unless the researchers in the field of Cellular Automata can overcome this bottleneck and develop fruitful applications, it will not be able to serve the human society of the twenty-first century, even though CAR has enormous potential to build a new class of models to address the problems human society faces in the current digital world.

In this context, I look forward to industrial/commercial applications of CAT from the current and future generations of researchers. In all probability, Data Science and Machine Learning (ML) framework to learn/analyze the evolving patterns of physical domain data (currently available in modern digital society) could be a good pointer for the development of an appropriate model for a specific application domain.

I conclude with an appeal to the global CA community to spread the discipline of CAR to different countries of the globe so that the new generation of researchers are encouraged to come forward with innovative and 'out of box' thinking of CAT that would attract researchers of other disciplines to develop industrial/commercial applications of their interest.

Kolkata, India<br>March 2022

P Pal Chaudhuri

# Preface

This volume contains three invited papers and fourteen accepted papers that were selected for presentation at the First Asian Symposium on Cellular Automata Technology, 2022 (ASCAT 2022), to be held during March 3–5, 2022. The symposium is officially hosted by Indian Institute of Engineering Science and Technology, Shibpur (IIEST, Shibpur), India, through video conferencing mode. The event is sponsored by *Scheme for Promotion of Academic and Research Collaboration* (SPARC), an initiative of the Ministry of Human Resource Development, Government of India, under the project titled "Exploring Cellular Automata Model for Hardware Security" with Sukanta Das as Indian PI, Jarkko Kari as Foreign PI, Biplab K Sikdar as Indian Co-PI, and Hannu Tenhunen and Pasi Liljeberg as Foreign Co-PIs.

Cellular automata as an emergent area of natural computing have been explored by a divergent group of researchers. There are a number of international conferences and workshops that consider cellular automata as the primary theme and deal with various aspects of cellular automata, their theoretical, modeling and application development issues. Since late 1980s, cellular automata have been used to devise solutions for technological problems, particularly in the domain of Very Large Scale Integration (VLSI) Design and Test, Pattern Recognition and Classification, Security and Image Processing. The aim of this symposium is to focus on the technology development issue. In our view, ASCAT is the first event of this kind. Since technology cannot get matured without a solid theoretical base, this symposium targets theoreticians and engineers to exchange their views and ideas.

ASCAT 2022 is proud to have Stephen Wolfram, Jarkko Kari, Mihir K. Chakraborty, Kenichi Morita, Debdeep Mukhopadhyay and Craig S. Lent as invited speakers, who have kindly agreed to deliver talks on related topics. We are indebted to them. Kenichi Morita, Jarkko Kari and Debdeep Mukhopadhyay (along with his co-researchers) contributed invited articles in this volume. We are especially grateful to them.

We have received 32 submissions in this event, each of which is reviewed by 2–3 experts. Apart from the members of the Program Committee, four additional reviewers have reviewed the papers. We are thankful to all the reviewers for their effort. An online Program Committee meeting was held on December 16, 2021, to consider the reviews, and based on the detailed discussion 14 papers have been accepted for presentation in the symposium. We thank all the authors for their contributions and hard work that made the event possible.

We are very much grateful to Prof. Parimal Pal Chaudhruri who has kindly written the Foreword for this volume. Professor Pal Chaudhuri is pioneer in cellular automata technology research and has promoted cellular automata research in India. Presently, India has a footprint in cellular automata research for which we are indebted to Prof. Pal Chaudhuri.

During this ongoing pandemic, the Indian researchers have come forward to form the *Cellular Automata India* research group. The group has hosted two webinar series on various topics of cellular automata and a special lecture series in the memory of John Horton Conway. Eminent scientists around the globe have delivered invited talks and shared their research in this platform. It has taken the leading role to launch the symposium as a meeting place for the international intelligentsia.

No international event like ASCAT can succeed without a great Program Committee. We are grateful to all the members of the Program Committee who have kindly accepted our invitation and extended their support in organizing this event. Special thanks to the members of the Organizing Committee and to the members of the departments of Information Technology, and Computer Science and Technology, IIEST, Shibpur, for extending their support toward the event. We are thankful to Easychair, the popular conference management system, for allowing us to manage the submissions. Our sincere thanks to the officials of Springer who agreed to publish the proceedings of this event as part of their Springer AISC Series. Finally, we extend our gratitude to the Indian Institute of Engineering Science and Technology, Shibpur, for giving us permission to host this symposium as an event of this institute.

Sukanta Das
Genaro J. Martinez
*Program Co-chairs*

Biplab K. Sikdar
Jarkko Kari
*General Co-chairs*

Kolkata, India
March 2022

# Contents

# About the Editors

**Sukanta Das** currently works as Associate Professor and Head, Department of Information Technology in Indian Institute of Engineering Science and Technology, Shibpur (formerly known as Bengal Engineering and Science University, Shibpur). He received his Ph.D. in 2007 from Bengal Engineering and Science University, Shibpur (currently known as IIEST, Shibpur) in Computer Science and Technology. He has been doing research in the different aspects of cellular automata for more than twenty years. His research area includes cellular automata: theory and applications. He is also Member of IFIP Working Group 1.5 (on "Cellular Automata and Discrete Complex Systems") and several programming committees for international conferences on cellular automata. He has more than 70 articles under his name. Currently, he is exploring the computational abilities of cellular automata, the non-uniform cellular automata, (partial) number conservation in cellular automata, and chaos and randomness in cellular automata.

**Genaro J. Martinez** is Full-Time Researcher at the National Polytechnic Institute in Mexico City. He is Active Member of several institutions and laboratories around the world, such as: the International Unconventional Computing Laboratory at the University of the West of England in the UK; the Foundation of Computer Science Laboratory at the Hiroshima University in Japan; the Centre for Chaos and Complex Networks at the City University of Hong Kong in China; the Unconventional Algorithms and Computing Lab at the Technical University of Ostrava in Czech Republic; the Laboratoire de Recherche Scientifique in France; and at the Institut des Systèmes Complexes en Normandie in France. He is Editorial Member of the Journal of Cellular Automata, Complex Systems, Physica D, International Journal of Bifurcation and Chaos, Artificial Life, Nonlinear Dynamics, Chaos, Information Sciences, International Journal of Natural Computing Research, IEEE Transactions on Circuits and Systems, and the Mathematical Reviews, American Mathematical Society. He together with Prof. Adamatzky had published the book "Designing Beauty: The Art of Cellular Automata." Also, he is leading two computational paradigms in cellular automata theory: colliders cellular automata and computation by competing patterns with cellular automata.

# Invited Papers

# Computing in a Simple Reversible and Conservative Cellular Automaton

**Kenichi Morita**

**Abstract** We show how to construct reversible computers in a simple reversible and conservative two-dimensional cellular space. We introduce the framework of an *elementary square partitioned cellular automaton* (ESPCA), and use a particular ESPCA with the hexadecimal ID number "01caef", which is denoted by ESPCA $P_0$ for short in this paper. A cell of ESPCA $P_0$ consists of four parts each of which has two states 0 and 1. The class of ESPCAs is the simplest subclass of square partitioned cellular automata, and each of their local functions is described by only six local transition rules. ESPCA $P_0$ is not only reversible, but also conservative in the sense that the total number of state 1's is conserved throughout its evolution process. We show that a space-moving pattern called a *glider* exists in $P_0$. Colliding a glider with another pattern called a *blinker*, several interesting phenomena appear. We give a method of constructing reversible Turing machines (RTMs) using such phenomena. We first implement a *reversible logic element with memory* (RLEM), rather than a reversible logic gate, using only three of these phenomena. Then, we compose reversible Turing machines out of RLEMs. In this systematic and modularized way, we can construct any RTM out of only two small patterns in a simple reversible and conservative cellular space.

**Keywords** Reversible cellular automaton · Elementary square partitioned cellular automaton · Reversible Turing machine · Reversible logic element with memory

---

Currently Professor Emeritus of Hiroshima University.

---

K. Morita (✉)
Hiroshima University, Higashi-Hiroshima 739-8527, Japan
e-mail: km@hiroshima-u.ac.jp

 
S. Das and G. J. Martinez (eds.), *Proceedings of First Asian Symposium on Cellular Automata Technology*, Advances in Intelligent Systems and Computing 1425,
https://doi.org/10.1007/978-981-19-0542-1_1

# 1 Introduction

Reversibility is one of the fundamental microscopic laws in physics, and thus it is important to investigate how computing is efficiently performed by utilizing reversible phenomena. Here, we study this problem using a new model of two-dimensional reversible cellular automaton. The model is an elementary square partitioned cellular automaton (ESPCA). A cell of an ESPCA is divided into four parts, and each part has two states 0 and 1. The next state of a cell is determined by the four adjacent parts of its four neighboring cells. A local transition function of an ESPCA can be described by only six local transition rules, and hence it is very simple. In addition, the framework of partitioned cellular automata (PCAs) is known to be useful for designing reversible CAs.

In this study, we newly introduce a particular reversible ESPCA with the hexadecimal ID number "01caef", which is denoted by ESPCA $P_0$ for short in this paper. ESPCA $P_0$ is conservative, as well as reversible, in the sense that the total number of the state 1's is conserved throughout its evolving process. It is an analogous concept to various conservation laws in physics. Here, we show that a reversible logic element with 1-bit memory (RLEM), rather than a reversible logic gate, is realizable in $P_0$. Once an RLEM is constructed, reversible Turing machines are systematically composed out of it.

ESPCAs were first studied in [8], and it was shown that the Fredkin gate, a universal reversible logic gate, is realized in reversible and conservative ESPCAs 02c5df and 02c5bf. However, it is not known whether an RLEM can be directly implemented in these ESPCAs. Note that, though an RLEM can be composed of Fredkin gates [2], the resulting circuit becomes very complex since the signal timing must be exactly adjusted at each gate.

On the other hand, an elementary *triangular* partitioned cellular automaton (ETPCA) with an ID number 0347 was studied in [3]. It is a reversible and non-conservative ETPCA. Since an ETPCA is a three-neighbor PCA, its local function is described by only four local transition rules, and thus simpler than that of an ESPCA. ETPCA 0347 has a space-moving pattern called a glider, and shows very interesting behavior [3]. It is shown that an RLEM is directly implemented using a glider and a few other small patterns [5]. By this, reversible Turing machines can be systematically constructed in ETPCA 0347.

In this paper, we investigate ESPCA $P_0$, and show that an RLEM is realizable using only two kinds of small patterns and three kinds of phenomena in $P_0$. Useful patterns are a space-moving pattern called a glider, and a periodic pattern called a blinker. Colliding a glider with a blinker appropriately, right-turn and U-turn of the glider, and shifting of the blinker is performed. Using these three phenomena, we compose a particular RLEM in $P_0$. We then construct reversible Turing machines out of RLEMs. By this, we can see that reversible computers can be constructed using only two small patterns and three phenomena observed in $P_0$.

## 2 A New Model of Reversible and Conservative Cellular Automaton $P_0$

A 4-neighbor *square partitioned cellular automaton* (SPCA) is a two-dimensional CA whose cell is divided into four parts as in Fig. 1a. The next state of a cell is determined depending on the present states of the four adjacent parts of the neighboring cells as shown in Fig. 1b.

**Definition 1** A 4-neighbor *square partitioned cellular automaton* (SPCA) is formally defined by

$$P = (\mathbb{Z}^2, (T, R, B, L), ((0, -1), (-1, 0), (0, 1), (1, 0)), f).$$

Here, $\mathbb{Z}^2$ is the set of all points with integer coordinates where cells are placed. The items $T$, $R$, $B$ and $L$ are non-empty finite sets of states of the top, right, bottom and left parts of a cell. The set of states of a cell is thus $Q = T \times R \times B \times L$. The quadruple $((0, -1), (-1, 0), (0, 1), (1, 0))$ is a *neighborhood*. The item $f : Q \to Q$ is a *local (transition) function*.

If $f(t, r, b, l) = (t', r', b', l')$ holds for $(t, r, b, l), (t', r', b', l') \in Q$, this relation is called a *local transition rule* of $P$. It is also indicated as in Fig. 1b. The local function $f$ is thus defined by a set of local transition rules.

**Definition 2** Let $P = (\mathbb{Z}^2, (T, R, B, L), ((0, -1), (-1, 0), (0, 1), (1, 0)), f)$ be an SPCA. A *configuration* of $P$ is a function $\alpha : \mathbb{Z}^2 \to Q$. The set of all configurations of $P$ is denoted by $\mathrm{Conf}(P)$, i.e., $\mathrm{Conf}(P) = \{\alpha \mid \alpha : \mathbb{Z}^2 \to Q\}$. Let $\mathrm{pr}_T : Q \to T$ be the *projection function* that satisfies $\mathrm{pr}_T(t, r, b, l) = t$ for all $(t, r, b, l) \in Q$. The projection functions $\mathrm{pr}_R : Q \to R$, $\mathrm{pr}_B : Q \to B$ and $\mathrm{pr}_L : Q \to L$ are defined similarly. The *global function* $F : \mathrm{Conf}(P) \to \mathrm{Conf}(P)$ of $P$ is defined as the one that satisfies the following.

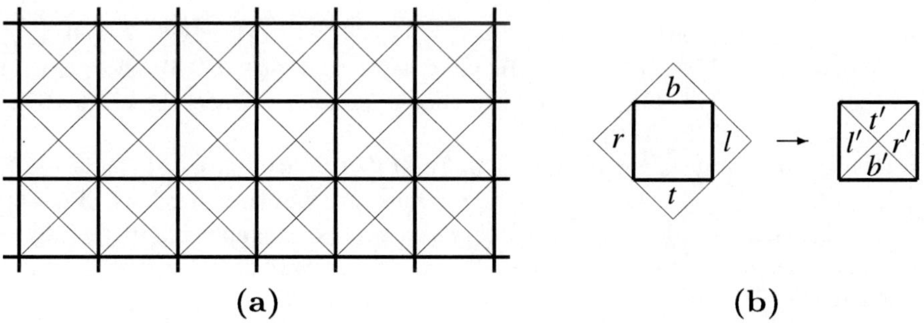


**Fig. 1** **a** Cellular space of a 4-neighbor square partitioned cellular automaton (SPCA), and **b** its local transition rule $f(t, r, b, l) = (t', r', b', l')$

$$\forall \alpha \in \mathrm{Conf}(P), \forall (x, y) \in \mathbb{Z}^2:$$
$$F(\alpha)(x, y) = f(\mathrm{pr}_T(\alpha(x, y-1)), \mathrm{pr}_R(\alpha(x-1, y)), \mathrm{pr}_B(\alpha(x, y+1)), \mathrm{pr}_L(\alpha(x+1, y)))$$

**Definition 3** An SPCA $P$ is called *reversible* if its global function is injective.

The next Lemma shows that injectivity of the global function is equivalent to injectivity of the local function [2, 6]. By this, we can easily obtain a reversible CA, since it is sufficient to design a PCA whose local function is injective.

**Lemma 1** *Let $P$ be an SPCA. Its global function $F$ is injective if and only if its local function $f$ is injective.*

Here we define the simplest subclass of SPCAs such that it is rotation-symmetric, and each of four parts of a cell has only two states. It is called an *elementary* SPCA as in the case of a one-dimensional *elementary cellular automaton* (ECA) [10]. We first define the notion of rotation-symmetry.

**Definition 4** Let $P = (\mathbb{Z}^2, (T, R, B, L), ((0, -1), (-1, 0), (0, 1), (1, 0)), f)$ be an SPCA. The SPCA $P$ is called *rotation-symmetric* (or *isotropic*) if the following conditions (1) and (2) holds.

(1) $T = R = B = L$
(2) $\forall (t, r, b, l), (t', r', b', l') \in T \times R \times B \times L:$
$f(t, r, b, l) = (t', r', b', l') \Rightarrow f(r, b, l, t) = (r', b', l', t')$.

**Definition 5** Let $P = (\mathbb{Z}^2, (T, R, B, L), ((0, -1), (-1, 0), (0, 1), (1, 0)), f)$ be an SPCA. We say $P$ is an *elementary triangular partitioned cellular automaton* (ESPCA), if $T = R = B = L = \{0, 1\}$, and it is rotation-symmetric.

Since an ESPCA is rotation-symmetric, its local function $f : \{0, 1\}^4 \to \{0, 1\}^4$ is defined by only six local transition rules, which are described by the following six values.

$$f(0, 0, 0, 0),\ f(0, 0, 1, 0),\ f(0, 0, 1, 1),\ f(1, 0, 1, 0),\ f(0, 1, 1, 1),\ f(1, 1, 1, 1)$$

Here, $f(0, 0, 1, 0),\ f(0, 0, 1, 1),\ f(0, 1, 1, 1) \in \{0, 1\}^4$. However, $f(1, 0, 1, 0) \in \{(0, 0, 0, 0), (0, 1, 0, 1), (1, 0, 1, 0), (1, 1, 1, 1)\}$ and $f(0, 0, 0, 0), f(1, 1, 1, 1) \in \{(0, 0, 0, 0), (1, 1, 1, 1)\}$, since it is rotation-symmetric. Hence, there are $16^3 \times 4 \times 2^2 = 65,536$ ESPCAs in total.

Reading the 4-bit values of $f(0, 0, 0, 0),\ f(0, 0, 1, 0),\ f(0, 0, 1, 1),\ f(1, 0, 1, 0), f(0, 1, 1, 1), f(1, 1, 1, 1)$ as six binary numbers, we can express an ESPCA by a 6-digit hexadecimal identification number $uvwxyz$. For example, if $f(0, 0, 1, 0) = (t, r, b, l)$, then $v = 2^3 t + 2^2 r + 2^1 b + 2^0 l$. An ESPCA with the ID number $uvwxyz$ is denoted by ESPCA $uvwxyz$.

Figure 2 shows a pictorial representation of the six local transition rules of ESPCA 01caef. It is easy to see that the local function of ESPCA 01caef is injective. Therefore it is a reversible ESPCA. Since we investigate this particular ESPCA in this paper, we denote it by $P_0$ for short in the following.

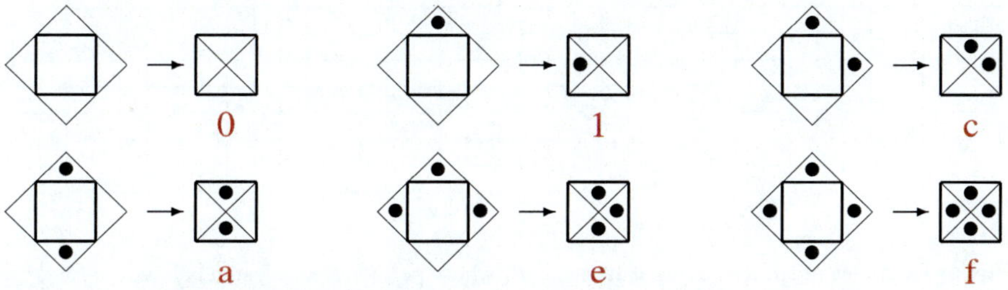


**Fig. 2** Local function defined by the six local transition rules of a particular reversible and conservative ESPCA 01caef, which is denoted by $P_0$ hereafter

An ESPCA is called *conservative* if the number of particles (i.e., state 1) is conserved in each local transition rule. It is an analog of various conservation laws in physics. We can see that ESPCA $P_0$ is conservative.

## 3 Useful Patterns and Phenomena in $P_0$

We investigate the reversible and conservative ESPCA $P_0$ to find useful patterns and phenomena that can be used to compose reversible Turing machines.

### 3.1 Useful Patterns in $P_0$

A *pattern* is a finite segment of a configuration (see [2] for its precise definition). A *periodic pattern* is one such that the same pattern appears at the same position after some time steps. The periodic pattern given in Fig. 3 is called a *blinker*. It is of period 2. Though there are many kinds of periodic patterns in $P_0$, a blinker is particularly useful among them as we shall see in Sect. 3.2

A *space-moving pattern* is one such that the same pattern appears at a different position after some time steps. In $P_0$ there are many kinds of space-moving patterns of various periods. In fact, if we start from a random-like pattern, then we often

**Fig. 3** Blinker in $P_0$

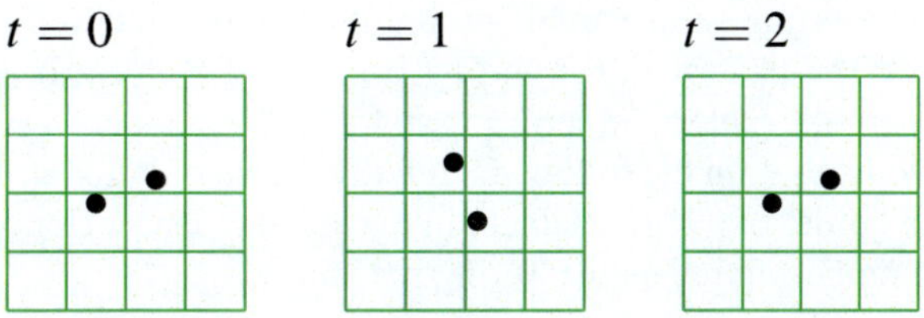

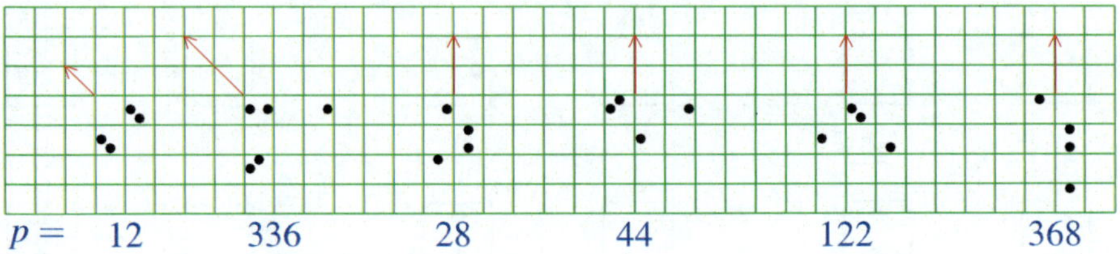


**Fig. 4** Examples of space-moving patterns in $P_0$ whose periods are indicated below

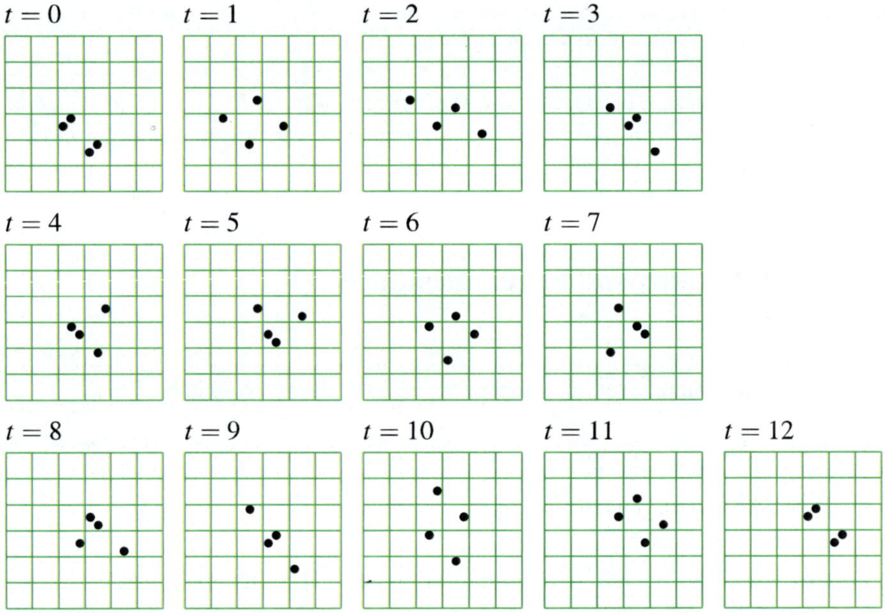


**Fig. 5** Glider in $P_0$

observe that space-moving patterns appear. Some examples of them are shown in Fig. 4.

The pattern having the shortest period among the space-moving patterns so far known is called the *glider* (Fig. 5). It travels one cell diagonally in 12 steps. It will be used as a signal when constructing reversible Turing machines, since it has an interesting and desirable properties as described in Sect. 3.2 .

The pattern shown in Fig. 6 is called a *block*. It is a *stable pattern*, which is a periodic pattern of period 1, and thus does not change its pattern if no other pattern touches it. In the following, it will be used only for writing comments and indicating a border of a logic element in the cellular space. Hence, it has no functional role for composing reversible Turing machines.

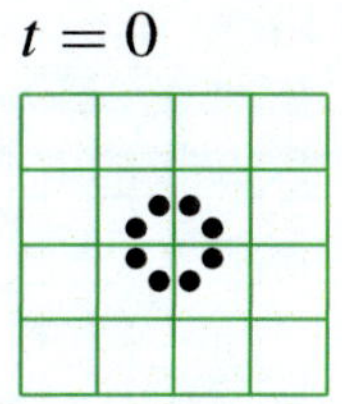


**Fig. 6** Block in $P_0$

## 3.2 Three Useful Phenomena in $P_0$

We show three experiments of colliding a glider with a blinker. Generally, if we collide a space-moving pattern with a periodic pattern in $P_0$, it first shows a complex behavior whose duration time is rather long. But, since $P_0$ is reversible and conservative, it finally gives two or more patterns consisting of at least one space-moving pattern and at most one periodic pattern. As we shall see below, in each of the three experiments, the products of the interaction are again a glider and a blinker, though its intermediate evolution process is complex. Therefore, these processes are useful for controlling a glider and a blinker.

The first experiment is given in Fig. 7. Colliding a glider with a blinker in this manner, a right-turn of a glider is realized.

The second experiment is in Fig. 8. By this, a glider makes a U-turn. It is used to test if a blinker exists at a specified position.

The third experiment is in Fig. 9. By this, the position of the blinker is shifted by 6 cells, and the glider makes a right-turn. Using this phenomenon, a kind of memory device is realized, where the memory states are kept by the positions of the blinker. It is also used to test if a blinker exists at a specified position.

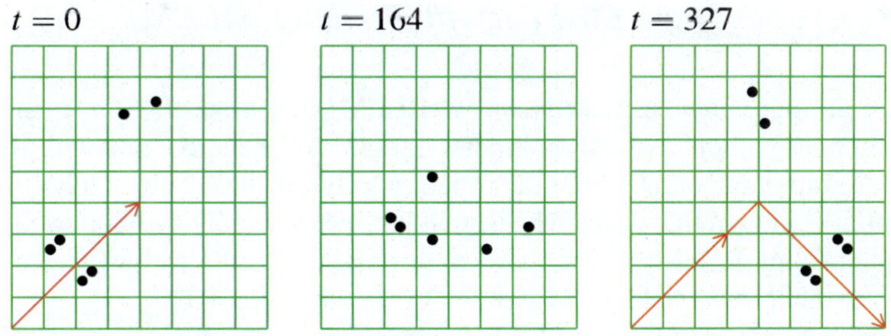


**Fig. 7** Right-turn of a glider in $P_0$

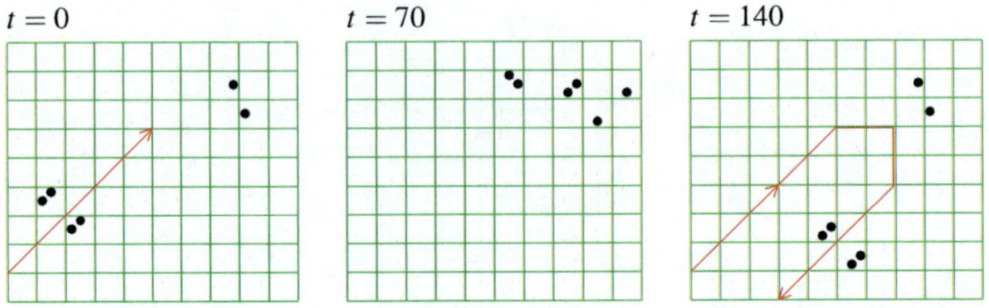


**Fig. 8** U-turn of a glider in $P_0$

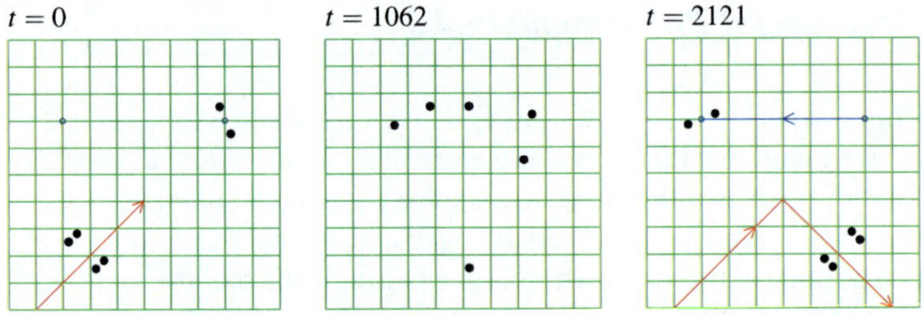


**Fig. 9** Shifting a blinker by a glider in $P_0$

# 4 Making Reversible Turing Machines in $P_0$

Using only three phenomena shown in Sect. 3.2, we can compose reversible Turing machines (RTMs). Here we employ a reversible logic element with memory (RLEM) [2, 7] rather than a reversible logic gate as a logical primitive. By this, construction of RTMs is greatly simplified.

## 4.1 Reversible Logic Element with Memory (RLEM)

A *reversible logic element with memory* (RLEM) is a kind of a finite automaton having output symbols as well as input symbols. A 2-state RLEM with the same numbers of input and output symbols is particularly important, since it is simple yet powerful. It is known that every RLEM that has three or more symbols is *universal*, which means any RLEM is composed only of it [2]. In the following we use the 2-state 4-symbol universal RLEM No. 4-31 for composing RTMs.

The move function of RLEM 4-31 is represented in a graphical form as in Fig. 10. Two rectangles in the figure correspond to the two states 0 and 1. Solid and dotted lines show the input-output relation. If an input signal goes through a dotted line, then the state does not change (Fig. 11a). If a signal goes through a solid line, then the state changes (Fig. 11b).

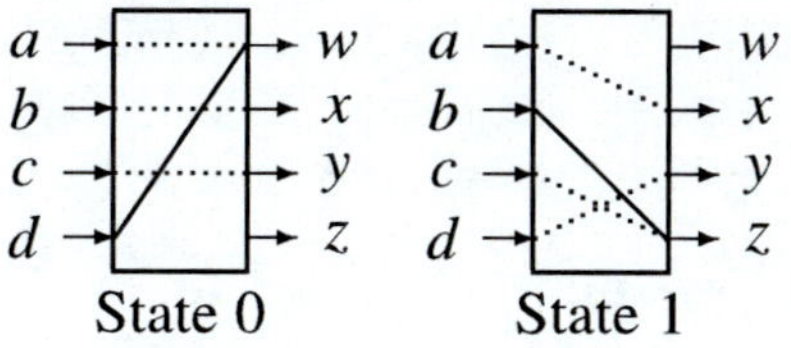


**Fig. 10** A four-symbol two-state RLEM 4-31

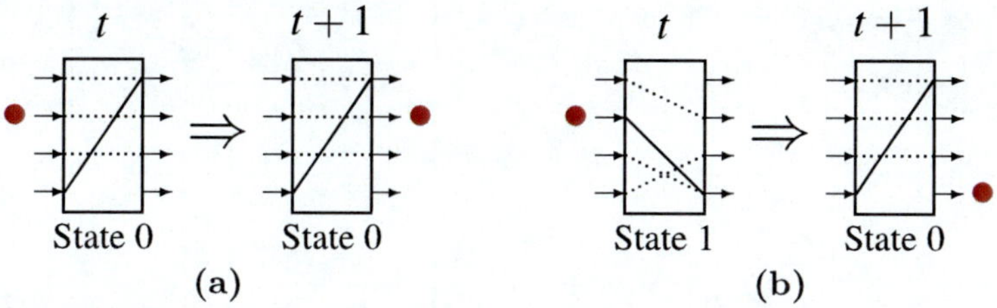


**Fig. 11** Operations of RLEM 4-31. **a** A case where a signal goes through a dotted line, and **b** a case where it goes through a solid line

**Fig. 12** A two-symbol two-state RLEM 2-2

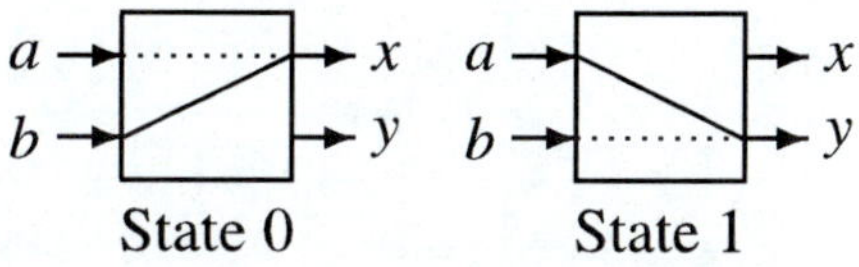


We first show that RLEM 2-2 (Fig. 12) can be realized in the cellular space of ESPCA $P_0$. Although RLEM 2-2 is not universal [2], it is implemented rather simply. Therefore, it is convenient for explaining how we can design a pattern that simulates an RLEM in $P_0$.

Figure 13 shows a pattern that simulates RLEM 2-2. There are seven blinkers in this pattern. One is used as a *position marker* for keeping the memory state 0 or 1, and six are used for turning a signal. Small two circles near the center of the pattern show possible positions of the marker. If the marker is at the left (right, respectively) position, we regard that the RLEM is in the state 0 (1).

First, consider the case where it is in the state 0 and a glider is given to the input port $b$ as in Fig. 13. The glider first turns right at the upper right blinker. Then, at P it shifts the marker to the position of the right circle, and it turns right. By this, the state changes from 0 to 1. Finally, it further turns right at the upper left blinker, and goes out from the output port $x$.

Next, consider the case where it is in the state 0 and a glider is given to the input port $a$. The glider first turns right at the lower left blinker. Then it goes straight ahead at Q without shifting the position marker. It means that the glider knows that the state is 0. It further makes four right-turns, and finally goes out from the output port $x$. It should be noted that at P two different signal paths of the two cases are merged

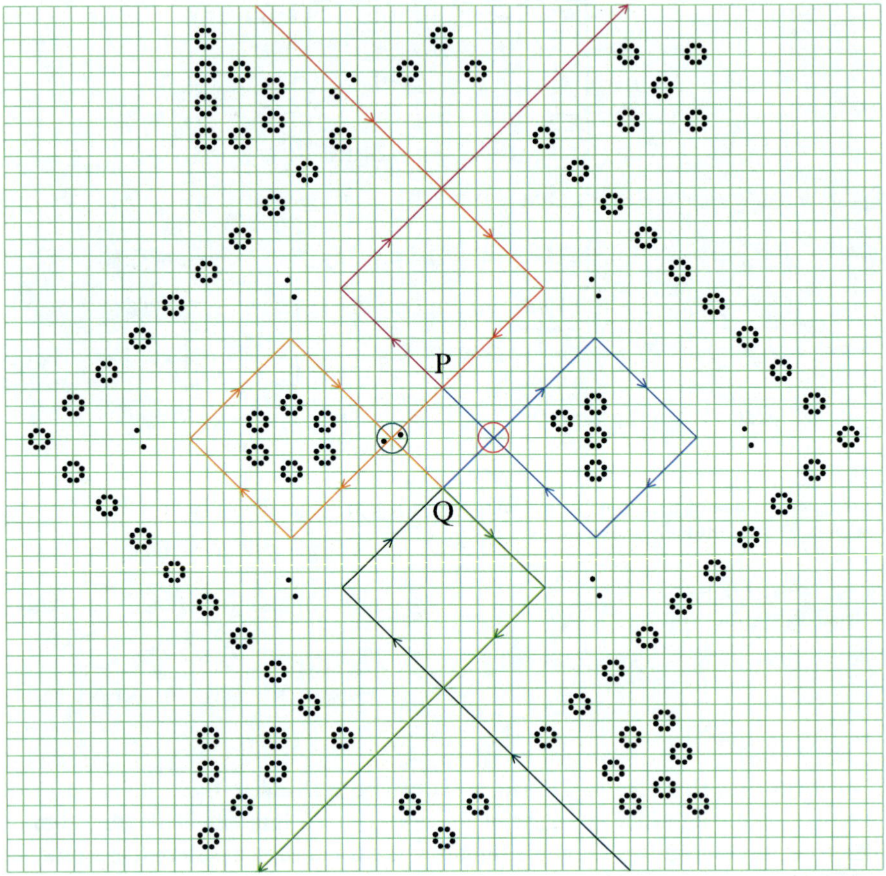


**Fig. 13** RLEM 2-2 implemented in $P_0$

into one. When we want to reversibly merge two paths into one, knowing the state is necessary.

The remaining two cases are similar, and thus we can see the pattern correctly simulates RLEM 2-2. Note that in the case of RLEM 2-2, only two phenomena, which are the right-turn of a glider (Fig. 7) and the shifting of a blinker (Fig. 9), are used.

A pattern that simulates RLEM 4–31 is given in Fig. 14. An outline of its design is explained below.

We use three markers to keep the state 0 or 1, and thus there are three pairs of small circles around the center of Fig. 14, which show possible positions of the markers. If the three markers are all placed in the left (right, respectively) circles of the three pairs, we assume that the state is 0 (1). Since there are eight cases of the combination of states and symbols, we must prepare a sufficient number of access paths leading to the markers to perform operations of testing the state, changing the

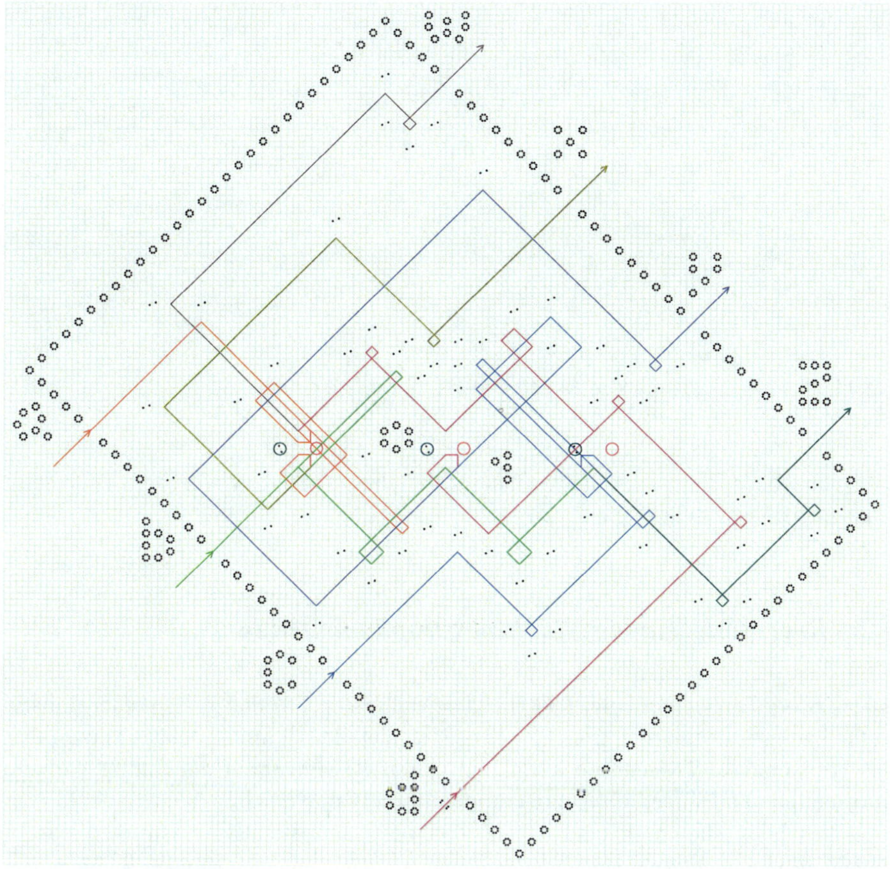

**Fig. 14** RLEM 4-31 implemented in $P_0$

state, and merging signal paths. By this reason, we need three position markers. In this construction, we also use the U-turn operation (Fig. 8) for testing the state and merging signal paths.

Figure 14 shows the case where the state is 0 and a glider is given to the input port $d$. In this case, the glider first collides with the right position marker, and shifts it to the right. Then the glider shifts the central and the left position markers to the right. By this, the state changes from 0 to 1. Finally, the glider goes out from the output port $w$.

Next, consider the case where the state is 0 and a glider is given to the input port $a$. The glider tests whether the state is 1 at the left position marker. But, since the state is 0 in this case, the glider moves straight ahead at this position, and finally goes out from the output port $w$. Merging the two paths into one, which leads to $w$, is performed at the left position marker.

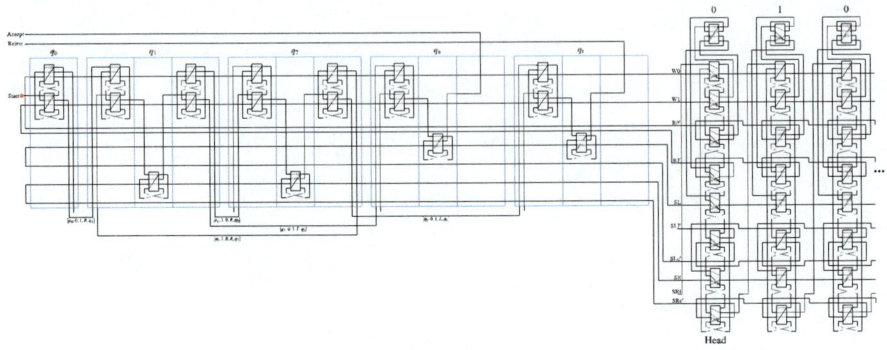


**Fig. 15** RTM $T_{\text{parity}}$ composed of RLEM 4-31 [7]

Since other cases are similar, we can see that the pattern shown in Fig. 14 correctly simulates RLEM 4-31.

## 4.2 *Reversible Turing Machines Implemented in $P_0$*

A *reversible Turing machine* (RTM) is a deterministic TM that is also deterministic to the backward time direction (see [2] for its detailed definition). It is known that any (irreversible) TM can be simulated by an RTM without generating garbage information [1]. Hence, RTMs are computationally universal.

It has been shown that any RTM can be systematically constructed out of RLEM 4-31 [7]. Figure 15 gives the circuit that simulates a simple RTM $T_{\text{parity}}$, which has the set of quintuples $\{[q_0, 0, 1, R, q_1], [q_1, 0, 1, L, q_{\text{acc}}], [q_1, 1, 0, R, q_2], [q_2, 0, 1, L, q_{\text{rej}}], [q_2, 1, 0, R, q_1]\}$. It is the following RTM. Assume a symbol string $0\,1^n\,0$ $(n = 0, 1, \ldots)$ is given as an input. Then, $T_{\text{parity}}$ halts in the accepting state $q_{\text{acc}}$ if and only if $n$ is even, and all the read symbols are complemented. Note that Fig. 15 is a slightly improved version of a circuit given in [7]. Detailed explanations on the circuit are omitted here.

Putting copies of the pattern of RLEM 4-31 given in Fig. 14 at the positions corresponding to the RLEMs in Fig. 15, and connecting them appropriately, we have a complete configuration of ESPCA $P_0$ that simulates $T_{\text{parity}}$. Figure 16 shows a configuration of another RTM $T_{\text{power}}$ that accepts the language $\{1^n \mid n = 2^k, k \in \{0, 1, 2, ...\}\}$ simulated on the CA simulator Golly [9]. Whole computing processes of RTMs embedded in $P_0$ can be seen on Golly using its emulator [4].

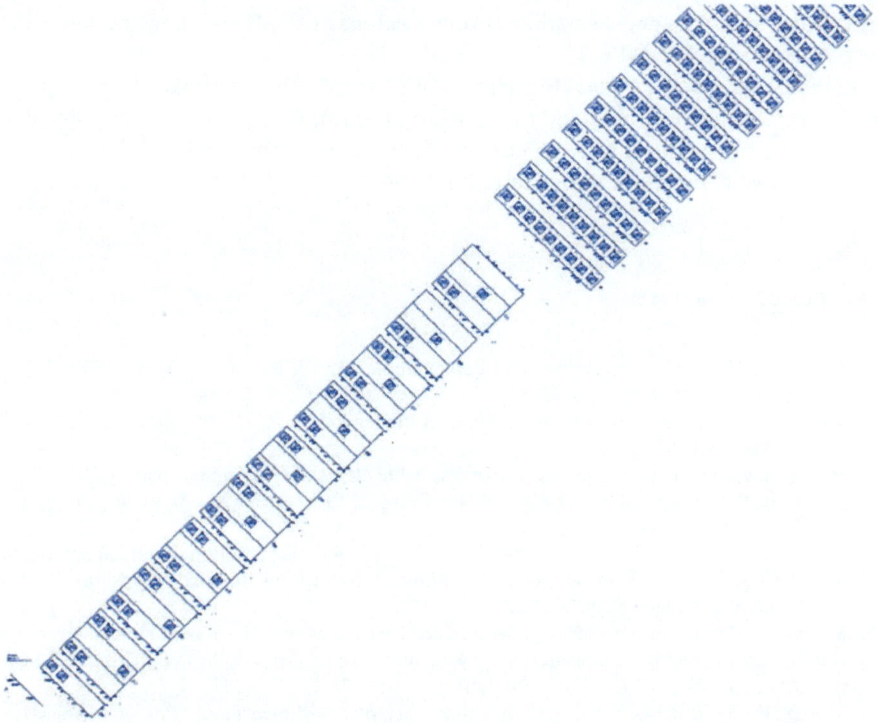

**Fig. 16** RTM $T_{\mathrm{power}}$ in $P_0$ simulated on *Golly* [4]

## 5 Concluding Remarks

We investigated how reversible computers are realized in a reversible and conservative cellular space of $P_0$ having a very simple local transition function (Fig. 2). We observe that for composing reversible Turing machines (RTMs) it is sufficient to have (i) a space-moving pattern acting as a signal, (ii) a periodic pattern for signal turning, and (iii) a periodic pattern acting as a position marker, which satisfy the following conditions.

1. The moving direction of a space-moving pattern can be controlled by a pattern for signal turning.
2. A space-moving pattern can know whether a marker exsists at a specified position or not. In other words, trajectory of a space-moving pattern can be branched by an interaction of the space-moving pattern and the marker.
3. The position of a marker can be shifted by colliding a space-moving pattern.

If we have three patterns that satisfy the conditions 1–3, we can in principle compose a universal reversible logic element with memory (RLEM) from them in the cellular space. Then, reversible Turing machines (RTMs) are constructed using the pattern for the RLEM.

In the ESPCA $P_0$, a glider (Fig. 5) acts as a signal, while a blinker (Fig. 3) plays both roles for signal turning and of a position marker. Figures 7, 8 and 9 show that they satisfy the conditions 1–3. Thus, in $P_0$, reversible computers are thus realized using only two patterns—a glider and a blinker—and three kinds of their interactions.

## References

1. Bennett, C.H.: Logical reversibility of computation. IBM J. Res. Dev. **17**, 525–532 (1973). https://doi.org/10.1147/rd.176.0525
2. Morita, K.: Theory of Reversible Computing. Springer, Tokyo (2017). https://doi.org/10.1007/978-4-431-56606-9
3. Morita, K.: A universal non-conservative reversible elementary triangular partitioned cellular automaton that shows complex behavior. Nat. Comput. **18**(3), 413–428 (2019). https://doi.org/10.1007/s11047-017-9655-9
4. Morita, K.: Data set for simulating a reversible elementary square partitioned cellular automaton with the ID number 01caef on Golly. Hiroshima University Institutional Repository (2021). http://ir.lib.hiroshima-u.ac.jp/00051576
5. Morita, K.: How can we construct reversible Turing machines in a very simple reversible cellular automaton? In: Yamashita, S., Yokoyama, T. (eds.) Proceedings of RC 2021. LNCS, vol. 12805, pp. 3–21 (2021). https://doi.org/10.1007/978-3-030-79837-6_1
6. Morita, K., Harao, M.: Computation universality of one-dimensional reversible (injective) cellular automata. Trans. IEICE **E72**, 758–762 (1989). http://ir.lib.hiroshima-u.ac.jp/00048449
7. Morita, K., Suyama, R.: Compact realization of reversible Turing machines by 2-state reversible logic elements. In: Ibarra, O.H., Kari, L., Kopecki, S. (eds.) Proceedings of UCNC 2014. LNCS, vol. 8553, pp. 280–292 (2014). https://doi.org/10.1007/978-3-319-08123-6_23
8. Morita, K., Ueno, S.: Computation-universal models of two-dimensional 16-state reversible cellular automata. IEICE Trans. Inf. Syst. **E75-D**, 141–147 (1992). http://ir.lib.hiroshima-u.ac.jp/00048451
9. Trevorrow, A., Rokicki, T., Hutton, T., et al.: Golly: an open source, cross-platform application for exploring Conway's Game of Life and other cellular automata (2005). http://golly.sourceforge.net/
10. Wolfram, S.: A New Kind of Science. Wolfram Media Inc. (2002)

# Algorithms for Group Cellular Automata

**Jarkko Kari**

**Abstract** We consider subshifts and cellular automata in the setup where the state set is a finite group. The group does not need to be commutative. A subshift that is also a subgroup is called a group shift, and we call a cellular automaton on a group shift a group cellular automaton if it is also a group homomorphism. Group cellular automata generalize the much studied concept of additive cellular automata into non-commutative groups. The set of space-time diagrams of a group cellular automaton is a group shift, so we can apply classical results by Bruce Kitchens and Klaus Schmidt on group shifts to analyze group cellular automata. In particular, we can effectively construct the limit set and the trace subshifts of any group cellular automaton. We can then algorithmically decide many properties concerning the cellular automaton that are in general undecidable. The talk is based on a joint work with Pierre Béaur.

**Keywords** Group cellular automata · Group shifts · Symbolic dynamics · Decidability

## 1 Introduction

This talk concerns cellular automata whose state set is a finite group $\mathbb{G}$, and the cellular automaton is a group homomorphism $\mathbb{X} \longrightarrow \mathbb{X}$ where $\mathbb{X} \subseteq \mathbb{G}^{\mathbb{Z}^d}$ is a *group shift*, i.e., a subshift that is a subgroup of $\mathbb{G}^{\mathbb{Z}^d}$. We call these *group cellular automata*. The concept is a generalization of additive cellular automata that are group cellular automata over commutative groups, and typically on the full shift. Additive cellular automata, in turn, generalize the concept of linear cellular automata that are linear functions over a commutative ring structure. We demonstrate that group cellular automata can be effectively analyzed, as many structural and dynamical properties that are undecidable for general cellular automata are decidable for group cellular automata. Properties considered include injectivity, surjectivity, equicontinuity, sensitivity and

J. Kari (✉)
Department of Mathematics and Statistics, University of Turku, Turku, Finland
e-mail: jkari@utu.fi



S. Das and G. J. Martinez (eds.), *Proceedings of First Asian Symposium on Cellular Automata Technology*, Advances in Intelligent Systems and Computing 1425,
https://doi.org/10.1007/978-981-19-0542-1_2

nilpotency of the cellular automaton. Non-transitivity is shown to be semi-decidable, while its decidability status remains open. We also establish that injectivity always implies surjectivity, that transitivity implies mixingness, that non-sensitivity implies equicontinuity, and that jointly periodic points are dense in the limit set. The talk is based on a joint work with Béaur [3].

## 2 Preliminaries

### 2.1 Subshifts and Cellular Automata

Let $A$ be a finite state set and $d \geq 1$ a dimension. A $d$-dimensional *configuration* over the state set $A$ is a coloring $c : \mathbb{Z}^d \longrightarrow A$ of the infinite $d$-dimensional grid by states. We denote the set of all such configurations by $A^{\mathbb{Z}^d}$, using the standard notation $X^Y$ for the set of functions $Y \longrightarrow X$. A translation $\tau$ by $\mathbf{n} \in \mathbb{Z}^d$ shifts a configuration so that the cell $\mathbf{n}$ is pulled to the cell $\mathbf{0}$, that is, the state of a cell $\mathbf{u} \in \mathbb{Z}^d$ in the translated configuration $\tau(c)$ is the same as the state of the cell $\mathbf{u} + \mathbf{n}$ in $c$. The configuration $c$ is called *strongly periodic* if for some $k > 0$ holds that $c(\mathbf{n} + k\mathbf{m}) = c(\mathbf{n})$ for all $\mathbf{n}, \mathbf{m} \in \mathbb{Z}^d$, i.e., if $c$ is invariant under translations by $k$ cells in any coordinate direction.

A finite *pattern* $p : D \longrightarrow A$ assigns states to a finite *shape* $D \subseteq \mathbb{Z}^d$. We say that a configuration $c$ contains the pattern $p$ if $\tau(c)|_D = p$ for some translation $\tau$. A *subshift* is a set of configurations defined by forbidding some finite patterns. More precisely, let $P$ be a set of finite patterns. Then $\mathcal{X}_P$, the subshift defined by $P$, consists of all those $c \in A^{\mathbb{Z}^d}$ that do not contain any $p \in P$. If $P$ is finite then the corresponding subshift $\mathcal{X}_P$ is a *subshift of finite type (SFT)*. Note that subshifts are precisely the translation invariant sets of configurations that are closed in the standard *prodiscrete* topology of $A^{\mathbb{Z}^d}$ (See, for example, [6] for details).

Let $\mathcal{X}_P \subseteq A^{\mathbb{Z}^d}$ be a subshift. A *cellular automaton (CA)* on $\mathcal{X}_P$ is a function $F : \mathcal{X}_P \longrightarrow \mathcal{X}_P$ with a finite *neighborhood* $D \subseteq \mathbb{Z}^d$ and a *local rule* $f : A^D \longrightarrow A$ such that for any $c \in A^{\mathbb{Z}^d}$ the configuration $F(c)$ has in any cell $\mathbf{n} \in \mathbb{Z}^d$ the state $f(\tau(c)|_D)$, where $\tau$ is the translation by $\mathbf{n}$. Quite often we considers cellular automata on the *full shift* $\mathcal{X}_P = A^{\mathbb{Z}^d}$.

A (bi-infinite) *trajectory* of a CA $F : \mathcal{X}_P \longrightarrow \mathcal{X}_P$ is a bi-infinite sequence $\ldots, c_{-2}, c_{-1}, c_0, c_1, c_2, \ldots$ of configurations $c_i \in \mathcal{X}_P$ such that $c_{i+1} = F(c_i)$ holds for all $i \in \mathbb{Z}$. Stacking configurations of a trajectory consecutively along a new dimension makes a *space-time diagram*, i.e., a $(d + 1)$-dimensional configuration that has in cell $(n_1, \ldots, n_d, i) \in \mathbb{Z}^{d+1}$ the state $c_i(n_1, \ldots, n_d)$, for all $(n_1, \ldots, n_d) \in \mathbb{Z}^d$ and $i \in \mathbb{Z}$. The set of the space-time diagrams of $F$ is then a $(d + 1)$-dimensional subshift $\mathrm{ST}(F)$, obtained by forbidding on the projections in the first $d$ dimensions the forbidden patterns in $P$ (thus forcing the projections to be elements of $\mathcal{X}_P$), and by forbidding on two consecutive such projections patterns that indicate an incorrect application of the local rule $f$ of the CA. Clearly, with this construction, if $\mathcal{X}_P$ is an

SFT then also $\mathrm{ST}(F)$ is an SFT. Moreover, we can effectively construct $\mathrm{ST}(F)$, i.e., a finite set of forbidden patterns that defines $\mathrm{ST}(F)$, given the local rule of $F$ and the forbidden set $P$ that defines the space $\mathcal{X}_P$.

Space-time diagrams of cellular automata capture information about their asymptotic dynamics, i.e., dynamics on their *limit sets*. The limit set $\Omega_F$ of a CA $F$ contains the configurations $c$ such that $F^{-n}(c)$ is non-empty for all $n \in \mathbb{N}$, which is easily seen equivalent to $c$ appearing in a trajectory (Recall that in our terminology trajectories and space-time diagrams are bi-infinite in the temporal dimension). Algorithms to analyze $(d+1)$-dimensional SFTs then provide algorithms to analyze asymptotic dynamics of $d$-dimensional cellular automata. Unfortunately, already 2-dimensional SFTs are generally beyond algorithmic treatment due to the undecidability of the domino problem. However, restricting the setup to SFTs with an additional algebraic structure helps, as seen in the following section.

## 2.2 Group Shifts and Group Cellular Automata

In group shifts and group cellular automata the state set is a finite group $\mathbb{G}$. Note that the group does not need to be commutative. The set $\mathbb{G}^{\mathbb{Z}^d}$ of configurations inherits the group structure as a infinite cartesian power of $\mathbb{G}$: the group operation are applied cell-wise so that for all $c_1, c_1 \in \mathbb{G}^{\mathbb{Z}^d}$ we have $(c_1c_2)(\mathbf{n}) = c_1(\mathbf{n})c_2(\mathbf{n})$ for all $\mathbf{n} \in \mathbb{Z}^d$. We call a subshift $\mathbb{X} \subseteq \mathbb{G}^{\mathbb{Z}^d}$ a *group shift* if it is a subgroup of $\mathbb{G}^{\mathbb{Z}^d}$. We call a cellular automaton $F : \mathbb{X} \longrightarrow \mathbb{X}$ on a group shift $\mathbb{X}$ a *group cellular automaton* if it is also a group homomorphism.

One can easily see that the subshift $\mathrm{ST}(F) \subseteq \mathbb{G}^{\mathbb{Z}^{d+1}}$ of space-time diagrams of a group cellular automaton $F$ is a $(d+1)$-dimensional group shift. This means that algorithms for $(d+1)$-dimensional group shifts provide algorithms for $d$-dimensional group cellular automata. This is the approach of the present talk, based on the following two classical results on group shifts.

**Theorem 1** *([12]) Every group shift is of finite type.*

Based on this result, every group shift has a finite presentation in terms of finitely many forbidden patterns. This is a presentation assumed in all algorithmic results. The second theorem states that for any finite pattern $p$ that appears in some configuration of a group shift $\mathbb{X}$ there exists a strongly periodic configuration in $\mathbb{X}$ that also contains $p$:

**Theorem 2** *([12]) Strongly periodic configurations are dense in group shifts.*

With these two results we get, for example, an algorithm to determine if a given finite pattern $p$ appears in a given group shift $\mathbb{X}$. This is a classical reasoning due to Wang [18]: Using compactness and the given finitely many forbidden patterns that define $\mathbb{X}$ one has a semi-algorithm to recognize negative cases (patterns $p$ that do not appear in any configuration of $\mathbb{X}$), and the density of strongly periodic configurations in SFT $\mathbb{X}$ provides a semi-algorithm for the positive cases ($p$ that appear in $\mathbb{X}$).

**Corollary 1** *There is an algorithm to decide if a given finite pattern p appears in a given group shift* $\mathbb{X}$.

Corollary 1 in turn implies an algorithm to decide if two given group shifts are identical: We have $\mathcal{X}_Q \subseteq \mathcal{X}_P$ iff no $p \in P$ appears in any configuration of $\mathcal{X}_Q$. This can be tested by Corollary 1 if the subshifts are group shifts. Testing $\mathcal{X}_P \subseteq \mathcal{X}_Q$ is analogous.

**Corollary 2** *There is an algorithm to decide if two given group shifts are identical.*

Our main technical contribution is an algorithm to construct any lower dimensional projection of a given group shift [3]. The projections are group shifts as well, and thus they have finite presentations (finite forbidden pattern sets) that our algorithm outputs. In particular, when applied to the subshift $\mathrm{ST}(F)$ of space-time diagrams of a group CA $F$, we obtain algorithms to construct the limit set and the *trace subshifts* of $F$. Recall that the the limit set $\Omega_F$ is the projection of $\mathrm{ST}(F)$ in the first $d$ dimensions. A trace subshift of $F$, on the other hand, is a one-dimensional group shift that records possible pattern sequences in trajectories. Let $D \subseteq \mathbb{Z}^d$ be finite. The $D$-trace extracted from a trajectory $\dots, c_{-1}, c_0, c_1, \dots$ is the sequence $\dots, c_{-1}|_D, c_0|_D, c_1|_D, \dots$ of patterns of $\mathbb{G}^D$. The $D$-trace subshift $\mathrm{Tr}_D(F)$ of $F$ contains the $D$-traces of all trajectories. It is a one-dimensional subshift over the state set $\mathbb{G}^D$, which is also a group (a cartesian power of $\mathbb{G}$). The $D$-trace subshift is a projection of $\mathrm{ST}(F)$ in the last, temporal dimension, and thus it a group shift that can be effectively constructed.

**Theorem 3** ([3]) *Given a group shift* $\mathbb{X} \subseteq \mathbb{G}^{\mathbb{Z}^d}$ *and a group cellular automaton* $F : \mathbb{X} \longrightarrow \mathbb{X}$ *one can effectively construct the limit set* $\Omega_F$*, which is a* $d$*-dimensional group shift over the group* $\mathbb{G}$*. Also, for any given finite* $D \subseteq \mathbb{Z}^d$ *one can effectively construct the* $D$*-trace subshift* $\mathrm{Tr}_D(F)$ *of* $F$*, which is a one-dimensional group shift over the group* $\mathbb{G}^D$.

Once the limit set and the trace subshifts are available, one can analyze many aspects of the cellular automaton.

## 3 Algorithms

### 3.1 Injectivity and Surjectivity

A cellular automaton $F : \mathcal{X} \longrightarrow \mathcal{X}$ on a subshift $\mathcal{X}$ is

- *injective* if $c \neq e \Longrightarrow F(c) \neq F(e)$,
- *surjective* if $F(\mathcal{X}) = \mathcal{X}$.

Injectivity and surjectivity are decidable properties on one-dimensional SFTs $\mathcal{X}$ [2], but they are undecidable already on two-dimensional full shifts $\mathcal{X} = A^{\mathbb{Z}^2}$ [9]. Injectivity and surjectivity of linear cellular automata over commutative rings are known to be decidable [17], and the following argumentations generalize this to all group cellular automata.

- Surjectivity: A CA $F : \mathbb{X} \longrightarrow \mathbb{X}$ is surjective if and only if $\Omega_F = \mathbb{X}$. For group cellular automata, the limit set $\Omega_F$ can be effectively constructed (Theorem 3) and the equivalence of two group shifts $\Omega_F$ and $\mathbb{X}$ can be effectively tested (Corollary 2).
- Injectivity: A homomorphism $F : \mathbb{X} \longrightarrow \mathbb{X}$ is injective if and only if the kernel

$$\ker(F) = F^{-1}(\mathbf{1}_{\mathbb{G}}^{\mathbb{Z}^d})$$

is the singleton set $\{\mathbf{1}_{\mathbb{G}}^{\mathbb{Z}^d}\}$, where $\mathbf{1}_{\mathbb{G}}$ is the identity element of group $\mathbb{G}$. The kernel can be effectively constructed by simply forbidding all neighborhood patterns that the local rule maps to a state different from $\mathbf{1}_{\mathbb{G}}$, as well as the patterns in $P$ used to define the group shift $\mathbb{X} = \mathcal{X}_P$. The equality of $\ker(F)$ and $\{\mathbf{1}_{\mathbb{G}}^{\mathbb{Z}^d}\}$ can be tested since both are group shifts.

Thus we have established the following result.

**Theorem 4** *([3]) It is decidable if a given group CA $F : \mathbb{X} \longrightarrow \mathbb{X}$ is injective/surjective.*

We can also note that all group shifts are *surjunctive*. Recall that a subshift $\mathcal{X}$ is called surjunctive if every injective CA $\mathcal{X} \longrightarrow \mathcal{X}$ is also surjective. Full shifts on $\mathbb{Z}^d$ in any dimension $d$ are surjunctive and, more generally, any strongly irreducible SFT is surjunctive [8]. But there exist (even one-dimensional) SFTs that are not surjunctive.

It is straightforward to see that if strongly periodic configurations are dense in a subshift then the subshift is surjunctive. Indeed, there are finitely many strongly periodic configurations with any fixed periods. An injective cellular automaton $F$ preserves the periods so that $F$ is an injective self-map on these finite sets, implying surjectivity on the sets as well. Thus every strongly periodic configuration has a (strongly periodic) pre-image. If the strongly periodic configurations are dense then every configuration of the subshift has a pre-image, meaning that the CA is surjective. With this argument the surjunctivity of all group shifts follows from Theorem 2.

**Theorem 5** *([3]) Every group shift $\mathbb{X} \subseteq \mathbb{G}^{\mathbb{Z}^d}$ is surjunctive.*

## *3.2 Density of Periodic Points*

Configuration $c \in \mathcal{X}$ is (temporally) *periodic* for a CA $F : \mathcal{X} \longrightarrow \mathcal{X}$ if there exists $n \geq 1$ such that $F^n(c) = c$. If $c$ is also strongly periodic then we call it *jointly periodic* (i.e., jointly for space and time). CA $F : \mathcal{X} \longrightarrow \mathcal{X}$ has *dense jointly periodic points (DJPP)* if every pattern $p$ that appears in some configuration of $\mathcal{X}$ appears also in some jointly periodic configuration.

It is a notorious open problem whether every surjective cellular automaton on a full shift (even in one-dimension) has DJPP. This is known to be true under some additional constraints: a one-dimensional surjective CA on a mixing SFT has DJPP if it is closing [5] or has equicontinuity points [4]; any one-dimensional surjective group CA has DJPP [5], as do all additive CA over the cyclic group $\mathbb{G} = \mathbb{Z}_m$ in any dimension $d$ [15]. We can generalize the latter results using Theorem 2, and show that every surjective group cellular automaton $F : \mathbb{X} \longrightarrow \mathbb{X}$ on any group shift $\mathbb{X}$ in any dimension has DJPP. We reason as follows. Any pattern $p$ that appears in some configuration of $\mathbb{X}$ appears in a trajectory under a surjective $F : \mathbb{X} \longrightarrow \mathbb{X}$. Thus $p$ appears in the projection of a space-time diagram into the first $d$ dimensions. As the subshift $\mathrm{ST}(F)$ of space-time diagrams is a group shift and thus has a dense set of strongly periodic configurations (Theorem 2), pattern $p$ appears in some strongly periodic space-time diagram, i.e., in a jointly periodic configuration under $F$.

**Theorem 6** *([3]) Every surjective group CA has a dense set of jointly periodic points.*

## 3.3 *Nilpotency and Periodicity*

A CA $F : \mathcal{X} \longrightarrow \mathcal{X}$ is

(i) *nilpotent* if $\Omega_F$ is a singleton set,
(ii) *eventually periodic* if $F^{n+p} = F^n$ for some $n$ and $p \geq 1$,
(iii) *periodic* if $F^p$ is the identity map, for some $p \geq 1$.

All three properties are undecidable for general one-dimensional CA [1, 10, 11]. However, for any group CA $F : \mathbb{X} \longrightarrow \mathbb{X}$ over any group shift in any dimension all three properties can be effectively decided.

(i) The identity element $\mathbf{1}_{\mathbb{G}}^{\mathbb{Z}^d}$ is a fixed point of every group CA and thus in the limit set $\Omega_F$. Consequently, a group CA $F$ is nilpotent if and only if $\Omega_F = \{\mathbf{1}_{\mathbb{G}}^{\mathbb{Z}^d}\}$. As group shifts $\Omega_F$ and $\{\mathbf{1}_{\mathbb{G}}^{\mathbb{Z}^d}\}$ can be effectively formed, and the equivalence of group shifts is decidable (Corollary 2), the nilpotency of $F$ can be effectively checked.
(ii) A CA $F$ is eventually periodic if and only if $F$ restricted to its limit set is periodic, that is, there is $p \geq 1$ such that $F^p(c) = c$ holds for all $c \in \Omega_F$. But this is equivalent to any of the trace subshifts being finite. Fix, for example, $D = \{\mathbf{0}\}$. As the trace subshift $\mathrm{Tr}_D(F)$ of a group CA is an SFT and can be constructed, and as the finiteness of a one-dimensional SFT is decidable, we can effectively check eventual periodicity of $F$.
(iii) A CA $F$ is periodic if and only if it is injective and eventually periodic. We have seen how both these properties are decidable, so also periodicity is decidable.

**Theorem 7** *([3]) It is decidable if a given group CA is nilpotent/periodic/eventually periodic.*

## 3.4 Sensitivity and Equicontinuity

An *equicontinuity point* of a CA $F : \mathcal{X} \longrightarrow \mathcal{X}$ is a configuration $c$ whose forward trajectory is tracked arbitrarily closely by the trajectory of any initial configuration that is sufficiently close to $c$. Let $\mathrm{eq}(F)$ be the set of equicontinuity points of $F$, so $c \in \mathrm{eq}(F)$ iff

$$\begin{aligned}
&\forall \text{ finite } E \subseteq \mathbb{Z}^d \\
&\quad \exists \text{ finite } D \subseteq \mathbb{Z}^d \\
&\qquad \forall e \in \mathcal{X} \,:\, e|_D = c|_D \quad \Longrightarrow \quad \forall t \geq 0 \,:\, F^t(e)|_E = F^t(c)|_E.
\end{aligned}$$

A cellular automaton $F : \mathcal{X} \longrightarrow \mathcal{X}$ is called *equicontinuous* if $\mathrm{eq}(F) = \mathcal{X}$. Equicontinuous cellular automata have stable dynamics. In contrast, $F$ is called *sensitive* to initial conditions if

$$\begin{aligned}
&\exists \text{ finite } E \subseteq \mathbb{Z}^d \\
&\quad \forall c \in \mathcal{X} \\
&\qquad \forall \text{ finite } D \subseteq \mathbb{Z}^d \\
&\qquad\quad \exists e \in \mathcal{X} \,:\, e|_D = c|_D \text{ and } \exists t \geq 0 \,:\, F^t(e)|_E \neq F^t(c)|_E.
\end{aligned}$$

A sensitive CA has no equicontinuity points (as is seen by swapping the first two lines in the four line expression above). We therefore have that

$$\begin{aligned}
F \text{ is equicontinuous} &\Longleftrightarrow \mathrm{eq}(F) = \mathcal{X}, \\
F \text{ is sensitive} &\longrightarrow \mathrm{eq}(F) = \emptyset,
\end{aligned}$$

The second implication is not an equivalence: there are multidimensional non-sensitive CA with $\mathrm{eq}(F) = \emptyset$ [16]. Also, there are CA with $\emptyset \subsetneq \mathrm{eq}(F) \subsetneq X$.

In the algebraic setting the situation is simpler. There is a known dichotomy for linear CA on the full shift in any dimension: such a CA is either equicontinuous or sensitive [14]. Analogous dichotomy holds also for one-dimensional additive CA on the full shift [7]. We can generalize the dichotomy result to arbitrary group cellular automata, on any group shift in any number of dimensions.

**Theorem 8** *([3]) For a group CA $F : \mathbb{X} \longrightarrow \mathbb{X}$ exactly one of the following two possibilities holds:*

- $\mathrm{eq}(F) = \mathbb{X}$ *and $F$ is equicontinuous*
- $\mathrm{eq}(F) = \emptyset$ *and $F$ is sensitive*

Concerning decidability, sensitivity and equicontinuity are undecidable properties in general, even for bijective one-dimensional CA over the full shift [11, 13]. In the algebraic setting, both properties are decidable for one-dimensional additive CA on the full shift [7].

Based on the fact that a group CA is equicontinuous if and only if it is eventually periodic we obtain the following as a corollary to Theorem 7.

**Theorem 9** *([3]) It is decidable if a given group CA $F : \mathbb{X} \longrightarrow \mathbb{X}$ is sensitive/equicontinuous.*

## 3.5 Mixing Properties

A cellular automaton $F : \mathcal{X} \longrightarrow \mathcal{X}$ is *transitive* if

$$\begin{aligned} &\forall \text{ finite } D \subseteq \mathbb{Z}^d \\ &\quad \forall c, e \in \mathcal{X} \\ &\qquad \exists c', e' \in \mathcal{X} \ : \ c'|_D = c|_D \text{ and } e'|_D = e|_D, \\ &\qquad\quad \exists t \geq 0 \ : \ F^t(c') = e'. \end{aligned}$$

It is lmixing if this condition holds for all sufficiently large $t$. In other words, in transitive CA there are trajectories between arbitrarily small neighborhoods of any two configurations.

The following facts are known in the algebraic setting: An additive CA on the full shift is transitive if and only if it is mixing [7], and transitivity (and hence also mixingness) is decidable for one-dimensional additive CA on the full shift [7].

Note that $F$ is transitive (mixing) if and only if its $D$-trace subshift $\mathrm{Tr}_D(F)$ is transitive (mixing, respectively) for all finite $D \subseteq \mathcal{X}$. As the $D$-trace subshifts are one-dimensional group shifts and can be effectively constructed, one can use properties of one-dimensional group shifts and the decidability of transitivity for one-dimensional SFTs to obtain the following result.

**Theorem 10** *([3]) A group CA $F : \mathbb{X} \longrightarrow \mathbb{X}$ is transitive if and only if it is mixing. There is a semi-algorithm to test non-transitivity of a given group CA $F : \mathbb{X} \longrightarrow \mathbb{X}$.*

The decidability status of transitivity remains open. A straightforward approach does not work as one needs to verify the transitivity of $D$-trace subshifts for all (infinitely many) finite $D \subseteq \mathbb{Z}^d$. In contrast, for the semi-decidability of non-transitivity one just needs to find one $D$ such that the $D$-trace subshift is non-transitive.

## References

1. Aanderaa, S., Lewis, H.: Linear sampling and the ∀∃∀ case of the decision problem. J. Symb. Logic **39**(3), 519–548 (1974)
2. Amoroso, S., Patt, Y.N.: Decision procedures for surjectivity and injectivity of parallel maps for tessellation structures. J. Comput. Syst. Sci. **6**(5), 448–464 (1972)
3. Béaur, P., Kari, J.: Decidability in group shifts and group cellular automata. In: Esparza, J., Král', D. (eds.) 45th International Symposium on Mathematical Foundations of Computer Science, MFCS 2020. Volume 170 of LIPIcs, pp. 12:1–12:13. Schloss Dagstuhl - Leibniz-Zentrum für Informatik (2020)
4. Blanchard, F., Tisseur, P.: Some properties of cellular automata with equicontinuity points. Annales de l'I.H.P. Probabilité et statistiques **36**(5), 569–582 (2000)

5. Boyle, M., Kitchens, B.: Periodic points for onto cellular automata. Indagationes Mathematicae **10**, 483–493 (1999)
6. Ceccherini-Silberstein, T., Coornaert, M.: Cellular Automata and Groups. Springer Monographs in Mathematics. Springer, Berlin Heidelberg (2010)
7. Dennunzio, A., Formenti, E., Grinberg, D., Margara, L.: Additive cellular automata over finite abelian groups: Topological and measure theoretic properties. In: Rossmanith, P., Heggernes, P., Katoen, J.-P. (eds.) 44th International Symposium on Mathematical Foundations of Computer Science, MFCS 2019
8. Fiorenzi, F.: Cellular automata and strongly irreducible shifts of finite type. Theor. Comput. Sci. **299**(1), 477–493 (2003)
9. Kari, J.: Reversibility and surjectivity problems of cellular automata. J. Comput. Syst. Sci. **48**(1), 149–182 (1994)
10. Kari, J.: The nilpotency problem of one-dimensional cellular automata. SIAM J. Comput. **21**(3), 571–586 (1992)
11. Kari, J., Ollinger, N.: Periodicity and immortality in reversible computing. In: Ochmański, E., Tyszkiewicz, J. (eds.) Mathematical Foundations of Computer Science 2008, MFCS 2008, volume 5162 of Lecture Notes in Computer Science, pp. 419–430. Springer (2008)
12. Kitchens, B., Schmidt, K.: Automorphisms of compact groups. Ergodic Theory Dyn. Syst. **9**(4), 691–735 (1989)
13. Lukkarila, V.: Sensitivity and topological mixing are undecidable for reversible one-dimensional cellular automata. J. Cellular Automata **5**(3), 241–272 (2010)
14. Manzini, G., Margara, L.: A complete and efficiently computable topological classification of d-dimensional linear cellular automata over $Z_m$. Theor. Comput. Sci. **221**(1), 157–177 (1999)
15. Margara, L.: On some topological properties of linear cellular automata. In: Kutyłowski, M., Pacholski, L., Wierzbicki, T. (eds.) Mathematical Foundations of Computer Science 1999, MFCS 1999, volume 1672 of Lecture Notes in Computer Science, pp. 209–219. Springer (1999)
16. Sablik, M., Theyssier, G.: Topological dynamics of 2d cellular automata. In: Beckmann, A., Dimitracopoulos, C., Löwe, B. (eds.) Logic and Theory of Algorithms, 4th Conference on Computability in Europe, CiE 2008, volume 5028 of Lecture Notes in Computer Science, pp. 523–532. Springer (2008)
17. Sato, T.: Decidability for some problems of linear cellular automata over finite commutative rings. Inform. Process. Lett. **46**(3), 151–155 (1993)
18. Wang, H.: Proving theorems by pattern recognition—II. Bell Syst. Tech. J. **40**(1), 1–41 (1961)

# Revisiting Logic Obfuscation Using Cellular Automata

**Akashdeep Saha, Hrivu Banerjee, Rajat Subhra Chakraborty, and Debdeep Mukhopadhyay**

**Abstract** Logic obfuscation has evolved as a promising countermeasure against IP piracy. The Finite State Machine (FSM) is often obfuscated in a sequential circuit using suitable strategies. One such strategy proposed to obfuscate each state transition of the FSM using a class of non-group additive cellular automata (CA) called $D1 * CA$ and $D1 * CA_{dual}$. This CA-based obfuscation strategy conceals the FSM states, providing high testability hence eliminating the requirement of any scan-based Design-for-Testability techniques. However, utilizing the information leaked by the implemented FSM observable externally, an end-to-end attack strategy (named ORACALL) was proposed. It could extract the secret key for each transition of the CA-based obfuscated FSM along with the CA state encodings of the FSM states. In this work, we investigate the root cause of the success of ORACALL on a CA-based obfuscation strategy. Utilizing those findings, we propose a couple of mitigation techniques by appending non-linearity to the existing CA structure along with a slight modification of the $D1 * CA$ rule vector. Experimental validation proves that these simple yet effective countermeasures could thwart ORACALL while preserving the elegance of the underlying structure of the CA-based obfuscation technique with minimal overhead.

**Keywords** Logic locking · Cellular automata · ORACALL · Non-linearity

A. Saha · R. S. Chakraborty · D. Mukhopadhyay (✉)
Indian Institute of Technology, Kharagpur, West Bengal, India
e-mail: debdeep@cse.iitkgp.ac.in

A. Saha
e-mail: akashdeep@iitkgp.ac.in

R. S. Chakraborty
e-mail: rschakraborty@cse.iitkgp.ac.in

H. Banerjee
Indian Institute of Engineering Science and Technology Shibpur,
Howrah, West Bengal, India
e-mail: banerjeehrivu21@gmail.com



S. Das and G. J. Martinez (eds.), *Proceedings of First Asian Symposium on Cellular Automata Technology*, Advances in Intelligent Systems and Computing 1425,
https://doi.org/10.1007/978-981-19-0542-1_3

## 1 Introduction

Over the past few years, the majority of the integrated circuit (IC) manufacturing enterprises are shifting to fabless operation, consequently exporting their designs to offshore foreign foundries. It helps them save resources like the cost of IC manufacturing and the time to market. However, as a flip side, these foreign laboratories are often untrusted and indulge in various malicious activities like IP piracy, design counterfeiting, and illegal over-production. A few known counter-measures are logic locking [1–3], IC metering [4], watermarking [5], camouflaging [6], etc. As mentioned earlier, logic locking has emerged as the most effective countermeasure against these threats. *Logic locking* can be defined as the act of interweaving an additional key-based logic into the original circuit. The resultant circuit behaves correctly upon applying the secret key or a key sequence. The secret key is often stored in a secure non-volatile memory from which it is read to unlock the circuit on power-up.

An adversary always tries to extract the secret key maliciously to pirate or tamper with the IC design. One such adversarial attack is the Boolean satisfiability-based (SAT) attack [7]. It was successful in unlocking all the existing logic locking techniques then. The attack works by eliminating the incorrect keys from the entire keyspace. The key/s left behind after pruning the keyspace are the correct key/s. To execute a SAT-based attack, an adversary requires (i) a functional chip acquired from the market and (ii) a reverse-engineered locked netlist obtained from a locked GDSII design database. The attack finds out the distinguishing input patterns (DIPs), eliminating incorrect key/s. It is to be noted that a DIP can be defined as an input pattern that produces different outputs (at most, one of which is correct) upon the application of different keys. Over the years, logic locking has evolved to produce SAT-based attack resilient techniques. These techniques can be broadly categorized into two classes, namely (i) low output corruptibility based constructions [8, 9] and (ii) SAT-Hard constructions [10, 11]. Low output corruptibility-based constructions utilize the concept of single-point function to corrupt the output of a single (or very few) input. Hence, searching for the DIPs to eliminate all incorrect keys using a SAT solver reduces to brute force search depending on the key complexity. On the other hand, the SAT-Hard constructions force the SAT solver to take exponential time to solve the successive SAT iterations, thus nullifying SAT-based attacks. The techniques SARLock and Anti-sat achieve SAT-based attack resiliency utilizing single-point functions.

However, the low output corruptibility of these schemes (single incorrect input-output pair for a wrong key to be precise) makes them vulnerable to bypass [12] and removal attacks [13]. The bypass attack produces a fully functional design by attaching an additional "bypass" circuitry that rectifies the single incorrect input-output pair. Likewise, the removal attack could unlock the obfuscated design by isolating and eliminating the protection logic from the locked netlist.

Stripped-functionality logic locking (SFLL-HD$^h$) [1] was proposed to mitigate the aforementioned attacks. It maintained a higher output corruption using a Hamming distance ($h$) based comparator between the applied key and the input to produce several incorrect input-output pairs. However, the functional analysis attack (FALL) [14] exploits the structural traces of the locked SFLL-HD$^h$ netlist to extract the hard-coded correct key.

The techniques mentioned above are primarily prevalent in the combinational setting. Securing sequential circuits largely focuses on obfuscating the state transition graph of the FSM. The basic idea is to insert additional states into the existing FSM state transition graph. The FSM enters a valid state and becomes functional upon applying the correct key sequence. Otherwise, the FSM gets lost in a labyrinth of invalid states from which it might be challenging to traverse back to a valid state. However, a simple attack strategy could be to bypass these additional authentication states to reach an original FSM state directly. This strategy was exploited to devise several attack strategies [15, 16] to mitigate this FSM obfuscation scheme successfully. The SAT-based attack can also be extended to sequential circuits if the scan-based DfT architecture is present.

A Cellular Automaton (CA) is a system of "cells" with specific characteristics, elaborated in detail in the following sections. A CA-based obfuscation strategy using non-group additive CA has been introduced in [17]. It proposes to obfuscate each transition of the FSM using a single-bit secret key. The original FSM states have been realized using a special type of CA—$D1 * CA$ and its counterpart $D1 * CA_{dual}$ [18]. This technique further proposes to obfuscate the state transition logic using XOR/XNOR key gates [2]. Hence, this obfuscation strategy comprising a combination of CA and XOR/XNOR key gates is expected to offer protection against various logic-locking attacks. Another advantage of using the aforementioned CA-based obfuscation is that it eliminates the requirement of scan-based DfT [18]. Hence, various attacks utilizing the scan-based DfT are by default mitigated. A sequence of externally observable leftmost CA cell values and the primary output are validated with the corresponding fault-free counterpart for testing purposes.

The work in [19] proposes an end-to-end attack algorithm (named ORACALL) that could successfully extract the secret key along with the CA state encodings of each FSM state. This work exploits the information leaked from the externally observable CA-cell values (unique for each state) utilized in its high fault coverage testing technique. This attack [19] also demonstrated that the secret key for the combinational interrupt logic circuit could also be revealed from the externally observable sequence.

In this work, we propose a countermeasure by modifying the CA-based obfuscation such that the aforementioned vulnerability no longer exists and the ORACALL attack could be prevented. The basic idea of the countermeasure is to incorporate non-linearity into the $D1 * CA$ structure and modify the rule vector or the characteristic matrix of the existing $D1 * CA$. However, these modifications do not hamper the inherent properties of the underlying $D1 * CA$. The experimental evaluation suggests that the ORACALL attack could be mitigated since the externally observable leftmost CA cell sequence of the modified $D1 * CA$ does not leak the state bits of the encoded CA states.

The rest of the paper is organized as follows: Sect. 2 throws light on the basics of CA. Sect. 3 elaborates the CA-based FSM obfuscation technique [17] and the

ORACALL attack on the same [19]. In Sect. 4, we introduce the proposed countermeasures utilizing non-linearity. Experimental validation of the countermeasures is provided in Sect. 5. We conclude in Sect. 6.

## 2 Background

In this section, we discuss in detail the $D1 * CA$ and its properties.

### 2.1 The Cellular Automata (CA)

A CA consists of a distinct lattice of cells [18]. Each cell has a storage element (usually a D flip-flop) and a subsequent state logic, a combinational circuit that calculates the next state of the existing CA cell from a linear combination of the present state of its two neighbors and itself. The rule of the CA determines the next state logic. If the rule of the CA consists of only XOR/XNOR functions, it is called an additive CA, whereas a nonadditive CA consists of AND/OR functions in its rule. All the states belong to a cycle created during the state transition in a group CA. A non-group CA consists of both cyclic and acyclic. A $n$-cell CA, can be represented by a vector as $X_t = [q_0(t), \cdots, q_{n-1}(t)]^T$, where $q_i(t)$ represents the $i$th cell at $t$th time instance. Therefore, the state of the $i$th cell at $(t+1)$th time instance (i.e., the next state) can be represented as $q_i(t+1) = f[q_{i-1}(t), q_i(t), q_{i+1}(t)]$. Here $f$ is the state transition function, usually known as the *rule of the CA*. The equation relates the next state and the present state of CA as follows:

$$X_{t+1} = T \cdot X_t$$

$T$ is a Boolean matrix with $n \times n$ dimensions. The $i$th row specifies the neighborhood dependency of the $i$th cell; the operation involved is a usual matrix multiplication with modulo-2 addition. $T$ is called the characteristic matrix of the CA, as shown below for the rule vector <90, 102, 102, $\cdots$, 102>.

$$T = \begin{pmatrix} 0\,1\,0\,0 \cdots 0\,0 \\ 0\,1\,1\,0 \cdots 0\,0 \\ 0\,0\,1\,1 \cdots 0\,0 \\ \vdots \quad\;\; \cdots \;\; \vdots \\ 0\,0\,0\,0 \cdots 1\,1 \\ 0\,0\,0\,0 \cdots 0\,1 \end{pmatrix}$$

## 2.2 A Non-group Additive CA: $D1 * CA$

A class of non-group additive CA [18] of *depth-one* is $D1 * CA$.[1] The next state logic of $D1 * CA$ is given by the rule vector <90, 102, 102, $\cdots$, 102>, where:

$$\textbf{Rule 90}: q_i(t+1) = q_{i-1}(t) \oplus q_{i+1}(t) \quad \textbf{Rule 102}: q_i(t+1) = q_i(t) \oplus q_{i+1}(t)$$
$$\textbf{Rule 165}: q_i(t+1) = \overline{q_{i-1}(t) \oplus q_{i+1}(t)} \quad \textbf{Rule 153}: q_i(t+1) = \overline{q_i(t) \oplus q_{i+1}(t)}$$

$D1 * CA_{dual}$ is obtained by complementing the XOR function in the next state logic for any CA cell of the existing $D1 * CA$. $D1 * CA_{dual_A}$ has been obtained by complementing the leftmost CA cell of the $D1 * CA$. It can be represented by the vector rule <165, 102, 102, $\cdots$, 102>. Likewise, $D1 * CA_{dual_B}$ is obtained by complementing the third leftmost CA cell. It can be represented by the vector rule <90, 102, 153, $\cdots$, 102>. The state transitions of 4-cell $D1 * CA$ and $D1 * CA_{dual_A}$ are illustrated in Fig. 1(a) and (b), respectively. Fig. 2 highlights the overall architecture.

## 2.3 Properties of $D1 * CA$

This section highlights a few $D1 * CA$ properties that has been used in embedding the FSM [17]. The elaborate proof for each property is present in [18].

1. All $D1 * CA$ reachable states are cyclic states.
2. All $D1 * CA$ cyclic states have an even number of 1's in their binary representation, i.e., they are *even-weighted* states.
3. A $D1 * CA$ having number of cells ranging from $(2^{k-1} + 2)$ to $(2^k + 1)$ has a *maximum transition cycle length* of $m = 2^k$.
4. For a $n$-cell $D1 * CA$ $(n \geq 3)$, let $m$ be the maximum transition cycle length in the state transition diagram, then the set of transition cycle lengths present is exactly the set of factors of $m$.

**Proposition 1** *Starting from any reachable (even-weighted) state $S_r$, if the machine makes $m$ or any multiple of $m$ number of $D1^*CA$ mode transitions, it returns back to state $S_r$.*

The proposition follows from Properties 1, 2 and 4.

[1] *depth* here is the minimum number of state transitions required from an unreachable state to reach the cyclic state

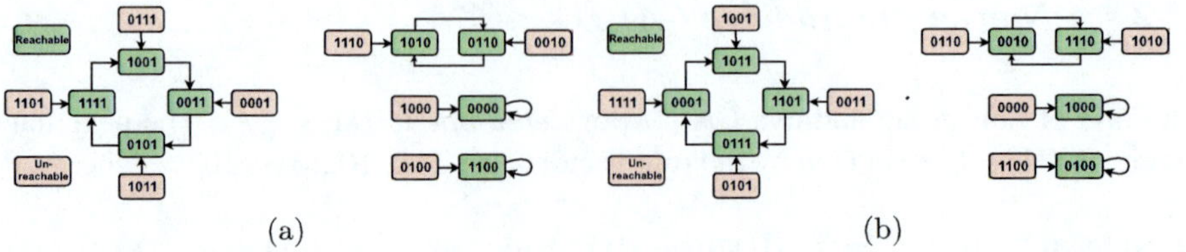


**Fig. 1** State transition graph of a 4-cell **a** $D1 * CA$ and **b** $D1 * CA_{dual_A}$

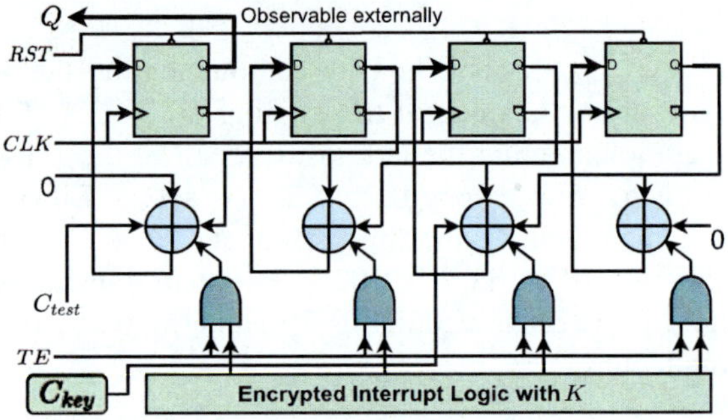


**Fig. 2** The overall architecture of an obfuscated CA-based FSM [19], which can be used for realising $D1 * CA$, $D1 * CA_{dual_A}$ and $D1 * CA_{dual_B}$ mode of transitions, depending on the combination of $TE$, $C_{test}$ and $C_{key}$ inputs

## 2.4 FSM Synthesis using $D1 * CA$

The $D1 * CA$ has been utilized as the basic building block [18] in realizing the FSM [17]. A $n$-state FSM requires a $\lceil \log_2 n \rceil$ cell $D1 * CA$. Some FSM state transitions can be mapped directly into CA state transitions, while others cannot. An additional combinational logic termed the *Interrupt Logic* has to be utilized for mapping these unmapped states. The synthesized circuit consists of two main blocks: (1) the CA block [18], and (2) a combinational logic consisting of *Interrupt Logic*; and the primary output logic. The output of the *Interrupt Logic* is known as the *Interrupt Vector* ($IV$), which is an input to the XOR gates associated with the state transitions of the CA cells.

For the state transitions directly realizable using the CA state transitions, the $IV$ value is zero. The CA is said to run in an autonomous mode in such a situation. Some $IV$ bits are set to '1', thus forcing them to realize the required transition in the CA. The CA runs in *Interrupt* mode for a non-zero $IV$ vector. The following equation gives the $IV$ value corresponding to the FSM transition $S_i \rightarrow S_k$ is:

$$IV_{i \rightarrow k} = S_j \oplus S_k$$

where $S_i$ is the initial state and $S_j$ is the *autonomous* successor of the FSM state $S_i$.

# 3 FSM Obfuscation using Cellular Automata

## 3.1 The Obfuscation Technique

The authors in [17] have utilized the aforementioned FSM synthesis scheme to execute both $D1 * CA$ and $D1 * CA_{dual_B}$ state transitions in a FSM, based on the choice of *Control Input* $C_{key}$ (Fig. 2). $C_{key} = 0$ realizes a $D1 * CA$ mode transition, while $C_{key} = 1$ induces a $D1 * CA_{dual_B}$ mode transition in the FSM. In a locked IC, the $C_{key}$ for each transition is the secret key and is supplied from a look-up table that is stored in the tamper-proof memory inside the chip.

The two possibilities for implementing the FSM transition $S_i \rightarrow S_k$ using the CA structure, where the states $S_i$ (the present FSM state) and $S_k$ are encoded to states $F^i$ and $F^k$ using the CA, respectively, are as follows: (a) using $D1 * CA$ ($C_{key} = 0$) such that $IV = F^k \oplus F^j$, or (b) using $D1 * CA_{dual_B}$ ($C_{key} = 1$) such that $IV = F^k \oplus F^l$. CA states $F^j$ and $F^l$ are termed the *autonomous successors* of $F^i$ in $D1 * CA$ and $D1 * CA_{dual_B}$, respectively. Table 1 summarizes the $IV$ values for two possible realizations of the transition $S_i \rightarrow S_k$.

The $IV$ is generated based on the mode of operation of the CA (either $D1 * CA$ or $D1 * CA_{dual_B}$), which in turn is controlled by the $C_{key}$. The $IV$ for the $i$th transition can be represented as:

$$IV_i = \alpha(PI_i, S_i, C_{key_i})$$

where the $IV$ values are generated based on the $PI_i$ (the primary input), $S_i$ (the present state) and the correct key $C_{key_i}$. The combinational *Interrupt Logic* circuit is also encrypted with the XOR/XNOR-based logic locking scheme [2]. Hence, the encrypted $IV$ can be represented as:

$$IV_i = \alpha(PI_i, S_i, C_{key_i}, K)$$

where the $k$-bit correct key $(K)$ is used to encrypt the *Interrupt Logic*.

**Table 1** The two possible $IV$ values for transitions

| FSM transition | Mode | CA realization | $IV$ | $C_{key}$ |
|---|---|---|---|---|
| $S_i \rightarrow S_k$ | *Interrupt* | $D1 * CA$ | $F^k \oplus F^j$ | 0 |
| | | $D1 * CA_{dual_B}$ | $F^k \oplus F^l$ | 1 |

### 3.1.1 Testing the FSM

The synthesized FSM operates in two modes: (a) *FSM Mode* and (b) *Test Mode*. In *FSM Mode*, the state transitions occur based on $C_{key}$ and $IV$ whereas, the *Test Mode* masks the $IV$ and transition is performed in *Autonomous Mode* using $D1 * CA$ and $D1 * CA_{dual_A}$. The testing strategy for the synthesized FSM makes use of the following unique state transition property [18]:

**Property 1** If $S_i$ is any random starting state, having an even/odd number of '1's, then in fault-free scenario, an Autonomous Mode run of $(m + 1)$ clock cycles in $D1 * CA_{dual_A}/D1 * CA$, followed by another Autonomous Mode run of $(m - 1)$ clock cycles in $D1 * CA/D1 * CA_{dual_A}$ should return the FSM back to the same starting state $S_i$, where $m$ is the maximum transition cycle length of the underlying $D1 * CA$-based machine.

Utilizing this property, any FSM state transition $S_i \rightarrow S_k$ can be verified by executing $S_i \rightarrow S_k$ in *FSM mode* in 1 clock cycle followed by an execution of *Autonomous Mode* transitions for $2m$ clock cycles (Property 1). The output sequence of leftmost CA cell (externally observable) and the primary output is observed during these $(2m + 1)$ clock cycles. The synthesized FSM is assumed to be correct if these two sequences match their fault-free counterpart. This testing approach provides high fault coverage [18] while eliminating the design overhead and security vulnerability associated with the DfT techniques.

## 3.2 *The ORACALL Attack Methodology*

The Lemma 1 in [19] proved that sequence $Q_{S,m+2}$ is unique for each CA state. By exploiting it, an adversary can extract the state encodings of a transition [19]. S/he can then calculate the two possible $IVs$, $IV_1$ (for $C_{key} = 0$) and $IV_2$ (for $C_{key} = 1$). Comparing $IV_1$ and $IV_2$ with the correct $IV$ ($IV_k$), reveals the actual $C_{key}$. Algorithms 1 and 2 in [19] summarizes the ORACALL attack upon the two variants (Case 1 and Case 2 in the following subsection) of CA-guided obfuscation technique. We leave out the complete description for brevity and refer interested readers to [19].

### 3.2.1 Adversary Model

The ORACALL attack requires an activated chip acting as the oracle and a locked design netlist ($N$). The adversary has the capability to extract the state elements and the combinational state-transition logic. S/he can also simulate the same with desired input and keys.

### 3.2.2 Case 1: The Interrupt Logic Unencrypted

Let us assume that each FSM transition $S_i \rightarrow S_j$ is encoded to CA states $F^i$ and $F^j$, respectively. The secret key ($C_{key}$) for each transition is to be extracted. It derives the state encodings of $F^i$ (CA encoded present state), from $Q_{S_i,m+2}$. The FSM transition $S_i \rightarrow S_j$ is executed with an appropriate primary input ($PI$).

Likewise, $F^j$ can be obtained from $Q_{S_j,m+2}$. The correct $IV$ ($IV_k$) for the transition $S_i \rightarrow S_j$ can be extracted by simulating the $IV$ netlist with the $PI$ and the encoding of $F^i$. Using $F^i$ encoding, its $D1 * CA$ and $D1 * CA_{dual_B}$ successors are calculated. Followed by calculation of the two $IVs$. Fig. 3 highlights the $C_{key}$ extraction process for the transition $S_0 \rightarrow S_1$. If $IV_k$ is equal to $IV_1$ (in Fig. 3), it is inferred that the $C_{key}$ is equal to 0 else the correct $C_{key}$ is 1 for the transition $S_i \rightarrow S_j$.

### 3.2.3 Case 2: Encrypted Interrupt Logic

The correct key ($K$) for the encrypted *Interrupt Logic* ($IV$) circuit is derived using SAT-based attack [7]. As already stated, the SAT-based attack requires a functional chip and a locked netlist ($N$). However, in this case, the unlocked circuit (oracle) for the locked $IV$ netlist is not present. Hence, two oracles $B_1$ and $B_2$ (Fig. 3 inset) are constructed, which outputs $IV_1$ and $IV_2$, respectively. These oracles have the same initial state as the state transition and are simulated with $PI$ for that corresponding state transition. Using the locked $IV$ netlist ($N$) with present state same as that of the oracles), two keys $K_1$ and $K_2$ are produced, only one of which is correct.

To distinguish between $K_1$ and $K_2$, another FSM transition (say $S_m \rightarrow S_n$) is considered. The two possible $IVs$ for this transition are obtained following the aforementioned procedure. Further, simulating two instances of $N$ with $F^m$ and $K_1$, and with $F^m$ and $K_2$, respectively, two $IVs$ are produced. Only one matches with the correct $IV$ for $S_m \rightarrow S_n$. Therefore, the key that produces the correct $IV$ for $S_m \rightarrow S_n$ is identified as the correct key $K$.

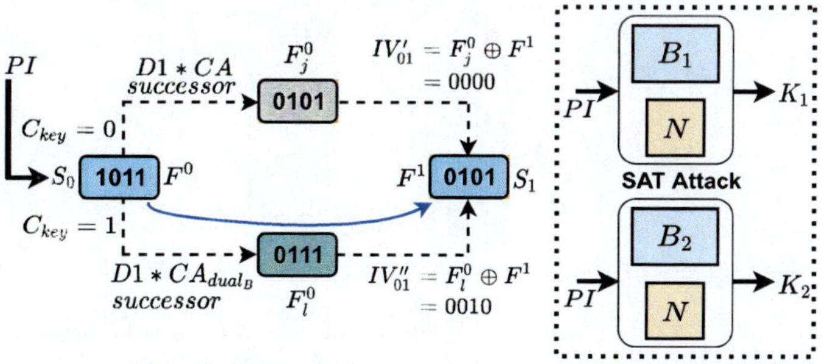


**Fig. 3** An illustration of the ORACALL attack methodology [19] on a state transition where the extracted state encodings are used to find out the two possible $IVs$ and subsequently reveal $K$ and $C_{key}$

## 4 Incorporating Non-linearity to Thwart ORACALL

The underlying $D1 * CA$ structure that has been utilized in obfuscating the FSM is linear since the next state can be obtained from the present state simply by matrix multiplication with a modulo-2 addition with the characteristic matrix. Further, each CA state ($S$) maps to a unique sequence $Q_{S,m+2}$ from which $S$ can be derived. Then $S$ is exploited to develop the subsequent portions of the ORACALL attack. Now, if $Q_{S,m+2}$ ceases to leak the state encodings of $S$, the ORACALL attack could be thwarted. *In this work, we harness this simple yet effective approach to successfully devise a couple of countermeasures (Case 1 and Case 2) against ORACALL.*

Initially, we propose to introduce non-linearity into the next state equation (Sect. 2.1) such that a non-linear CA structure is produced [20]. However, introducing non-linearity without hampering the structure of the linear $D1 * CA$ is difficult. We address this challenge by utilizing the well-established and cryptographically robust S-Box functions. S-Boxes are reversible non-linear Boolean functions, an integral component of block ciphers. This countermeasure has been elaborated in Case 1. Precisely, we have utilized a cascade of two well studied S-Box constructions—the PRESENT cipher S-Box ($\alpha$) [21] and CA-based S-Box ($\beta$) [22] to provide non-linearity. The modified next state equation can be represented as follows:

$$X_{t+1} = \alpha\beta T\beta^{-1}\alpha^{-1} \cdot X_t$$

We propose to use a cascade of two S-Boxes since we found that though a single layer of S-Box function ($\alpha^{-1}$ or $\beta^{-1}$) provides the required non-linearity and could prevent the information leakage, it still leaked information in some cases about the state encoding bits, however with a diminished probability compared to the original $D1 * CA$ in [17].

Finally, we propose to modify the rule vector of the $D1 * CA$ structure while using a single S-Box structure as the second countermeasure thwarting the ORACALL attack. This is elaborated in Case 2. The new characteristic matrix of the modified rule vector can be represented as:

$$T^* = \begin{pmatrix} 0\,1\,0\,0\,0 \cdots 0\,0 \\ 0\,1\,1\,1\,0 \cdots 0\,0 \\ 0\,0\,1\,1\,1 \cdots 0\,0 \\ \vdots \qquad\quad \cdots \quad \vdots \\ 0\,0\,0\,0\,0 \cdots 1\,1 \\ 0\,0\,0\,0\,0 \cdots 0\,1 \end{pmatrix}$$

Another reason for choosing the S-Boxes as the required non-linear function here is their reversibility (i.e., $\alpha^{-1}$ and $\beta^{-1}$ exist), such that the crucial properties of $D1 * CA$ (Sect. 2.3) are still satisfied. The non-linearity of both CA-based and PRESENT cipher S-Box is 4 [22].

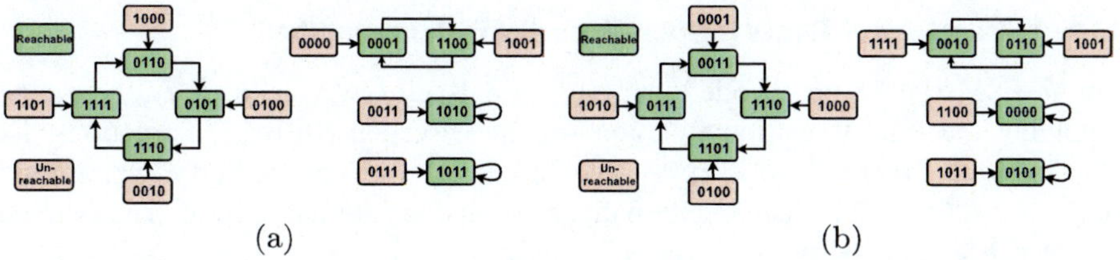


**Fig. 4** State transition graph of a 4-cell modified $D1 * CA$ with **a** characteristic matrix $T_1 = \alpha\beta T\beta^{-1}\alpha^{-1}$ and **b** characteristic matrix $T_r$

An important property of the CA-based obfuscation is that if $X$ is a reachable cyclic state, then after $m$ transitions, it should traverse back to $X$. Therefore, $T^m.X = X$. This property should not be affected by introducing the non-linearity or modification of the characteristic matrix. Let us assume that $Y$ be a reachable cyclic state in the modified $D1 * CA$ setting. The characteristic matrix for the modified $D1 * CA$ is $T_1 = \alpha\beta T\beta^{-1}\alpha^{-1}$. The non-linearity of $T_1$ is 2. Now, for $m$ transitions using $T$ of $D1 * CA$, we get:

$$\begin{aligned} T_1^m \cdot X &= (\alpha\beta T\beta^{-1}\alpha^{-1})^m \cdot X \\ &= \alpha\beta T\beta^{-1}\alpha^{-1}\alpha\beta T\beta^{-1}\alpha^{-1} \cdots (m\ times) \cdots T\beta^{-1}\alpha^{-1} \cdot X \\ &= \alpha\beta T^m\beta^{-1}\alpha^{-1} \cdot X \\ &= X \end{aligned}$$

Hence, the cyclic property remains intact for the Case 1 construction, even for the modified next state transition function, as illustrated in Fig. 4a. The same holds for the proposed Case 2 construction with $T^*$ in Fig. 4b.

Likewise, the sequence $Q_{S,m+2}$, which was unique for each CA state in $D1 * CA$ and revealed the CA state encodings, is no longer unique for the modified $D1 * CA$ and does not leak the CA state encodings (Cases 1 and 2). From the characteristic matrix ($T$) of $D1 * CA$, Rule 102 can be generalized according to the formula proposed in [19]. Thus, the first $n-$bits in the sequence $Q_{S,m+2}$ leak the $n-$bit encoding the CA state.

The state transition function of the modified cascaded non-linear $D1 * CA$ as elaborated in Case 1, can be described as follows: the CA state is initially substituted using the PRESENT S-Box ($\alpha$ function) and then by the CA-based S-Box ($\beta$ function). This is followed by multiplication with the characteristic matrix ($T$), and the obtained product is again substituted with inverse S-Box mapping of CA-based S-Box ($\beta^{-1}$) and then with PRESENT S-Box ($\alpha^{-1}$).

The state transition function of the modified rule $D1 * CA$ structure, as elaborated in Case 2, replaces $T$ in the aforementioned structure with $T^*$ while using a single S-Box function to impede ORACALL attack. Now, we move on to the cases illustrating the utilities of the two constructions.

**Case 1: Cascaded S-Boxes for non-linearity with unmodified T**

Let an FSM state $S$ be encoded using using a 4-cell CA state $F = f_1, f_2, f_3, f_4$. Applying the state transition function for the cascaded S-Box constructions, let $(T_1(F)) = Z$, where $Z = z_1, z_2, z_3, z_4$. Utilizing the testing property, where only the leftmost CA cell is visible externally, $z_1$ is visible externally for $Z$. The relation between $F$ and the output bit $z_1$ is as follows:

$$z_1 = f_2 f_3 f_4 \oplus f_3 f_4 \oplus f_4 \oplus f_1 f_2 f_3 \oplus f_2 f_3 \oplus f_3$$

It is to be noted that the output $z_1$ does not yield the input, or even if $f_1$ bit is revealed, all possible values of $f_2$, $f_3$ and $f_4$ bits are still possible.

Now, if a single S-Box (say only PRESENT S-Box) was utilized instead of the cascaded approach, the resultant state transition function becomes $T_p = \alpha T \alpha^{-1}$. The non-linearity of $T_p$ is 2. The relation between $F$ and $Z$ using $T_p$ is:

$$z'_1 = f_2 f_3 f_4 \oplus f_1 f_3 f_4 \oplus f_3 f_4 \oplus f_1 f_4 \oplus f_4 \oplus f_2 f_3 \oplus f_1 f_2 \oplus f_2$$

The bit $f_1$ is already known. If $f_1 = 0$, the above equation reduced to $f_2 f_3 f_4 \oplus f_3 f_4 \oplus f_4 \oplus f_2 f_3 \oplus f_4$. So, all possible values of $f_2$, $f_3$ and $f_4$ are possible, which is expected. However, if $f_1 = 0$, $z'_1 = f_2 f_3 f_4 \oplus f_2 f_3$. This equation can leak all the state remaining bits $(f_2, f_3, f_4)$ if $z'_1 = 1$, which is not desired. To mitigate this problem, we propose to use a cascade of two S-Boxes. Though non-linearity remains the same, the leakage could be avoided.

**Case 2: Single S-Box for non-linearity with modified T**

In this countermeasure, we modify the existing rule vector of $D1 * CA$ while utilizing a single S-Box layer for providing non-linearity. The characteristic matrix for the modified rule is $T^*$. The state transition function for the modified rule and single S-Box is $T_r = \alpha T^* \alpha^{-1}$. The non-linearity of $T_r$ is 2. Let $T_r(F) = W$, where $W = w_1, w_2, w_3, w_4$. The relation between $F$ and $w_1$, which is externally observable is:

$$w_1 = f_1 f_2 f_3 \oplus f_3 f_4 \oplus f_2 f_4 \oplus f_1 f_4 \oplus f_2 \oplus f_2 f_3 \oplus f_1 f_3 \oplus f_1$$

From the above equation it is evident that even if $f_1$ is known from the first bit of $Q_{S,m+2}$, the observable bit $w_1$ does not leak information about any of remaining bits of the CA state encoding $F$. Next, we move on to the experimental validations of the proposed countermeasures that could mitigate the ORACALL attack.

## 5 Experimental Results

In this section, we provide the experimental validations of the proposed countermeasures. We have utilized four different non-linear 4 * 4 S-Box functions—the PRESENT cipher S-Box and three CA-based S-Boxes belonging to different S-Box

**Table 2** Overhead in terms of number of gates for various S-Box structures

| S-Box | PRESENT | CA (1, 4, 2) | CA (3, 3, 3) | CA (3, 4, 2) | Cascaded S-Box |
|---|---|---|---|---|---|
| Overhead | 84 | 66 | 80 | 91 | 71 |

**Table 3** The ORACALL attack can be mitigated using the two countermeasures. Here, ✗ implies that the ORACALL is entirely thwarted, whereas ✗* signifies that though ORACALL is mitigated, some information leakage remains

| Circuit | # I/O | # States | # Transitions | S-Box type | Non-linearity | ORACALL Attack |
|---|---|---|---|---|---|---|
| bbara | 4/2 | 10 | 60 | PRESENT | 2 | ✗* |
| cse | 7/7 | 16 | 91 | PRESENT | 2 | ✗* |
| bbara | 4/2 | 10 | 60 | CA (1, 4, 2) | 2 | ✗* |
| cse | 7/7 | 16 | 91 | CA (1, 4, 2) | 2 | ✗* |
| bbara | 4/2 | 10 | 60 | Cascaded | 2 | ✗ |
| cse | 7/7 | 16 | 91 | Cascaded | 2 | ✗ |
| bbara | 4/2 | 10 | 60 | PRESENT with T* | 2 | ✗ |
| cse | 7/7 | 16 | 91 | PRESENT with T* | 2 | ✗ |

classes $(1, 4, 2)$,[2] $(3, 3, 3)$ and $(3, 4, 2)$. The mapping equations were obtained using the Sage tool. The countermeasures were evaluated using circuits from the MCNC benchmark suite. The CA-based FSM obfuscation strategy and the ORACALL attack were performed using circuits from the MCNC benchmark. The modified state transition function and the state encodings have been automated using Python programming language. The ABC synthesis and verification framework tool [23] was utilized for further optimization purpose. The open-source SAT-based attack tool in [7] was utilized for performing the required SAT attack in ORACALL. The experiments were performed on an Ubuntu workstation having four core Intel i7-7700 CPU executing at 3.6 GHz equipped 16 GB of main memory.

Table 2 shows the incurred overhead in terms of additional gates owing to the S-Box function. Incurred overhead due to the countermeasures being minimal. The non-linearity of all these S-Boxes is 4. The Cascaded S-Box column shows diminished gate count compared to PRESENT and CA-based S-Box since we have used their cascaded mappings to create an equivalent circuitry. Table 3 shows the various methods of introducing non-linearity into the CA-based design. The first four rows show the effect of utilizing a single S-Box function in providing non-linearity. However, in these cases, there are possibilities of information leakage. However, the amount of leakage, which in the case of ORACALL could leak the entire state bit, is

[2] *Class* $(a, b, c)$ infers that the S-Box equation comprises of $a$ cubic terms, $b$ quadratic terms, and $c$ linear terms, respectively.

now diminished. Hence, obtaining the CA state encodings becomes challenging. In the subsequent rows in Table 3 show the ORACALL attack on cascaded S-Box structure (Case 1) and PRESENT S-Box with modified characteristic matrix (Case 2). In these cases, there is no information leakage. The non-linearity remains the same and the structure of the $D1 * CA$ is preserved (Fig. 4). They could successfully mitigate the ORACALL attack.

## 6 Conclusion

The ORACALL attack exposed severe vulnerabilities in the CA-based FSM obfuscation technique, which could leak the entire CA state encodings and the secret key. In this work, we have proposed two countermeasures that could mitigate the vulnerabilities while preserving the overall structure of the existing CA-based scheme. The proposed countermeasures are based on introducing non-linearity into the otherwise linear $D1 * CA$ to prevent information leakage. The countermeasures are simple yet efficient in mitigating the ORACALL attack at minimal overhead. An exciting direction for future work in the FSM obfuscation technique could be developing more such obfuscation techniques utilizing the rich structures of CA.

## References

1. Yasin, M., Sengupta, A., Nabeel, M. T., Ashraf, M., Rajendran, J., Sinanoglu, O.: Provably-secure logic locking: from theory to practice. In: Proceedings of the 2017 ACM SIGSAC Conference on Computer and Communications Security, pp. 1601–1618 (2017)
2. Roy, J.A., Koushanfar, F., Markov, I.L.: Ending piracy of integrated circuits. Computer **43**(10), 30–38 (2010)
3. Yasin, M., Mazumdar, B., Rajendran, J.J., Sinanoglu, O.: TTLock: Tenacious and traceless logic locking. In: 2017 IEEE International Symposium on Hardware Oriented Security and Trust (HOST), p. 166. IEEE Computer Society (2017)
4. Alkabani, Y., Koushanfar, F.: Active hardware metering for intellectual property protection and security. In: USENIX Security Symposium, pp. 291–306 (2007)
5. Kahng, A.B., Lach, J., Mangione-Smith, W.H., Mantik, S., Markov, I.L., Potkonjak, M., ... & Wolfe, G.: Watermarking techniques for intellectual property protection. In: Proceedings of the 35th Annual Design Automation Conference, pp. 776–781 (1998)
6. Rajendran, J., Sam, M., Sinanoglu, O., Karri, R.: Security analysis of integrated circuit camouflaging. In: Proceedings of the 2013 ACM SIGSAC Conference on Computer & Communications Security, pp. 709–720 (2013)
7. Subramanyan, P., Ray, S., Malik, S.: Evaluating the security of logic encryption algorithms. In: 2015 IEEE International Symposium on Hardware Oriented Security and Trust (HOST), pp. 137–143. IEEE (2015)
8. Xie, Y., Srivastava, A.: Anti-sat: mitigating sat attack on logic locking. IEEE Trans. Comput.-Aided Des. Integr. Circ. Syst. **38**(2), 199–207 (2018)
9. Yasin, M., Mazumdar, B., Rajendran, J. J., Sinanoglu, O. SARLock: SAT attack resistant logic locking. In: 2016 IEEE International Symposium on Hardware Oriented Security and Trust (HOST), pp. 236–241. IEEE (2016)

10. Saha, A., Saha, S., Chowdhury, S., Mukhopadhyay, D., Bhattacharya, B.B.: LoPher: SAT-hardened logic embedding on block ciphers. In: 2020 57th ACM/IEEE Design Automation Conference (DAC), pp. 1–6. IEEE (2020)
11. Kamali, H. M., Azar, K. Z., Homayoun, H., Sasan, A.: Full-lock: Hard distributions of sat instances for obfuscating circuits using fully configurable logic and routing blocks. In: Proceedings of the 56th Annual Design Automation Conference, pp. 1–6 (2019)
12. Xu, X., Shakya, B., Tehranipoor, M.M., & Forte, D.: Novel bypass attack and BDD-based tradeoff analysis against all known logic locking attacks. In: 2017 International Conference on Cryptographic Hardware and Embedded Systems, pp. 189–210. Springer, Cham (2017)
13. Yasin, M., Mazumdar, B., Sinanoglu, O., Rajendran, J.: Removal attacks on logic locking and camouflaging techniques. IEEE Trans. Emerg. Top. Comput. **8**(2), 517–532 (2017)
14. Sirone, D., Subramanyan, P.: Functional analysis attacks on logic locking. IEEE Trans. Inform. Forensics Sec. **15**, 2514–2527 (2020)
15. Meade, T., Zhao, Z., Zhang, S., Pan, D., Jin, Y.: Revisit sequential logic obfuscation: attacks and defenses. In: 2017 IEEE International Symposium on Circuits and Systems (ISCAS), pp. 1–4. IEEE (2017)
16. Fyrbiak, M., Wallat, S., Déchelotte, J., Albartus, N., Böcker, S., Tessier, R., Paar, C.: On the difficulty of FSM-based hardware obfuscation. In: IACR Transactions on Cryptographic Hardware and Embedded Systems, pp. 293–330 (2018)
17. Karmakar, R., Jana, S.S., Chattopadhyay, S.: A cellular automata guided obfuscation strategy for finite-state-machine synthesis. In: 2019 56th ACM/IEEE Design Automation Conference (DAC), pp. 1–6. IEEE (2019)
18. Chaudhuri, P.P., Chowdhury, D.R., Nandi, S., Chattopadhyay, S.: Additive Cellular Automata: Theory and Applications, vol. 43. Wiley (1997)
19. Saha, A., Banerjee, H., Chakraborty, R.S., Mukhopadhyay, D.: ORACALL: an oracle-based attack on cellular automata guided logic locking. IEEE Trans. Comput.-Aided Des. Integr. Circ. Syst. (2021)
20. Mukhopadhyay, D.: Design and analysis of cellular automata based cryptographic algorithms (2007). https://cse.iitkgp.ac.in/~debdeep/research/thesis.pdf
21. Bogdanov, A., Knudsen, L.R., Leander, G., Paar, C., Poschmann, A., Robshaw, M.J., Seurin, Y., Vikkelsoe, C.: PRESENT: an ultra-lightweight block cipher. In: International Workshop on Cryptographic Hardware and Embedded Systems, pp. 450–466. Springer (2007)
22. Ghoshal, A., Sadhukhan, R., Patranabis, S., Datta, N., Picek, S., Mukhopadhyay, D.: Lightweight and side-channel secure 4×4 s-boxes from cellular automata rules. In: IACR Transactions on Symmetric Cryptology, pp. 311–334 (2018)
23. Berkeley Logic Synthesis and Verification Group: ABC: a system for sequential synthesis and verification, release 10216 (2018). http://www.eecs.berkeley.edu/~alanmi/abc/

# Accepted Papers

# Grade Estimation of Mineral Resources: An Application of Cellular Automata

**Soumyadeep Paty and Supreeti Kamilya**

**Abstract** Mineral industry contributes a major part of economical growth to a nation. Therefore, it is important to understand the concentration measurement of minerals in different locations of earth's crust. The concentration refers to as the grade values of minerals. A block model is a simplified geometrical representation of a mineral deposit. There are several methods to find out the grade values of each block, such as, inverse distance method, copula, krigging etc. Krigging is a popular method to find out the grade values. However, number of mathematical calculations increase with the increasing number of sample points and thus the computational complexity becomes high. Here, we use a mathematical tool, *cellular automata (CAs)* where each cell is represented as a block. Using cellular automata, we here find out the grade values with much less computations. 2 dimensional CAs are used in this study where the local rule is the ordinary krigging estimator function using spherical variogram model. Also, in this study we deal with multiple horizontal slices of 3 dimensional mineral deposits. It can be seen that the CA based estimation is same as ordinary krigging based estimation, however, in a much simpler way.

**Keywords** Block model · Krigging · Cellular automata · Variogram model

## 1 Introduction

Contribution of mineral industry for an economic growth of a nation is undeniable. Mineral is a naturally occurring substance that is found in the earth's crust. To mention a few, hematite is a mineral of iron, bauxite is a mineral of aluminium, chalcopyrite is a mineral of copper, galena is a mineral of lead. Obtaining the metallic or non-metallic element involves two steps method: mining and mineral processing. To properly mine and process the mineral, prior *grade* estimation of mineral is required. Grade of a

S. Paty (✉)
Department of Mining Engineering, Kazi Nazrul University, Asansol, West Bengal, India

S. Kamilya
Department of Computer Science and Engineering, BIT Mesra, Ranchi, Jharkhand, India

 
S. Das and G. J. Martinez (eds.), *Proceedings of First Asian Symposium on Cellular Automata Technology*, Advances in Intelligent Systems and Computing 1425,
https://doi.org/10.1007/978-981-19-0542-1_4

mineral indicates the concentration of element (metal/non-metal) it contains and it is represented in percentage.

A *block model* is a simplified representation of a mineral deposit in a three dimensional space. In general, it consists of multiple equal sized cubical blocks that are arranged in a 3 dimensional space. Each block contains a grade value. It is not possible to directly determine the grade value of each and every block, before the mining operation. However, grade values of a small number of blocks can be identified by means of *geological exploration*. The grade value is considered as a spatial data where every data value is connected to a spatial location (or co-ordinates), i.e., they are spatially correlated. Therefore, it is possible to estimate the grade values of unknown blocks by the help of the small number of known blocks. Because, the known grade values carry information of the unknown ones [12]. Geostatistics is a discipline that deals with such spatially correlated data. The study on spatially correlated data has been popularized by Krige [5]. Later, the theory on regionalized variables (that deals with spatial data) has been mathematically formalized by Matheron [7, 8]. With the help of the mathematical formulation, the spatial data can be represented by regionalized structures and also can be analyzed by statistical methods. In 1990, Kim has proposed spatial characteristics of the regionalized variables [4]. According to the author, regionalized variables are highly variable in space, however, they are not completely independent.

There are different geostatistical methods to regionalize the spatial variables, such as correlation coefficient, covariance and variogram. Variogram is the most commonly used method to estimate mineral grades. A method has been proposed in [7] that estimates regionalized variables. It is based on known data points and variogram model. The method is commonly known as *Krigging* which has been named after Dan Krige. Krigging method involves a number of mathematical equation solving techniques that is computationally expensive. A simplified tool, *cellular automata (CAs)*, is here with us that can reduce the computational complexity while estimating grade values of minerals. A very few work has been reported that use cellular automata model in mineral industry [6]. In the paper [6], the authors have modelled the size (thickness) of a coal deposit. However, in our work, we estimate the grade of minerals using CAs.

A cellular automaton (CA) is a D-dimensional lattice of cells where each cell is in a particular state. A cell changes its state depending on a local function (or rule) and its neighbors. In 1940s, J. von Neumann, relying on the concept of Turing machine, became interested in the theory of self-reproductive automata and worked on the concept of self-reproductive machine [1, 2, 11]. After getting suggestions from Stanislaw Ulam, he used a regular lattice to mathematically model the self-reproduction, which is presently known as cellular automata [10].

In the present work, we use 2-dimensional cellular automata. The 3-D block of a block model is used as a layers of 2-D blocks in this work. Each block depends on its neighbouring blocks as per variogram model. Therefore, a 2-D CA becomes a suitable tool for the estimation purpose. The local rule, we use in this study, is the variogram function. The exploration data of a iron deposit is collected from a mine situated in the Eastern part of India. Our study is based on a mineral deposit that spreads 1160 m North–South and 1020 m East–West direction, at a depth of 140–340 m. The deposit is subdivided into 20 × 20 × 20 meter blocks. CA is applied on a single horizontal slice and the mineral grade for each block is calculated.

## 2 Preliminaries

### 2.1 Cellular Automata

Let us formally define a cellular automaton.

**Definition 1** A cellular automaton (CA) is defined as a quadruple $(\mathscr{L}, S, \mathscr{M}, f)$ where,

- $\mathscr{L} \subseteq \mathbb{Z}^D$ is $D$-dimensional cellular space, $\vec{v} \in \mathscr{L}$
- $S$ is the set of states a cell can assume
- $\mathscr{M} = (\vec{v_1}, \vec{v_2}, \ldots, \vec{v_m})$ is the *neighborhood vector* of each $\vec{v}$ of the lattice, and $(\vec{v} + \vec{v_i}) \in \mathscr{L}$,
- $f : S^m \to S$ is the transition function, called *local rule* of the CA.

**Definition 2** *(Configuration)* A configuration is a mapping that assigns states to the cells $c : \mathscr{L} \to S$.

This indicates that $S^{\mathscr{L}}$ is the set of all possible configurations. A CA can also defined by a function $G : S^{\mathscr{L}} \to S^{\mathscr{L}}$, which is called as global transition function.

In this study, we consider a finite size of CAs with $D = 2$. Neighborhood of a two dimensional CA are generally determined in two ways—(1) von Neumann neighborhood dependency and (2) Moore neighborhood dependency. CA proposed by John von Neumann is a two-dimensional CA with square grids.

Moore neighborhood [9] of two dimensional CAs follows 9-neighborhood dependency, where four non-orthogonal cells are additionally considered as neighbors.

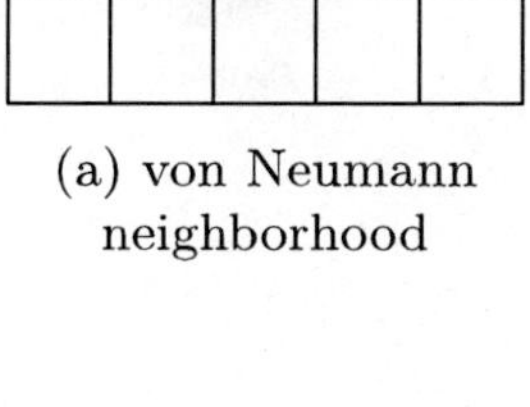

(a) von Neumann neighborhood

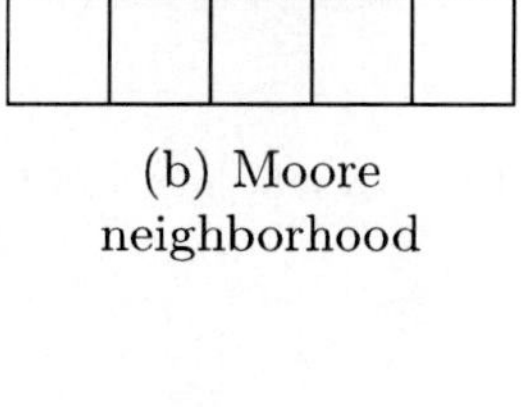

(b) Moore neighborhood

**Fig. 1** Neighborhood dependencies of two-dimensional CAs where the black cell is the cell under consideration, and the black cell changes its state depending on its neighbors marked in gray colors

Figure 1b shows Moore neighborhood dependency of a CA. This neighborhood structure has been utilized to design the famous *Game of Life*, a CA which was introduced by John Conway and popularized by Martin Gardner [3]. In this study also, we use Moore neighborhood of a 2-D CA.

## 2.2 Variogram Model and Krigging

Variogram model describes the relationship between spatially related data with distance. The semivariogram function has been originally defined in [7], which can be calculated as,

$$\gamma(h) = \frac{1}{2 * N(h)} \sum_{i=1}^{N(h)} [z(x_i) - z(x_i + h)]^2 \tag{1}$$

where $z(x_i)$ and $z(x_i + h)$ are the grade values at $x_i$ and $x_{i+h}$ locations respectively, which are $h$ distance apart from each other, $N(h)$ is the number of pairs $z(x_i)$ and $z(x_i + h)$.

A variogram model can be constructed using lag distance ($h$) and the corresponding semivariogram function $\gamma(h)$. Figure 2 shows a plot of variogram model.

A variogram model has several parameters, such as, range, nugget effect, sill, and spatial variance.

*Range* ($a$): Range is the distance within which spatial data are correlated and beyond which no correlation exists.

*Nugget effect* ($C_0$): Nugget effect shows the measurement error or micro scale variation of data. The value of the nugget effect can be calculated from $\gamma(h)$ when $h = 0$.

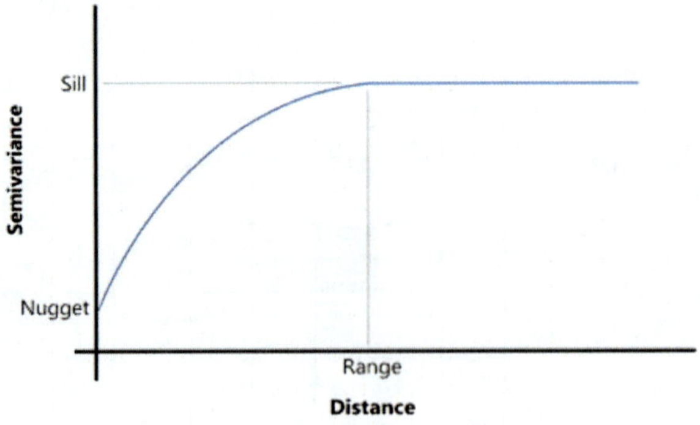


**Fig. 2** A variogram model

*Sill (C)*: Sill is the variogram value at $h$ = range.

*Spatial Variance* ($C_1$): It is the difference between sill and nugget effect.

Kriging is an estimation method for spatially related data. The data depends on their neighboring data. The estimation is done by the following equation.

$$z(x_0) = \sum_{i=1}^{n} \lambda_i z(x_i) \quad (2)$$

where, $z(x_0)$ is the estimated mineral grade at $x_0$, $\lambda_i$ is the weight of each data $i$. It is calculated from the covariance matrix [6].
$n$ is the number of samples and $x_i$ is the position of $i$th sample data.

For ordinary Kriging purpose, following $n + 1$ number of equations are used to calculate the weight $\lambda_i$.
$\gamma_{1,1} \cdot \lambda_1 + \gamma_{1,2} \cdot \lambda_2 + \cdots + \gamma_{1,n} \cdot \lambda_n + \mu = \gamma_{1,0}$. Similarly, for $n$th equation, we can write, $\gamma_{n,1} \cdot \lambda_1 + \gamma_{n,2} \cdot \lambda_2 + \cdots + \gamma_{n,n} \cdot \lambda_n + \mu = \gamma_{n,0}$. The $(n + 1)$th equation can be defined as, $\lambda_1 + \lambda_2 + \cdots + \lambda_n = 1$. Here, $\mu$ is the lagrangian parameter.

## 3 Estimation of Grade Values Using Cellular Automata

We now estimate the grade values of the minerals using a mathematical model, i.e., cellular automata. However, before proceeding to the new method, let us first see the conventional way of estimation for our collected data, by means of ordinary krigging.

Samples are collected from 71 boreholes. A spherical variogram model is constructed from the collected sample data. Figure 3 shows the spherical variogram model for our sample data.

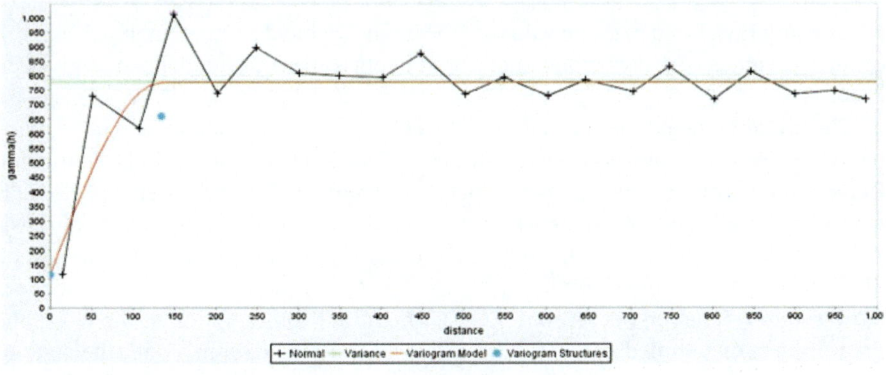


**Fig. 3** Experimantal variogram model of the sampled data

East Coordinate (meter)

2229 2249 2269 2289 2309 2329 2349 2369 2389 2409 2429 2449 2469 2489 2509 2629

North Coordinate (meter)

2323 2343 2363 2383 2403 2423 2443 2463 2483 2503 2523 2543 2563 2583 2603 2723

| | | | | | | | | | | | | | | |
|---|---|---|---|---|---|---|---|---|---|---|---|---|---|---|
| 60.25 | 60.67 | 61.60 | 61.64 | 61.39 | 61.33 | 60.55 | 60.98 | 59.39 | 58.17 | 56.86 | 55.38 | 53.96 | 52.94 | 52.68 |
| 59.06 | 59.42 | 60.37 | 60.92 | 61.59 | 61.57 | 61.75 | 61.60 | 61.26 | 60.15 | 59.98 | 58.89 | 57.21 | 56.02 | 55.97 |
| 57.66 | 58.11 | 58.92 | 60.31 | 61.26 | 61.70 | 62.46 | 61.96 | 62.12 | 62.67 | 62.61 | 62.28 | 60.30 | 59.27 | 57.87 |
| 56.81 | 57.41 | 58.46 | 59.81 | 61.05 | 62.26 | 63.12 | 62.81 | 62.89 | 63.22 | 63.82 | 63.71 | 62.54 | 60.98 | 59.60 |
| 56.26 | 57.04 | 58.49 | 59.83 | 61.29 | 62.60 | 63.57 | 63.48 | 63.44 | 63.67 | 64.02 | 64.47 | 63.74 | 62.49 | 62.00 |
| 58.24 | 58.24 | 58.99 | 60.18 | 61.45 | 62.37 | 62.58 | 62.39 | 62.04 | 62.10 | 62.51 | 62.44 | 64.14 | 63.46 | 62.57 |
| 59.43 | 59.54 | 60.11 | 60.81 | 61.63 | 62.12 | 61.47 | 60.49 | 60.64 | 60.69 | 61.04 | 61.74 | 63.11 | 63.99 | 62.44 |
| 60.63 | 60.54 | 61.07 | 61.64 | 61.86 | 61.41 | 60.53 | 59.53 | 59.42 | 59.22 | 60.11 | 60.81 | 61.96 | 64.01 | 64.15 |
| 62.10 | 62.05 | 62.16 | 62.38 | 61.64 | 60.74 | 59.72 | 58.69 | 57.83 | 58.15 | 58.72 | 59.94 | 60.47 | 61.99 | 63.97 |
| 61.81 | 61.85 | 61.90 | 61.68 | 61.29 | 60.01 | 58.91 | 57.89 | 57.27 | 57.03 | 58.21 | 59.39 | 60.87 | 61.86 | 63.97 |
| 61.47 | 61.44 | 61.02 | 61.00 | 60.85 | 59.93 | 59.00 | 58.06 | 57.46 | 57.54 | 58.65 | 59.78 | 61.05 | 63.11 | 63.63 |
| 61.04 | 60.95 | 60.45 | 60.44 | 60.25 | 59.65 | 58.82 | 58.39 | 58.25 | 58.30 | 58.72 | 60.31 | 61.48 | 62.49 | 63.01 |
| 60.57 | 60.09 | 60.13 | 60.17 | 59.84 | 59.39 | 58.52 | 58.57 | 59.07 | 59.14 | 59.69 | 60.19 | 62.24 | 62.35 | 63.00 |
| 60.24 | 59.89 | 59.88 | 59.77 | 59.52 | 59.08 | 58.59 | 59.26 | 59.88 | 60.48 | 61.68 | 61.86 | 62.11 | 62.81 | 62.56 |
| 60.30 | 59.83 | 59.74 | 59.61 | 59.40 | 59.74 | 59.71 | 60.23 | 60.81 | 60.88 | 61.86 | 62.05 | 61.84 | 62.33 | 62.58 |

**Fig. 4** A snap shot of krigging estimated values of a horizontal slice

Here, the range ($a$) of the variogram for grade value of mineral is 142.6 m, Sill value ($C$) is 662.5, and nugget effect ($C_0$) is 115.9. So, the spatial variance, $C_1$ (difference between sill and nugget effect) is 546.6. From this model, we can calculate the variogram for distance $h$ as follows.

$$\gamma(h) = \begin{cases} C_0 + C_1 \cdot \frac{h}{a}\left[1.5 - 0.5\left(\frac{h}{a}\right)^2\right] & \text{if } h \le a \\ C & \text{otherwise} \end{cases} \tag{3}$$

By using Eqs. 2 and 3, grade values of the unknown blocks are estimated. Krigging estimated values of a horizontal slice (with depth 280 m) of the deposit is given in Fig. 4.

The krigging method is computationally complex as it needs to solve a large number of equations. So, we now try to estimate the values using 2-D cellular automata. Cells of a CA can be considered as blocks of minerals. The cells interact with their neighboring cells and evolve depending on the local transition function. The transition function of the CA is defined as the ordinary krigging function. The new state of a cell $(i, j)$ is obtained by applying the krigging function, taking into account the variogram function (as given in Eq. 3). The initial configuration of the CA is the grade values of sampled data. If there exists no sampled data in a cell, the state of the cell is considered as 0. In the next steps of transition, the state of the cells may be changed from 0 to some grade value. Figure 5 shows the grade values after first

East Coordinate (meter)

2229 2249 2269 2289 2309 2329 2349 2369 2389 2409 2429 2449 2469 2489 2509 2629

North Coordinate (meter)

| | | | | | | | | | | | | | | | |
|---|---|---|---|---|---|---|---|---|---|---|---|---|---|---|---|
| 2323 | 59.77 | 59.77 | 59.77 | 59.77 | 59.77 | 59.77 | 59.77 | 59.77 | 59.77 | 59.77 | 59.77 | | | | |
| 2343 | 59.77 | 59.77 | 59.77 | 59.77 | 59.77 | 59.77 | 59.77 | 59.77 | 59.77 | 59.77 | 59.77 | | | | |
| 2363 | 59.77 | 58.11 | 58.92 | 59.80 | 60.26 | 60.70 | 60.46 | 61.96 | 62.12 | 59.77 | 59.77 | | | | |
| 2383 | 59.77 | 57.95 | 58.46 | 59.81 | 61.05 | 61.26 | 61.12 | 62.81 | 62.89 | 59.77 | 59.77 | | | | |
| 2403 | 59.77 | 57.04 | 58.49 | 59.83 | 61.29 | 62.60 | 62.57 | 61.48 | 60.44 | 59.77 | 59.77 | | | | |
| 2423 | 59.77 | 58.24 | 58.99 | 60.18 | 61.45 | 62.37 | 62.58 | 61.39 | 60.04 | 59.77 | 59.77 | | | | |
| 2443 | 59.77 | 59.84 | 60.11 | 60.81 | 61.63 | 62.12 | 61.47 | 60.49 | 60.64 | 59.77 | 59.77 | 60.44 | 60.44 | 60.44 | 60.44 |
| 2463 | 59.77 | 60.54 | 61.07 | 61.64 | 61.86 | 61.41 | 60.53 | 59.53 | 59.42 | 59.77 | 59.77 | 60.44 | 60.44 | 60.44 | 60.44 |
| 2483 | 59.77 | 62.05 | 62.16 | 62.38 | 61.64 | 60.74 | 59.72 | 58.69 | 57.83 | 59.77 | 59.77 | 61.85 | 62.45 | 61.18 | 62.61 |
| 2503 | 59.77 | 61.85 | 61.90 | 61.68 | 61.29 | 60.01 | 59.91 | 57.89 | 57.27 | 59.77 | 59.77 | 61.39 | 60.87 | 61.86 | 63.97 |
| 2523 | 59.77 | 61.44 | 61.02 | 61.00 | 60.85 | 59.93 | 59.00 | 58.06 | 57.46 | 59.77 | 59.77 | 61.08 | 61.05 | 63.11 | 63.63 |
| 2543 | 59.77 | 60.95 | 60.45 | 60.44 | 60.25 | 59.65 | 59.82 | 58.39 | 58.25 | 59.77 | 60.01 | 60.31 | 61.48 | 62.49 | 63.01 |
| 2563 | 59.77 | 60.09 | 60.13 | 60.17 | 59.84 | 59.39 | 59.82 | 58.57 | 59.07 | 60.58 | 61.05 | 60.19 | 62.24 | 62.35 | 63.00 |
| 2583 | 59.77 | 59.77 | 59.77 | 59.77 | 59.77 | 59.77 | 59.77 | 59.77 | 59.77 | 61.02 | 61.52 | 61.86 | 62.11 | 62.81 | 62.56 |
| 2603 2723 | 59.77 | 59.77 | 59.77 | 59.77 | 59.77 | 59.77 | 59.77 | 59.77 | 59.77 | 60.35 | 61.95 | 62.05 | 61.84 | 62.33 | 62.58 |

**Fig. 5** Grade values after the first stage of evolution of CA

evolution. In this figure, it can be observed that the grade values converge to 59.77 and 60.44. That is, the evolution will halt after the grade value reaches to these points.

After the first stage of evolution, we keep the values of neighboring cells of cell $(i, j)$ fixed and apply the transition rule for evolution. The fixed neighboring cells are shown by dotted lines in Fig. 5. Again, Fig. 6 shows the fixed cells and the cell values after second evolution. A similar method is applied in the next stages where the next neighboring cells of the fixed cells are kept constant. The transition rule is applied until we get fixed values of all the cells.

To obtain fixed values of all the cells, total 6 stages of evolution is required in our study. The final grade values are shown in Fig. 7.

It can be observed that the values obtained by using cellular automata is same as that of ordinary kriging method as the same variogram model is used in both the cases. However, it is easier to use cellular automata for such estimation. Moreover, the computation is much simpler in case of CAs compared to the kriging method. We use AMD dual core processor with 8 GB of RAM for the study. The language used here is C. It can be observed that, CA model executes faster compared to the kriging model.

After the estimation of grade values in horizontal 2-D slices, we now combine the multiple slices into a 3-D structure. The 3-D structure is shown in Fig. 8. The grade value of each block is represented by different colors.

East Coordinate (meter)

| North Coordinate (meter) | 2229–2249 | 2249–2269 | 2269–2289 | 2289–2309 | 2309–2329 | 2329–2349 | 2349–2369 | 2369–2389 | 2389–2409 | 2409–2429 | 2429–2449 | 2449–2469 | 2469–2489 | 2489–2509 | 2509–2629 |
|---|---|---|---|---|---|---|---|---|---|---|---|---|---|---|---|
| 2323–2343 | | | | | | | | | | | | | | | |
| 2343–2363 | | | | | | | | | | | | | | | |
| 2363–2383 | | | | | | | | | | | | | | | |
| 2383–2403 | | | | | | | | | | | | | | | |
| 2403–2423 | | 57.45 | 58.89 | 59.83 | 61.29 | 62.60 | 62.57 | 61.75 | | | | | | | |
| 2423–2443 | | 58.26 | 58.99 | 60.18 | 61.45 | 62.37 | 58.99 | 61.64 | | | | | | | |
| 2443–2463 | | 59.44 | 60.11 | **60.81** | **61.63** | **62.12** | 60.11 | 60.46 | | | | | | | |
| 2463–2483 | | 60.44 | 61.07 | **61.64** | **61.86** | **61.41** | 61.07 | 59.45 | | | | | | | |
| 2483–2503 | | 61.25 | 62.16 | **62.38** | **61.64** | **60.74** | 62.16 | 58.78 | | | | | | | |
| 2503–2523 | | 61.24 | 61.90 | 61.68 | 61.29 | 60.01 | 61.90 | 57.89 | | | | | | | |
| 2523–2543 | | 61.40 | 61.02 | 60.59 | 60.45 | 59.44 | 58.87 | 58.07 | | | 58.64 | 61.04 | 60.94 | 62.57 | 62.25 |
| 2543–2563 | | | | | | | | | | | 60.01 | 60.31 | 61.48 | 62.49 | 63.01 |
| 2563–2583 | | | | | | | | | | | 61.00 | 60.19 | **62.24** | **62.35** | **63.00** |
| 2583–2603 | | | | | | | | | | | 61.24 | 61.86 | **62.11** | **62.81** | **62.56** |
| 2603–2723 | | | | | | | | | | | 60.84 | 62.05 | **61.84** | **62.33** | **62.58** |

**Fig. 6** Grade values after the second stage of evolution of CA

East Coordinate (meter)

| North Coordinate (meter) | 2229–2249 | 2249–2269 | 2269–2289 | 2289–2309 | 2309–2329 | 2329–2349 | 2349–2369 | 2369–2389 | 2389–2409 | 2409–2429 | 2429–2449 | 2449–2469 | 2469–2489 | 2489–2509 | 2509–2629 |
|---|---|---|---|---|---|---|---|---|---|---|---|---|---|---|---|
| 2323–2343 | 60.25 | 60.67 | 61.60 | 61.64 | 61.39 | 61.33 | 60.55 | 60.98 | 59.39 | 58.17 | 56.86 | 55.38 | 53.96 | 52.94 | 52.68 |
| 2343–2363 | 59.06 | 59.42 | 60.37 | 60.92 | 61.59 | 61.57 | 61.75 | 61.60 | 61.26 | 60.15 | 59.98 | 58.89 | 57.21 | 56.02 | 55.97 |
| 2363–2383 | 57.66 | 58.11 | 58.92 | 60.31 | 61.26 | 61.70 | 62.46 | 61.96 | 62.12 | 62.67 | 62.61 | 62.28 | 60.30 | 59.27 | 57.87 |
| 2383–2403 | 56.81 | 57.41 | 58.46 | 59.81 | 61.05 | 62.26 | 63.12 | 62.81 | 62.89 | 63.22 | 63.82 | 63.71 | 62.54 | 60.98 | 59.60 |
| 2403–2423 | 56.26 | 57.04 | 58.49 | 59.83 | 61.29 | 62.60 | 63.57 | 63.48 | 63.44 | 63.67 | 64.02 | 64.47 | 63.74 | 62.49 | 62.00 |
| 2423–2443 | 58.24 | 58.24 | 58.99 | 60.18 | 61.45 | 62.37 | 62.58 | 62.39 | 62.04 | 62.10 | 62.51 | 62.44 | 64.14 | 63.46 | 62.57 |
| 2443–2463 | 59.43 | 59.54 | 60.11 | 60.81 | 61.63 | 62.12 | 61.47 | 60.49 | 60.64 | 60.69 | 61.04 | 61.74 | 63.11 | 63.99 | 62.44 |
| 2463–2483 | 60.63 | 60.54 | 61.07 | 61.64 | 61.86 | 61.41 | 60.53 | 59.53 | 59.42 | 59.22 | 60.11 | 60.81 | 61.96 | 64.01 | 64.15 |
| 2483–2503 | 62.10 | 62.05 | 62.16 | 62.38 | 61.64 | 60.74 | 59.72 | 58.69 | 57.83 | 58.15 | 58.72 | 59.94 | 60.47 | 61.99 | 63.97 |
| 2503–2523 | 61.81 | 61.85 | 61.90 | 61.68 | 61.29 | 60.01 | 58.91 | 57.89 | 57.27 | 57.03 | 58.21 | 59.39 | 60.87 | 61.86 | 63.97 |
| 2523–2543 | 61.47 | 61.44 | 61.02 | 61.00 | 60.85 | 59.93 | 59.00 | 58.06 | 57.46 | 57.54 | 58.65 | 59.78 | 61.05 | 63.11 | 63.63 |
| 2543–2563 | 61.04 | 60.95 | 60.45 | 60.44 | 60.25 | 59.65 | 58.82 | 58.39 | 58.25 | 58.30 | 58.72 | 60.31 | 61.48 | 62.49 | 63.01 |
| 2563–2583 | 60.57 | 60.09 | 60.13 | 60.17 | 59.84 | 59.39 | 58.52 | 58.57 | 59.07 | 59.14 | 59.69 | 60.19 | 62.24 | 62.35 | 63.00 |
| 2583–2603 | 60.24 | 59.89 | 59.88 | 59.77 | 59.52 | 59.08 | 58.59 | 59.26 | 59.88 | 60.48 | 61.68 | 61.86 | 62.11 | 62.81 | 62.56 |
| 2603–2723 | 60.30 | 59.83 | 59.74 | 59.61 | 59.40 | 59.74 | 59.71 | 60.23 | 60.81 | 60.88 | 61.86 | 62.05 | 61.84 | 62.33 | 62.58 |

**Fig. 7** The final grade values using cellular automata

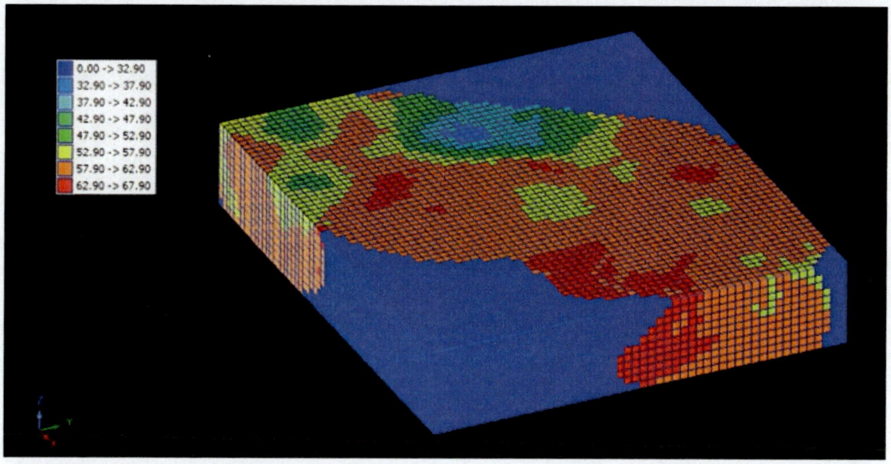


**Fig. 8** The final 3-D structure of Mineral Structure. Color represents the grade values of the deposit

## 4 Conclusion

This study represents a cellular automata based grade estimation of mineral deposits. A 2 dimensional CA has been used for the estimation. Each cell has been represented as a block of mineral. Krigging is a popular way of such block estimation, however, it needs a larger number of mathematical calculation. Use of cellular automata makes it more simpler. The outcome of grade values using krigging based cellular automata is same as that of using ordinary krigging method. In this study, we have used 2 dimensional horizontal slices of 3 dimensional block model and combined it to 3 dimensional representation. In future, the study can be done using three dimensional cellular automata.

## References

1. Burks, A.W.: Essays on Cellular Automata. University of Illinois Press (1970)
2. Das, S.: Theory and Applications of Nonlinear Cellular Automata In VLSI Design. Ph.D. thesis, Bengal Engineering and Science University, Shibpur, India (2007)
3. Gardner, M.: The fantastic combinations of John Conway's new solitaire game 'Life'. Sci. Am. **223**, 120–123 (1970)
4. Kim, Y.C.: Introductory Geostatistics and Mine Planning. University of Wollongong, Key Centre for Mines (1991)
5. Krige, D.G.: A statistical approach to some basic mine valuation problems on the Witwatersrand. J. S. Afr. Inst. Min. Metall. **52**, 201–203 (1951)
6. Martins, A.C., Nader, B., De Tomi, G.: A novel application of cellular automata for the evaluation and modelling of mineral resources. Int. J. Min. Miner. Eng. **3**(4), 303–315 (2011)
7. Matheron, G.: Principles of geostatistics. Econ. Geol. **58**(8), 1246–1266 (1963)

8. Matheron, G.: The theory of regionalized variables and its applications, vol. 5. Paris: École National Supérieure des Mines, p. 211 (1971)
9. Moore, E.F.: Machine models of self-reproduction. In: Proceedings of Symposia in Applied Mathematics, vol. 14, pp. 17–33. American Mathematical Society (1962)
10. Ulam, S.: On some mathematical problems connected with patterns of growth of figures. In: Proceedings of Symposia in Applied Mathematics, vol. 14, pp. 215–224. American Mathematical Society (1962)
11. von Neumann, J.: The Theory of Self-reproducing Automata, Burks, A.W. ed. University of Illinois Press, Urbana and London (1966)
12. Zhang, J., Yao, N.: The geostatistical framework for spatial prediction. Geo-spat. Inf. Sci. **11**(3), 180–185 (2008)

# Shape Generation Using Cellular Automata with Clustering

**Caleb Zeeman and Lynette van Zijl**

**Abstract** When using cellular automata with clustering in the Lumer-Faieta ant sorting algorithm, the method used to drop the cells in a cluster affects the geometric shape that the cluster will form. We investigate various algorithms for dropping strategies that can be used to generate pre-defined geometric shapes.

**Keywords** Cellular automata · Clustering · Shape generation

## 1 Introduction

Cellular automata with clustering (CCA) is a hybrid form of cellular automata (CA), where adjacent cells with the same attributes are allowed to join into a super-cell named a cluster [1, 9, 10]. The cluster information is stored extraneous to the cellular automaton itself. When a cell in a cluster is updated via the CCA rule, all the cells in the cluster are synchronously updated to the same attribute value. Both Adams [1] and Smal [9] showed that CCA typically require fewer time evolutions to achieve a goal than is the case with CA.

In this article, we consider the application of CCA to shape generation. The use of CA in shape generation has been considered before [3, 4, 8, 12], but due to the single cell model, the shape formation is not easy to achieve. Delormeo et al. [4], for example, used signals in grouped cellular automata based on the mathematical definition of the shapes, whilst Chavoya and Duthen [3] evolved a shape from a single cell in a 2D grid, by using genetic algorithms to update lookup tables for the CA. In the case of CCA, we show that simple geometric shapes such as circles or

This research was supported by the National Research Foundation, grant number 133051.

C. Zeeman (✉) · L. van Zijl
Department of Computer Science, Stellenbosch University, Stellenbosch, South Africa
e-mail: 21018030@sun.ac.za

L. van Zijl
e-mail: lvzijl@sun.ac.za

 
S. Das and G. J. Martinez (eds.), *Proceedings of First Asian Symposium on Cellular Automata Technology*, Advances in Intelligent Systems and Computing 1425,
https://doi.org/10.1007/978-981-19-0542-1_5

squares can quite easily be explicitly forced in CCA. The CCA model used for the shape formation is the standard Lumer and Faieta (LF) ant sorting model [7]. In the LF model, an ant picks up a single object, and drops a single object. The CCA model is identical, except that an ant picks up a cluster of objects, and drops a cluster of objects. This requires a dropping strategy, as the objects in the cluster have to be dropped in open cells in the immediate vicinity of the target cell. We design three families of dropping strategies, and show how these determine the geometrical shape of the cluster, based on different algorithms for the dropping strategies.

Section 2 gives a brief overview of CA, CCA, and the LF ant sorting model. Algorithms for different dropping strategies are investigated in Sect. 3, with a discussion on how these strategies affect the shape that the clusters form. We conclude in Sect. 4.

## 2 Background

It is assumed that the reader is familiar with the theoretical background to CA, such as in [11].

### *2.1 Cellular Automata*

A CA is a discrete system that consists of a $k$-dimensional array of automata that execute simultaneously. Automata in the array communicate by means of a CA rule. Typically, this rule concerns the automata that are in close proximity in the array – that is, an automaton is affected locally by the other automata in its immediate neighbourhood. The CA execute in concrete time steps, where the state of the automaton in cell $c_i$ at time step $t + 1$, is determined by the state of both $c_i$ and its neighbours at time step $t$, via the CA rule. These time steps are known as the evolution of the CA. The array can be considered to be bounded on all sides (null boundary CA), or to be connected on opposing sides (periodic boundary CA). In this article, the generation of geometric shapes in two dimensions are considered, and hence only two-dimensional grids with periodic boundaries are investigated.

The idea of a neighbourhood in CA is defined by means of a distance $r$, and geometrical adjacency of cells. For example, the well-known Von Neumann and Moore neighbourhood for a cell at coordinates $(a, b)$ in the grid, is defined respectively as

$$N^r_{(a,b)} = \{(i, j) : |i - a| + |j - b| \leq r\} . \tag{1}$$

and

$$N^r_{(a,b)} = \{(i, j) : |i - a| \leq r, |j - b| \leq r\} .$$

Thus, any cell in the Von Neumann neighbourhood of the cell $(a, b)$ for distance $r = 1$, is geometrically adjacent to it; that is, the cells share a common border.

Likewise, any cell in the Moore neighbourhood of cell $(a, b)$ for $r = 1$, is either geometrically adjacent to cell $(a, b)$ or it is geometrically adjacent to another cell which is adjacent to $(a, b)$. Note that cells that are diagonally connected are not considered to be adjacent. Hence, to include cells that are diagonal to $(a, b)$ in the Moore neighbourhood, a connected path of adjacent cells from $(a, b)$ to the target cell, can be constructed. The idea of neighbourhoods and adjacency of cells are important in the CCA shape generation, and is discussed in more detail in Sects. 2.2 and 2.4.

## 2.2 *Cellular Automata with Clustering*

Two-dimensional CA with clustering was first proposed by Smal [9], and applied to a pattern layout optimisation problem. The idea of biological cellular clustering is well-known, but normal CA models function only on a single cell behaviour. Deutsch et al. [5] point out the collective behaviour in biological multicellular organisms, and in particular the issue of collective migration, which reflects the idea of clusters as applied in CCA.

Informally, a cluster is a collection of adjacent cells with the same attribute values. Thus, a cluster has no predefined shape or size, and it becomes necessary to redefine the idea of adjacency for clusters. Otherwise, it would not be possible to determine the neighbourhood of a cluster. The definitions from [1] are briefly summarised (see Fig. 1).

Two cells $c(i, j)$ and $c(u, v)$ in a grid are properly adjacent if they share a common geometric border. That is, $|u - i| + |v - j| \leq 1$ (see Eq. (1) above). Consider a graph with the cells as nodes, and an edge between two nodes only if the corresponding cells are properly adjacent. Properly adjacent cells are then also properly connected. This enables the definition of a cluster as a set of properly connected cells. Given any two clusters $A$ and $B$, the clusters are properly adjacent if there exists at least one cell $a(i, j)$ in $A$ and at least one cell $b(k, m)$ in $B$, such that $a(i, j)$ is properly adjacent to $b(k, m)$. For example, the Von Neumann neighbourhood of distance $r$ for a cluster $A$ is given by

$$\begin{aligned}N_A^r =\{B : \exists c_{ij} \in B \text{ such that } |i - a| + |j - b| \leq r \text{ and}\\ \exists\ c_{km} \notin A \text{ and } c_{ab} \in A \text{ such that}\\ |a - k| = 1 \text{ and } |b - m| = 0 \text{ or}\\ |a - k| = 0 \text{ and } |b - m| = 1\}.\end{aligned}$$

Similarly, for the Moore neighbourhood of a cluster, one has:

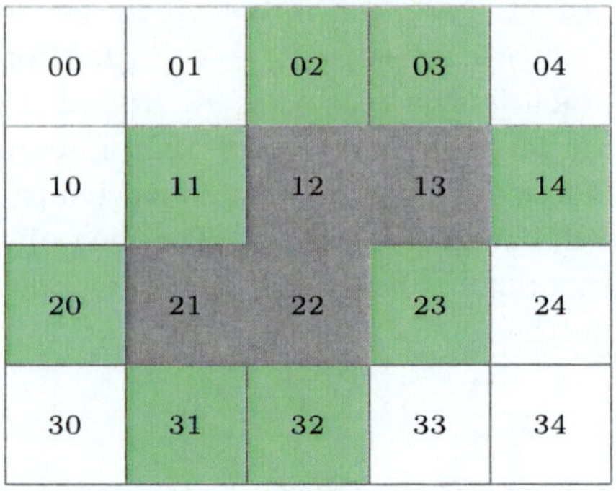


**Fig. 1** Von Neumann neighbourhood for cluster $\{c_{12}, c_{13}, c_{21}, c_{22}\}$

$$
\begin{aligned}
N_A^r = \{B : \exists c_{ij} \in B \text{ such that } |i-a| \leq r \text{ and } |j-b| \leq r, \text{ and} \\
\exists\, c_{km} \notin A \text{ and } c_{ab} \in A \text{ such that} \\
|a-k| = 1 \text{ and } |b-m| = 0 \text{ or} \\
|a-k| = 0 \text{ and } |b-m| = 1\}.
\end{aligned}
$$

Figure 1 illustrates the concept of the neighbourhood of a cluster. Consider a cluster consisting of cells $\{c_{12}, c_{13}, c_{21}, c_{22}\}$. Its Von Neumann neighbourhood consists of all cells which are properly adjacent to at least one cell in the cluster. For example, $c_{20}$ must be in the Von Neumann neighbourhood, as it is properly adjacent to $c_{21}$. On the other hand, $c_{33}$ shares no common border with any of the cells in the cluster, and hence is not in the Von Neumann neighbourhood of the cluster.

## *2.3 Ant Sorting Using the Lumer and Faieta Model*

Lumer and Faieta [7] provided a CA model which reflects the sorting behaviour of ants, where corpses are organised into piles known as cemetries. This model is useful in practical applications ranging from optimisation problems to distributed control [6].

In the Lumer and Faieta (LF) model, a number of objects are initially randomly placed on a grid, where each cell can contain at most one object. Ants are then randomly placed across the grid. At each time step, these ants move randomly to an adjacent cell. Before it moves, if the current cell is not empty, the ant may pick up the object with a certain probability, and carry it to the adjacent cell. Likewise, if the ant is carrying an object, and the current cell is empty, the ant may drop the object with a certain probability and move unladen (without carrying an object) to the adjacent cell. The pickup and drop probabilities depend on the density of objects with the same attributes that are located in the neighbourhood of the current cell. This density is reflected by the so-called fitness function $f(o_i)$ of an object $o_i$. In the simplest form of the LF model, each object is assigned a certain colour, and the goal is then to sort objects of the same colour into one vicinity on the grid. Informally, if an ant encounters an object, it will pick up the object if the surrounding neighbourhood does not contain many other objects with the same attribute. Likewise, if an ant carries an object, the probability of dropping the object in a certain cell will increase if there are many objects with the same attribute in the immediate neighbourhood. For the mathematical details of the fitness function and the pickup and drop probabilities, see [7].

### 2.4 Ant Sorting Using Clustered Cellular Automata

Adams [1] provided an adaptation of the LF algorithm for CCA (see Algorithm 1). In principle, the main modification to the original LF algorithm is that an entire cluster of objects is picked up and dropped by an ant, instead of only individual objects. At the completion of the time step, any adjacent clusters with the same attributes are merged into a single cluster.

The dropping of a cluster requires some clarification. To drop a cluster, an ant moves into an empty cell. Since all the objects in the cluster cannot be dropped into this single cell, the cluster must be distributed across empty cells in the surrounding area. The dropping strategy determines how these extra empty cells are located, and the resulting shape of the cluster on the grid varies depending on the dropping strategy employed.

If there are not enough empty cells found to fully drop the cluster, the original cluster is split into two clusters. The cells already dropped form one cluster, and the other cells still being carried are joined into a new cluster, which remains as the ant's payload. The ant will then continue to search for a new dropping location.

In addition, note that when a cluster $c_i$ is dropped directly adjacent to another existing cluster $c_j$ (typically with the same attributes), the cells in $c_i$ are dropped into empty cells next to $c_j$. This may lead to the formation of joined partial shapes during the time evolution of the CCA in the shape generation process.

## 3 Shape Generation

In order to generate a given geometric shape in the ant sorting model, objects in a cluster need to be dropped so that the required shape forms on the grid. Note that geometric shapes can be formed by dropping the objects in relation to the current cell, in a specified order. For example, to form a straight horizontal line of width one, the objects in the cluster need to be dropped to the right (or the left) of the current cell. Likewise, to form a perpendicular line of width one, the objects need to be dropped to the top (or bottom) of the current cell. However, in the ant sorting model, clusters are typically dropped adjacent to other clusters with the same attributes. This means that not all the cells around the current cell are necessarily empty, and therefore the

**Algorithm 1:** The modified Lumer-Faieta algorithm for clustering

```
for every object o_i do
    place o_i randomly on grid;
for all ants do
    place ants at randomly selected site;
calculate clusters;
for t = 1 to t_max do
    for all ants do
        if (ant unladen) and (site occupied by object o_i in cluster c_j) then
            compute f(o_i) and p_pickup(o_i);
            select random real number R between 0 and 1;
            if R ≤ p_pickup(o_j) then
                pick up cluster c_j;
        else if (ant carrying cluster c_j) and (site empty) then
            compute f(o_i) and p_drop(o_j);
            select random real number R between 0 and 1;
            if R ≤ p_drop(o_j) then
                drop cluster c_j;
        move to randomly selected neighboring unoccupied site if available,
        else do nothing;
```

dropping strategy needs to search for the 'best' open cell to drop an object. The method of determining the best cell can influence the eventual shape that forms.

Hence, any dropping strategy must identify the current cell, the neighbourhood to be explored for an empty cell, the order in which the neighbourhood will be explored, and a method to define the best empty position in which to drop the next object from the cluster.

Three families of dropping strategies are introduced, namely, fitness-based strategies, queue-based strategies, and priority queue-based strategies. Within each family, the same general dropping technique will be used, but different neighbourhoods and different exploration orders of the neighbourhoods are considered. For each geometric shape, a strategy is suggested, and visual output is shown to verify the correctness.

## 3.1 *Fitness-Based Strategies*

Dropping strategies in this family follow the same general approach: while there are still cells remaining in the cluster to be dropped, some neighbourhood around the current cell is analysed using some fitness function. This function will return an empty cell in the neighbourhood if one exists. The next object in the cluster is dropped at the empty cell found, and the process is repeated until there are no more objects to drop. The fitness-based strategies therefore depend on the neighbourhood

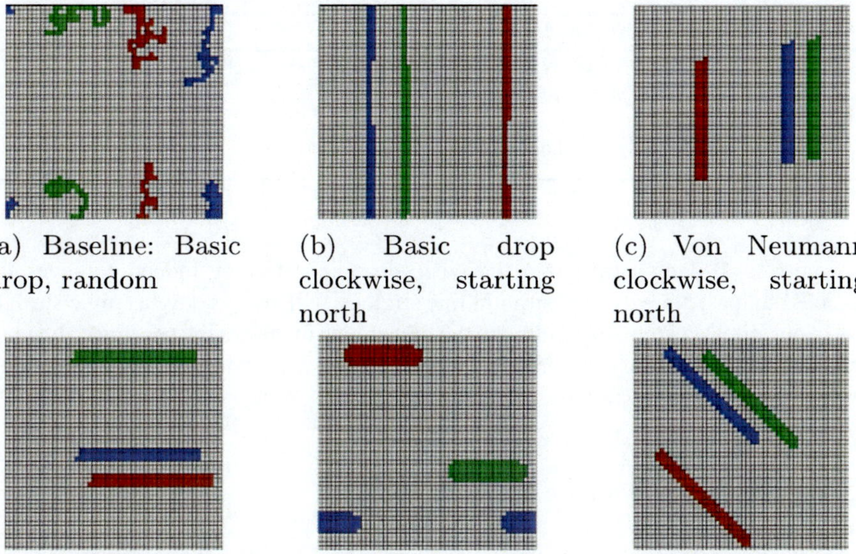

(a) Baseline: Basic drop, random

(b) Basic drop clockwise, starting north

(c) Von Neumann clockwise, starting north

(d) Von Neumann anti-clockwise, starting west

(e) Moore, clockwise, starting north

(f) Moore, clockwise, starting north-east

**Fig. 2** Results of the fitness-based strategies

chosen, and the fitness function used. The size of the neighbourhood directly affects how computationally expensive the particular dropping strategy will be. To maintain the association with the LF model, the fitness function used in these experiments is $f(o_i)$, as it reflects the idea of clustering similar cells together.

The three strategies discussed below are the basic drop, the Von Neumann drop and the Moore drop. The basic drop simply returns any valid cell, while the latter two strategies use a fitness function to determine the best cell for a drop, using the Von Neumann or Moore neighbourhoods respectively. The results from the fitness-based strategies are given in Fig. 2.

### 3.1.1 Basic Dropping Strategy

This is the most basic of all the dropping strategies. A comparative baseline is first established – if the fitness function is applied to return a random empty cell in the neighbourhood, no discernible shape should be generated. To force the simplest of shapes, a straight line can be generated by exploring the neighbourhood in a fixed order and returning the first empty cell. The high-level implementation is given in Algorithm 2.

**Algorithm 2:** Basic drop method

```
/* c_i the cluster being dropped */ p ←find empty neighbourhood cell, or null if none;
while p ≠ null and c_i not empty do
    drop next object in c_i at p;
    p ← find empty neighbourhood cell, or null if none;
```

Figure 2a, b demonstrate that the basic dropping strategy indeed achieves the intended shapes. Note that the straight line strategy will always have a single straight line in one direction (corresponding to the first choice in the order), varying only when colliding with other clusters. Only the Von Neumann neighbourhood was considered for the basic dropping strategy.

#### 3.1.2 Von Neumann/Moore Dropping Strategy

Given the thin rectangular shape produced by the basic drop, is it possible to generate thicker rectangular shapes? We show that the fitness function $f(o_i)$ can be used to produce a neighbourhood-wide drop. As opposed to the basic drop, the fitness function can be used to quantitatively identify the best cell for the next drop. Then cells are preferred when located geometrically nearest to other dropped cells. Recall that the LF fitness function for a cell $c_i$ is calculated by default over all the eight cells surrounding $c_i$.

In this strategy, the cell with the best fitness value in the Von Neumann (resp. Moore) neighbourhood of the current cell is chosen to drop the next object in the cluster. The best fitness value is defined as the highest object density $f(o_i)$. Note that if a second cell is found with the same fitness value as the current best fit, the current best fit is kept. Since the fitness function must be evaluated for all neighbours of all cells placed, this strategy is computationally more expensive than the basic strategy.

These two examples illustrate how the order of the neighbourhood exploration affects the shape of the cluster. In Fig. 2c, the Von Neumann neighbourhood is explored in the order [north, east, south, west]. This results in a three-wide vertical line forming for each cluster. In contrast, reversing the order to [west, south, east, north] provides three-wide horizontal lines instead (see Fig. 2d).

The starting point in the exploration also plays a role. In Fig. 2e, the Moore neighbourhood is explored in a clockwise direction, starting in the north. This produces thick horizontal clusters. In contrast, starting in the northeast corner and exploring in a clockwise direction, results instead in diagonals in a southeast orientation (see Fig. 2f). Thus, both the thickness of the cluster and the orientation of the cluster formed is dependent on the exploration order of the neighbourhood.

## 3.2 *FIFO Queue-Based Strategies*

In order to produce shapes that are not simply rectangular, one has to build the shape incrementally, with knowledge of cells already used. For example, one could form a diamond shape around a cell by dropping objects into the Von Neumann neighbourhood of the cell at increasing distances. All strategies in this family therefore make use of a first-in-first-out (FIFO) queue that keeps track of all empty cells in neighbourhoods explored so far. At the start of the drop, the initial cell is added to the queue. While there are still objects in the cluster to drop, a new cell is popped from the top of the queue, an object is dropped at this cell, and all empty cells in some neighbourhood around this new cell is added to the queue. This process is continued until all the objects in the cluster have been dropped, or the queue is empty.

Algorithm 3 shows a generic FIFO queue-based dropping strategy (adapted from [1]). Note that both the queue-based and the priority queue-based strategy implement a memory mechanism, somewhat reminiscent of higher order cellular automata [2].

**Algorithm 3:** Generic FIFO queue drop method

```
/*c_i the cluster being dropped*/
queue ← new Queue();
p ← position of the ant;
while p ≠ null and c_i still has objects to drop do
    o_j ← the next object to drop in c_i;
    aP ← all empty positions in some neighbourhood of p;
    queue.addAll(aP);
    drop object o_j at p;
    /*find next empty position*/
    if queue is empty then
        /*incomplete drop*/
        p ← null;
    else
        p ← queue.dequeue();
```

Since these dropping strategies do not use the fitness function when considering cells, all strategies in this family will often place cells directly adjacent to another cluster. When sorting via clustering, this is undesirable—as such, if the focus of the application is to sort data, it is better to use a fitness-based strategy such as the Von Neumann drop.

The shape formed by a FIFO queue-based strategy depends entirely on which neighbours are explored—and the order in which they are explored—when adding cells to the queue. Four examples of queue-based dropping strategies are given, with the results of each strategy shown in Fig. 3.

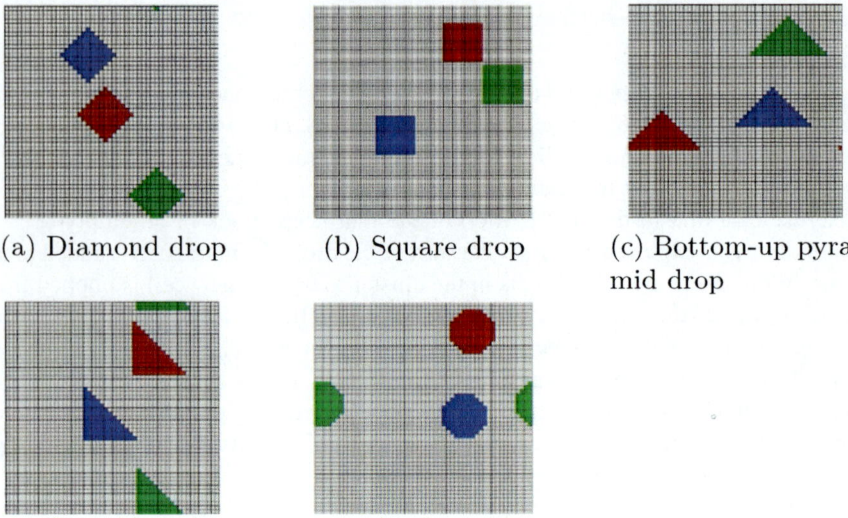

(a) Diamond drop (b) Square drop (c) Bottom-up pyramid drop

(d) Triangle drop, north-east (e) Circle drop

**Fig. 3** Results of the queue-based and priority queue-based strategies

### 3.2.1 Diamond Dropping Strategy

Since a Von Neumann neighbourhood is by definition a diamond-shaped area, the diamond dropping strategy is straightforward to implement. When adding new positions to the queue after dropping an object, the Von Neumann neighbourhood (explored clockwise or anti-clockwise) is used. Since a FIFO queue is being used, all positions in a Von Neumann neighbourhood of radius one from the centre are considered, then radius two, and so on. The final form is, therefore, that of a diamond. The first cell chosen will determine the starting point of the shape, and the rotation direction will only affect the shape if there are insufficient cells in the cluster to complete a full layer of the diamond.

### 3.2.2 Square Dropping Strategy

Again, producing a square using the queue-based method is easy. The square dropping strategy is similar to the diamond dropping strategy, except that the Moore neighbourhood is used when adding new cells to the queue. This results in the shape forming a full Moore neighbourhood of increasing radii around the centre (the starting point) of the drop. Since the Moore neighbourhood is inherently square-shaped, this results in a square forming around the starting point of the drop.

### 3.2.3 Pyramid Dropping Strategy

Noting that a pyramid is half of a diamond, a pyramid shape can be generated analogously to a diamond, both for top-down and bottom-up pyramids.

In the top-down approach, the starting coordinates of the drop will be the top of the pyramid. The neighbourhood used for exploration is chosen by picking one cardinal direction in which the pyramid will be generated, and adding the two adjacent diagonal directions. For example, [south, southeast, southwest] will cause a pyramid to form downwards, with the top being at the coordinates of the drop. Note that the cardinal direction should always be first in the exploration order.

In the bottom-up approach, the pyramid will form such that the starting coordinates of the drop will be the centre of the base of the pyramid. This is achieved by letting the neighbourhood used for exploration consist of any three cardinal directions. Exactly one of the three directions will be directly adjacent to both of the other directions: this direction denotes the direction in which the pyramid will grow. For example, if the neighbourhood [east, north, west] is chosen, then only north is adjacent to both east and west, so the pyramid will form with the base growing in an east-west direction, and the top directly north of the starting point.

### 3.2.4 Triangle Dropping Strategy

To generate right-angled triangles, the right-angled corner of the triangle is placed at the starting coordinate of the drop, and two cardinal directions are chosen. The direction of the two sides projecting from that corner depends on which directions are chosen. The general pattern for this dropping strategy is to pick the two adjacent sides surrounding the diagonal in which the triangle should grow. For example, a neighbourhood [north,east] will form a triangle to the northeast, while [east,south] will form a triangle to the southeast.

Note that the memory usage of all the queue-based strategies are identical, as each strategy builds the queue, and the shape is determined by the order in which the objects are added to the queue. The upper bound on the queue size is the number of objects on the grid. This upper bound can be reached if all the objects on the grid have the same colour.

## *3.3 Priority Queue-Based Strategies*

This family of dropping strategies follows the same general approach as the FIFO queue-based strategies, but uses a priority queue instead. A priority is associated with each item in the priority queue such that the item with the highest priority is at the front of the queue. This allows more complex features to be incorporated into the dropping strategy. The generic priority queue-based dropping strategy is given in

Algorithm 4, and the use of the priority queue is demonstrated in the circle dropping strategy.

**Algorithm 4:** Generic priority queue drop method

$priorityQueue \leftarrow$ new PriorityQueue();
$start \leftarrow$ coordinates of the starting point of the drop;
$p \leftarrow$ position of the ant;
**while** $p \neq$ *null and* $c_i$ *still has objects to drop* **do**
  $o_j \leftarrow$ the next object to drop in $c_i$;
  $priorities, positions \leftarrow$ all empty positions in *some* neighbourhood of $p$, where the priorities are generated by $f(start, positions)$;
  priorityQueue.addAll($priorities, positions$);
  drop object $o_j$ at $p$;
  /*find next empty position*/
  **if** *priorityQueue is empty* **then**
    /*incomplete drop*/
    $p \leftarrow$ null;
  **else**
    $p \leftarrow$ priorityQueue.dequeue();

### 3.3.1 Circle Dropping Strategy

As a geometrical shape, a circle is formed as a structure of a certain radius around a centre point. Therefore, the starting point of the drop is chosen to be the centre of the circle, and cells are dropped around the centre in order of distance. This causes a filled circle to form of radius one, then radius $\sqrt{2}$, and so on, until no more cells need to be dropped. See Fig. 3e.

The priority queue is used to store the coordinates of the starting point. For every coordinate added to the queue, the priority is the distance from the starting point to this new coordinate. When taking the next coordinate from the priority queue, the algorithm uses the coordinate that has the lowest distance from the starting point. The periodic boundary of the CCA has to be taken into account when calculating euclidian distances that span the boundary.

In summary, we proposed different strategies for the generation of different shapes. Rectangular directional shapes can be generated by the fitness-based strategies, and the size of the neighbourhood influences the computation cost. More complex shapes, such as squares or triangles, require the use of queue-based strategies. Here, the computational cost is less than that of the fitness-based strategies, at the cost of a increased memory requirement.

## 4 Conclusion

Geometric shape generation was successfully performed for CCA based on an ant sorting model. It was shown that different dropping strategies, with different exploration orders, can be used to drop clusters of different shapes. Fitness-based dropping strategies can be more expensive in terms of execution time depending on the given neighbourhood, whilst queue-based and priority-queue-based strategies are computationally less expensive, but requires extra memory usage for the queue.

The geometrical forms generated by the strategies discussed in this article are relatively simple shapes. An investigation into other dropping strategies, that generate more complex geometrical forms, is appropriate. As a first step, the effect of changing the fitness function to operate over a whole cluster, rather than over a single cell, should be investigated. Another avenue of investigation is to extend this work to generate three-dimensional shapes.

## References


1. Adams, R., van Zijl, L.: Ant sorting based on cellular automata with clustering. In: Proceedings of the Annual Conference of the South African Institute of Computer Scientists and Information Technologists. SAICSIT '18, pp. 29–38. Association for Computing Machinery, New York, NY, USA (2018)
2. Baas, N.A., Helvik, T.: Higher order cellular automata. Adv. Complex Syst. **8**, 169–192 (2005)
3. Chavoya, A., Duthen, Y.: A cell pattern generation model based on an extended artificial regulatory network. Biosystems **94**(1–2), 95–101 (2008)
4. Delorme, M., Mazoyer, J., Tougne, L.: Discrete parabolas and circles on 2D cellular automata. Theor. Comput. Sci. **218**(2), 347–417 (1999)
5. Deutsch, A., Nava-Sedeño, J.M., Syga, S., Hatzikirou, H.: BIO-LGCA: a cellular automaton modelling class for analysing collective cell migration. PLOS Comput. Biol. **17**(6), e1009066 (2021)
6. Dorigo, M., Di Caro, G., Stützle, T.: Special issue on ant algorithms. Future Gener. Comput. Syst. **16**(8), 851–956 (2000)
7. Lumer, E., Faieta, B.: Diversity and adaptation in populations of clustering ants. In: Proceedings of the Third International Conference on Simulation of Adaptive Behaviour: From Animals to Animats, pp. 499–508. MIT Press, Cambridge, MA, USA (1994)
8. Macedo, Y.P., Chaimowicz, L.: Improving procedural 2D map generation based on multi-layered cellular automata and Hilbert curves. In: 2017 16th Brazilian Symposium on Computer Games and Digital Entertainment, pp. 116–125. IEEE (2017)
9. Van Zijl, L., Smal, E.: Cellular automata with cell clustering. In: Proceedings of AUTOMATA 2008: Theory and Application of Cellular Automata, pp. 425–440. Luniver Press (2008)
10. Van Zijl, L.: Clustering in 1D binary elementary cellular automata. In: Proceedings of the 2015 Annual Research Conference on South African Institute of Computer Scientists and Information Technologists. SAICSIT '15, pp. 1–7. Association for Computing Machinery, New York, NY, USA (2015)
11. Wolfram, S.: Theory and Applications of Cellular Automata, vol. 1. World Scientific Singapore (1986)
12. Zhang, D., Choi, C., Kim, J., Kim, Y.M.: Learning to generate 3D shapes with generative cellular automata. In: 9th International Conference on Learning Representations, ICLR, poster presentation (March 2021). https://arxiv.org/pdf/2103.04130.pdf

# Evolution Patterns of Some One-Dimensional Non-uniform Cellular Automata

**Sreeya Ghosh and Sumita Basu**

**Abstract** This paper investigates patterns generated by some one-dimensional spatially non-uniform cellular automata. The non-uniform cellular automaton is taken to be composed of finite celled blocks where each block has a different local transition function and within any block all the cells follow same local transition function. The evolution patterns generated by identity, complementary, constant and shift functions and a combination of them as local transition functions are studied in this paper.

**Keywords** Non-uniform cellular automata · Local transition function · Left/right shift function

## 1 Introduction

A system generated by a lattice of individuals whose local intersections evolve a global pattern is called a cellular automaton(CA) ([9–11, 13, 16–18]). Each individual (called a *cell*) acts as a finite automaton which updates itself by a local rule. If the local rule is uniform then the CA is uniform/homogeneous otherwise the CA is said to be non-uniform/hybrid. Obviously non-uniform CA is a generalization of uniform CA.

In recent times researchers have worked on one-dimensional non-uniform CA ([1, 3, 14]).

Previously we have worked on the algebraic properties and evolution patterns of one-dimensional uniform(homogeneous) CA ([4–7]) as well as one-dimensional fuzzy CA being non-uniform over a period of time ([2]). In the present article, we have chosen our CA model to have one dimensional finite $(n_i | i \in \mathbb{Z})$ celled blocks

S. Ghosh (✉)
Department of Applied Mathematics, Calcutta University, Kolkata 700009, India
Indian Statistical Institute, Kolkata 700108, India

S. Basu
Bethune College, Kolkata 700006, India
Guest Faculty, Department of Mathematics, The Heritage College, Kolkata 700107, India



S. Das and G. J. Martinez (eds.), *Proceedings of First Asian Symposium on Cellular Automata Technology*, Advances in Intelligent Systems and Computing 1425,
https://doi.org/10.1007/978-981-19-0542-1_6

where each block has a different local transition function and within any block all the cells follow same local transition function. The local transition functions are restricted to identity, complementary, constant and shift functions and a combination of them.

The structure of the paper is as follows. In Sect. 2 basic concepts are introduced and some fundamental results are reported. Relation between local and global transitions of a non-uniform CA have been investigated and included in Sect. 3. Section 4 is devoted to the evolution patterns studied by us.

## 2 Basic Concepts

We give a formal definition of a one-dimensional Cellular Automaton which consists of cells placed on a linear grid.

**Definition 2.1** The memory elements of the automata placed on the grid line belong to a finite state set $Q$.

A **global configuration** is a mapping from the group of integers $\mathbb{Z}$ to the set $Q$ given by $C : \mathbb{Z} \to Q$.
The set $Q^{\mathbb{Z}}$ is the set of all global configurations where $Q^{\mathbb{Z}} = \{C | C : \mathbb{Z} \to Q\}$.

A mapping $\tau : Q^{\mathbb{Z}} \to Q^{\mathbb{Z}}$ is called a **global transition function.**

A **CA**(denoted by $\mathcal{C}_\tau^Q$) is a triplet $(Q, Q^{\mathbb{Z}}, \tau)$ where $Q$ is the finite state set, $Q^{\mathbb{Z}}$ is the set of all configurations, $\tau$ is the global transition function.

If for a particular CA, $|Q| = 2$ so that we can write $Q = \{0, 1\}$, then the CA is said to be a **binary CA** or a **Boolean CA**.

The states of CA are usually categorized as either quiescent or non-quiescent. A cell in a quiescent state remains in the same state if all of its neighbours are also in the same quiescent state. A configuration $C \in Q^{\mathbb{Z}}$ of a CA is said to be finite if only a finite number of cells are in non-quiescent state. Now a finite CA may be defined as follows:

**Definition 2.2** An infinite CA where only a finite number of consecutive cells are in the non-quiescent state is said to be a **finite CA**. All the other cells are in quiescent state(say $q_B \in Q$). So the positions of non-quiescent states, (say $q_i \in Q$, such that $i = 1, 2, ..., n \in \mathbb{N}$) may be described by a finite lattice as shown in *(Fig. A)*.

| | | | | | | | | | |
|---|---|---|---|---|---|---|---|---|---|
| ... | $q_B$ | $q_B$ | $q_1$ | $q_2$ | ... | $q_n$ | $q_B$ | $q_B$ | ... |

*Fig. A*
*A typical grid of a finite CA*

**Definition 2.3** For a finite CA the boundary cells maybe:

Periodic(or 'wrap') : The 'n' consecutive cells such that the $(n + 1)$th cell is identified with the 1st cell. Cells having non-quiescent states form a continuous ring in the one-dimensional case.

Fixed: States of all the cells of the CA except 'n' consecutive cells remain quiescent(fixed) throughout the simulation. If the cells with fixed states be in '0' state, then the CA is null boundaried.

Reflective : Cells to the left of the 1st cell obtain the state of the 1st cell and cells to the right of the $n$th cell obtain the state of the $n$th cell.

**Definition 2.4** For $i \in \mathbb{Z}, r \in \mathbb{N}$, let $S_i = \{i - r, ..., i - 1, i, i + 1, ..., i + r\} \subseteq \mathbb{Z}$. $S_i$ is the neighbourhood of the $i$th cell. r is the radius of the neighbourhood of a cell. It follows that $\mathbb{Z} = \bigcup_i S_i$

A restriction from $\mathbb{Z}$ to $S_i$ induces the following:

1. Restriction of $C$ to $c_i$ is given by $c_i : S_i \to Q$; and $c_i$ may be called **local configuration** of the $i$th cell.
2. Restriction of $Q^{\mathbb{Z}}$ to $Q^{S_i}$ is given by $Q^{S_i} = \{c_i | c_i : S_i \to Q\}$;and $Q^{S_i}$ may be called the **set of all local configurations** of the $i$th cell.

The mapping $\mu_i : Q^{S_i} \to Q$ is known as a **local transition function** for the $i$th automaton having radius $r$. So, $\forall i \in \mathbb{Z}$, $\mu_i(c_i) \in Q$. So,if the local configuration of the $i$th cell at time $t$ is denoted by $c_i^t$ then $\mu_i(c_i^t) = c_i^{(t+1)}(i)$.

**Definition 2.5** A local transition function $\mu$ is an **identity function** denoted by $\mu_e$ if, $\forall i \in \mathbb{Z}$, $\mu(c_i) = c_i(i)$

A global transition function $\tau_e$ is an **identity function** provided for all $C \in Q^{\mathbb{Z}}$, $\tau_e(C) = C$.

A CA is said to be an **identity Cellular Automaton** if the global transition function is an identity function.

**Definition 2.6** A local transition function $\mu$ is a **constant function** denoted by $\mu_q$ if, $\forall i \in \mathbb{Z}$, $\mu(c_i) = q$ for a particular state $q \in Q$.

A global transition function $\tau$ is a **constant function** provided for all $C \in Q^Z$, $\tau(C) = C^*$ for a particular constant configuration $C^* \in Q^{\mathbb{Z}}$.

A CA is said to be a **constant Cellular Automaton** if the global transition function is a constant function.

**Definition 2.7** A local transition function $\mu$ is an **m-place left shift function** denoted by $\mu_{Lm}$ where $m \in \mathbb{N}$ is finite, if the state of the $i$th automaton $c_i(i)$ shifts $m - place$ leftwards. So, $\forall i \in \mathbb{Z}$, $\mu_{Lm}(c_i) = c_i(i + m)$ where $c_i$ is the restriction of $C$.

A global transition function $\tau$ is an **m-place left shift function** denoted by $\tau_{Lm}$ where $m \in \mathbb{N}$ is finite, if $\forall i \in \mathbb{Z}$, $\tau_{Lm}(C)(i) = C(i + m)$.
A CA is said to be a **m-place left shift Cellular Automaton** if the global transition function is a m-place left shift function.

**Definition 2.8** A local transition function $\mu$ is an **m-place right shift function** denoted by $\mu_{Rm}$ where $m \in \mathbb{N}$ is finite, if the state of the $i$th automaton $c_i(i)$ shifts $m - place$ rightwards. So, $\forall i \in \mathbb{Z}$, $\mu_{Rm}(c_i) = c_i(i - m)$ where $c_i$ is the restriction of $C$.

A global transition function $\tau$ is an **m-place right shift function** denoted by $\tau_{Rm}$ where $m \in \mathbb{N}$ is finite, if $\forall i \in \mathbb{Z}$, $\tau_{Rm}(C)(i) = C(i - m)$.

A CA is said to be a **m-place right shift Cellular Automaton** if the global transition function is a m-place right shift function.

**Definition 2.9** For a binary CA $(Q, Q^{\mathbb{Z}}, \tau)$ if $C_1, C_2 \in Q^{\mathbb{Z}}$ such that $\tau(C_1) = C_2$ where

$$\forall i \in \mathbb{Z},\ C_1(i) = 0 \leftrightarrow C_2(i) = 1 \ and\ C_1(i) = 1 \leftrightarrow C_2(i) = 0$$

then $C_1$ is the complement of $C_2$ and vice versa. $\tau$ is said to be the **complementary transition function** and is denoted by $\tau^c$.

If a local transition function $\mu$ be such that, for any $i \in \mathbb{Z}$, and local configuration $c_i$,

$$c_i(i) = 0 \leftrightarrow \mu(c_i) = 1 \ \ and,\ \ c_i(i) = 1 \leftrightarrow \mu(c_i) = 0$$

then $\mu$ is said to be a local complementary function and is denoted by $\mu^c$.

# 3 Relation Between Local and Global Transition Function of a Non-uniform CA

In a uniform CA, the local transition function of each cell is identical. So the nature of evolution of the local transition function of any cell would reflect the nature of evolution of the entire grid of cells of the CA. However, the following proposition shows that such may not be the case for a non-uniform CA.

**Proposition 3.1** *If the global transition function* $(\tau)$ *of a finite celled non-uniform CA with periodic boundaries composed of finite blocks be invertible then the local transition function* $(\mu_i)$ *of the cells will be invertible.*

***Proof*** We know that $\forall i \in \mathbb{Z},\ C(i) = c_i(i)$ and $\tau(C)(i) = \mu_i(c_i)$. If $\tau$ be invertible then,

$$\tau^{-1}(\tau(C)(i)) = C(i) = c_i(i)$$

Therefore, $\exists\ \mu_i^{'}$ such that $\forall i \in \mathbb{Z},\ \mu_i^{'}(\mu_i(c_i)) = c_i$.

Hence the proposition follows. □

*Remark 3.1* However, converse may not be true is justified by the following example.

*Example 3.1* Let us consider a non-uniform CA with periodic boundaries having initial configuration $C_0 = \overbrace{1010}^{B_1}\ \overbrace{010}^{B_2}$ such that cells of block $B_1$ follow $\mu_{R1}$ and cells of $B_2$ follow $\mu_{L1}$. Now $(\mu_{R1})^{-1} = \mu_{L1}$. But for the cells of block $B_1$, $\mu_{L1}(\mu_{R1}(c_i))$ gives $\overbrace{1011}^{B_1}$ and not $\overbrace{1010}^{B_1}$.

Hence the CA is not invertible globally and the result follows.

## 4 Patterns of Evolution of a Non-uniform CA Under Some Transitions

The grid line of the non-uniform CA studied in this section consists of a number of blocks of cells. Each block of cells has a different local transition function so that the CA becomes non-uniform or hybrid. We have considered identity, complementary, constant and shift functions and a combination of them as local transition functions and studied their evolution patterns. If a non-uniform CA have a block of cells whose local transition function is complementary then the CA must be a binary CA.

We have also restricted our model so that the blocks are of any finite length and the states of the automata are finite. So it is obvious that the evolution pattern will either be stationary after a finite time or be cyclic. In the first case we calculated the time that elapses before the evolution becomes stationary and in the second case we found the period of the cycle.

Any CA mentioned in this section will mean a non-uniform (hybrid) CA as described above in which the initial configuration can be random in nature.

**Proposition 4.1** *A finite celled non-uniform CA with periodic boundaries composed of two blocks having two distinct transitions from the set of functions* $S = \{(\mu_e), (\mu^c), (\mu_q | q \in Q)\}$ *transition function is either stable or 2-period cyclic from the initial step or after the first step.*

***Proof*** The proof for the different cases are as follows :

- If a two blocked finite celled non-uniform CA follow two transition functions $\mu_e$ and $\mu^c$, the CA is 2-period cyclic from the initial step.
- If a two blocked finite celled non-uniform CA follow two transition functions $\mu_e$ and $\mu_q$, the CA is stable from initial step if all the cells which follow $\mu_q$ be at state $q$ initially; otherwise the CA is stable after the first step.
- If a two blocked finite celled non-uniform CA follow two transition functions $\mu^c$ and $\mu_q$, the CA is 2-period cyclic from initial step if all the cells which follow $\mu_q$ be at state $q$ initially; otherwise the CA is 2-periodic after the first step.

Hence the proposition. □

**Corollary 4.1** *The evolution of a three blocked finite celled non-uniform CA following all the three transition functions of S in any order whatsoever is 2-period cyclic.*

**Corollary 4.2** *A non-uniform CA with periodic boundaries composed of finite blocks having transition functions from S is 2-period cyclic if atleast one of the blocks have* $\mu^c$ *as local transition function; otherwise the CA is stable.*

*Remark 4.1* Evolution patterns of a non-uniform CA with fixed boundaries composed of finite blocks having transition functions from $S$ is similar to that with periodic boundaries.

**Proposition 4.2** *Let a non-uniform CA with periodic boundaries be composed of finite blocks of length* $k_1, k_2, k_3, k_4$ *in a row. The transition function of the* $k_1$ *celled block is* $\mu_e$ *or* $\mu_q$ *for some* $q \in Q$, $k_2$ *celled block is* $\mu_{Lm}$ *and* $k_4$ *celled block is* $\mu_{Rn}$ *such that for* $i = 1, 2, 3, 4$, *each* $k_i \geq max\{m, n(\in \mathbb{N})\}$.

*The CA will be stable after atmost*

1. $p + 1$ *steps if* $k_3$ *celled block follows* $\mu_q$
2. $p$ *steps if* $k_3$ *celled block follows* $\mu_e$

*where* $p = max\{\lambda, \nu\}$ *such that* $\lambda = \lceil \frac{k_2}{m} \rceil$ *and* $\nu = \lceil \frac{k_4}{n} \rceil$.

***Proof*** At each discrete time step $m$ left-most cell elements will be deleted from the block of $k_2$ cells and $m$ left-most elements from the block of $k_3$ cells will occupy the $m$ right-most position of the $k_2$ celled block. Similarly $n$ right-most cell elements from the block of $k_4$ cells will be deleted and $n$ right-most elements from the $k_3$ celled block will occupy the $n$ left-most position of the $k_4$ celled block.

The above process will continue until some element of the $k_3$ celled block reaches the left-most position of the $k_2$ celled block and reaches the right-most position of the $k_4$ celled block for the first time.

Henceforth the CA remains stationary.

Let $\lambda = \lceil \frac{k_2}{m} \rceil$, $\nu = \lceil \frac{k_4}{n} \rceil$ and $p = max\{\lambda, \nu\}$.

1. If $k_3$ celled block follows $\mu_q$, then the left-most position of the $k_2$ celled block is reached by $q$ after atmost $\lambda + 1$ steps and the right-most position of the $k_4$ celled block is reached by $q$ after atmost $\nu + 1$ steps. So, after $p + 1$ time steps each cell of $k_2, k_3, k_4$ celled blocks will be in the state $q$ and the CA becomes stable.
2. If $k_3$ celled block follows $\mu_e$, then the left-most position of the $k_2$ celled block is reached by some element of $k_3$ after atmost $\lambda$ steps and the right-most position of the $k_4$ celled block is reached by by some element of $k_3$ after atmost $\nu$ steps. The CA becomes stable after $p$ steps.

Hence the proposition. □

**Corollary 4.3** *A non-uniform CA composed of finite blocks of length* $k_1, k_2, k_3, k_4$ *in a row where* $k_1, k_2, k_4$ *celled blocks follow* $\mu^c, \mu_{Lm}, \mu_{Rn}$ *respectively, will be 2-period cyclic after atmost* $p + 1(p)$ *steps if* $k_3$ *celled block follows* $\mu_q(\mu_e)$

**Corollary 4.4** *The above proposition also holds true if the blocks* $k_1$ *and* $k_3$ *following same transition functions be infinite in length.*

*Remark 4.2* A non-uniform CA with periodic boundaries composed of three finite blocks of length either $k_1, k_2, k_3$ or $k_3, k_4, k_1$ in a row under transitions defined in the above proposition also evolves similar patterns as above.

*Remark 4.3* The above proposition also holds true for a finite celled non-uniform CA with fixed boundaries since the evolution of the $k_2$ and $k_4$ celled block is dependent on the elements of the $k_3$ celled block only.

**Proposition 4.3** *Let a non-uniform CA with boundaries fixed to $q_B \in Q$ be composed of two finite blocks of length $k_1$ and $k_2$ such that for $m < n(\in \mathbb{N})$ and some $r_1, r_2 \in \mathbb{N}$, $k_1 = mr_1$ and $k_2 = nr_2$.*

*The transition function of the $k_1(k_2)$ celled block is the shift function $\mu_{Lm}(\mu_{Ln})$.*

*The CA will become stationary with all cells acquiring $q_B$ state after atmost $r_1 + r_2$ steps*

***Proof*** At each discrete time step $m$ left-most cell elements will be deleted from the block of $k_1$ cells and $m$ left-most elements from the block of $k_2$ cells will occupy the $m$ right-most position of the $k_1$ celled block.

Consequently, since $m < n$, $(m+1)$th to $n$th left-most elements from the block of $k_2$ cells will be deleted and $n$ right-most positions of the $k_2$ celled block will be occupied by $q_B$ state.

Since $r_1 = \frac{k_1}{m}$ and $r_2 = \frac{k_2}{n}$, all the cells acquire $q_B$ state after atmost $r_1 + r_2$ steps. □

*Remark 4.4* Similar evolution patterns can be obtained for a non-uniform CA with fixed boundaries composed of two finite blocks of length $k_1$ and $k_2$ having transition function $\tau_{Rm}$ and $\tau_{Rn}$ respectively.

**Theorem 4.1** *Let a non-uniform CA with periodic boundaries be composed of two finite blocks of length $k_1$ and $k_2$. The transition function of the $k_1(k_2)$ celled block is the shift function $\mu_{Lm}(\mu_{Rn})$ such that $k_1, k_2 \geq max\{m, n(\in \mathbb{N})\}$.*

*The CA will be periodic from pth step where $p = max\{\lambda, \nu\}$ such that $\lambda = \lceil \frac{k_1 - n}{m} \rceil$ and $\nu = \lceil \frac{k_2 - m}{n} \rceil$.*

*Periodicity of the CA will be given by $h = \frac{m+n}{g.c.d.(m,n)}$*

***Proof*** Let the initial configuration of an $i$th cell $C_0(i)$ for $i = 1, 2, ..., k_1 + k_2$, be denoted by $C_0^i$

| ← | | $\mu_{Lm}$ | | | | | | | $\mu_{Rn}$ | | → |
|---|---|---|---|---|---|---|---|---|---|---|---|
| $C_0^1$ | ... | $C_0^{k_1-n+1}$ | ... | $C_0^{k_1-1}$ | $C_0^{k_1}$ | $C_0^{k_1+1}$ | $C_0^{k_1+2}$ | ... | $C_0^{k_1+m}$ | ... | $C_0^{k_1+k_2}$ |
| | ... | $a_1$ | ... | $a_{n-1}$ | $a_n$ | $b_1$ | $b_2$ | ... | $b_m$ | ... | |

We rename the element at each cell as :

$$
\begin{aligned}
C_0^i &= a_{i+n-k_1} \quad if\ 1 \leq i \leq k_1 \\
&= b_{i-k_1} \quad if\ k_1 < i \leq k_2
\end{aligned}
$$

At each discrete time step $m$ left-most cell elements will be deleted from the block of $k_1$ cells and $m$ left-most elements from the block of $k_2$ cells will occupy the $m$ right-most position of the $k_1$ celled block. Similarly $n$ right-most cell elements from the block of $k_2$ cells will be deleted and $n$ right-most elements from the $k_1$ celled block will occupy the $n$ left-most position of the $k_2$ celled block.

The left-most position of the $k_1$ celled block is reached by '$a_1$'or some '$b_i$'($i = 1, 2, ..., m$) for the first time after $\lambda$ steps where $\lambda m + n \geq k_1 \Rightarrow \lambda = \lceil \frac{k_1 - n}{m} \rceil$.

Similarly the right-most position of the $k_2$ celled block is reached by '$b_m$'or some '$a_j$'($j = 1, 2, ..., n$) for the first time after $\nu$ steps where $\nu = \lceil \frac{k_2 - m}{n} \rceil$.

After $p$th step where $p = max(\lambda, \nu)$ there will be a rotation of blocks '$a_1 a_2 ... a_n b_1 b_2 ... b_m$' in a cyclic form.

Hence each configuration will be repeated after $h$ steps where $h = \frac{m+n}{g.c.d.(m,n)}$. □

*Remark 4.5* The above proposition also holds true for a finite celled non-uniform CA with fixed boundaries since the evolution of the $k_1$ and $k_2$ celled block is dependent on the $n$ right-most elements of the $k_1$ celled block and $m$ left-most element of the $k_2$ celled block.

**Theorem 4.2** *Let a non-uniform CA with periodic boundaries be composed of s blocks for some finite $s \in \mathbb{N}$ where any $i^{th}$ ($i = 1, 2, ..., s$) block is composed of two finite sub-blocks of length $k_1^i$ and $k_2^i$.*

*The transition function of the $k_1^i$ ($k_2^i$) celled block is the shift function $\mu_{Lm^i}$ ($\mu_{Rn^i}$) such that $k_1^i, k_2^i \geq max\{m^i, n^i (\in \mathbb{N})\}$.*

*The CA will be periodic from pth step where $p = max\{p^1, p^2, ..., p^s\}$ such that $p^i = max\{\lambda^i, \nu^i\}$ and $\lambda^i = \lceil \frac{k_1^i - n^i}{m^i} \rceil$ and $\nu^i = \lceil \frac{k_2^i - m^i}{n^i} \rceil$.*

*Periodicity of the CA will be given by*

$$l.c.m. \left\{ \frac{m^1 + n^1}{g.c.d.(m^1, n^1)}, \frac{m^2 + n^2}{g.c.d.(m^2, n^2)}, ..., \frac{m^s + n^s}{g.c.d.(m^s, n^s)} \right\}$$

***Proof*** From the previous proposition we get that any $i^{th}$ ($i = 1, 2, ..., s$) block of the non-uniform CA will be periodic from the ($p^i$)th time step where

$p^i = max\{\lambda^i, \nu^i\}$ such that $\lambda^i = \lceil \frac{k_1^i - n^i}{m^i} \rceil$ and $\nu^i = \lceil \frac{k_2^i - m^i}{n^i} \rceil$ having periodicity $\frac{m^i + n^i}{g.c.d.(m^i, n^i)}$.

Hence it follows that the non-uniform CA will be periodic from $p$th time step where $p = max\{p^1, p^2, ..., p^s\}$ and the periodicity of the CA will be given by

$$l.c.m. \left\{ \frac{m^1 + n^1}{g.c.d.(m^1, n^1)}, \frac{m^2 + n^2}{g.c.d.(m^2, n^2)}, ..., \frac{m^s + n^s}{g.c.d.(m^s, n^s)} \right\}.$$

□

## 5 Conclusion

Evolution patterns of non-uniform CA of combinations of a few standard local transition functions have been studied in this paper. We could specify the exact pattern.

It can be observed that the evolution of a uniform CA can be treated as a special case of evolution of a non-uniform CA having a single block of cells which follow the same transition function.

The investigation on evolution patterns of a non-uniform CA may be extended to other linear transition functions. Extension of the length of the blocks from finite to infinite may also be worthwhile.

## References


1. Adak, S., Mukherjee, S., Das, S.: Reachability problem in non-uniform cellular automata. Inf. Sci. **543**, 72–84 (2021). Jan
2. Basu, S., Ghosh, S.: Fuzzy cellular automata model for discrete dynamical system representing spread of mers and covid-19 virus. In: Internet of Medical Things for Smart Healthcare, pp. 267-304. Springer (2020)
3. Dennunzio A., Formenti E., Provillard J.: Computational complexity of rule distributions of non-uniform cellular automata. In: Proceedings of the 6th International Conference on Language and Automata Theory and Applications, LATA'12 (2012). ISBN 978-3-642-28331-4
4. Ghosh S, Basu S: Evolution patterns of some boolean cellular automata having atmost one active cell to model simple dynamical systems. Bull. Cal. Math Soc. **108**(6) 449–464 (2016)
5. Ghosh S, Basu S: Evolution patterns of finite celled synchronous cellular automata having atmost one active cell. In: Proceedings of the 10th International Conference MSAST 2016 (December 2016)
6. Ghosh S, Basu S: Some algebraic properties of linear synchronous cellular automata (30 AUG. 2017). arXiv:1708.09751v1 [nlin CG]
7. Ghosh, S.: Evolutions of some one-dimensional homogeneous cellular automata. Complex Syst. **30**, 75–92 (2021)
8. Hedlund, G.A.: Endomorphisms and automorphisms of the shift dynamical system. Math. Syst. Theory **3**(4), 320–375 (1969)
9. Kari, J.. Theory of cellular automata: a survey. Theor. Comp. Sci. **334**, 3 33 (2005)
10. Nishio, H., Worsch, T.: Changing the neighbourhood of cellular automata: local structure, equivalence and isomorphism. J. Cell. Autom. **5**(3), 227–240 (2010)
11. Nishio H : Automorphism classification of cellular automata, Fundamenta Informaticae **104**(1), 125–140 (2010)
12. von Neumann, J.: Theory of Self-Reproducing Automata. University of Illinois Press, Illinois (1966). Edited and completed by A.W. Burks
13. Salo, V.: Groups and Monoids of Cellular Automata, Cellular Automata and Discrete Complex Systems, LNCS 9099, 17–45. Springer, Berlin Heidelberg (2015)
14. Sipper, M.: Co-evolving non-uniform cellular automata to perform computations. Phys. D **92**, 193–208 (1996)
15. Ulam S.M : On some mathematical problems connected with patterns of growth of figures. Proc. Symp. Appl Math. Am. Math. Soc. **14**, 215–224 (1962)
16. Voorhees, B.: Predecessor states for certain cellular automata evolutions. Commun. Math. Phys. **117**, 431–439 (1988)
17. Stephen, W.: Theory and Applications of Cellular Automata. World Scientific, Singapore (1986)
18. Wolfram S: A New Kind of Science. Wolfram Media (2002)

# Relationship of Two Discrete Dynamical Models: One-Dimensional Cellular Automata and Integral Value Transformations

**Sreeya Ghosh, Sudhakar Sahoo, Sk. Sarif Hassan, Jayanta Kumar Das, and Pabitra Pal Choudhury**

**Abstract** Cellular Automaton(CA) and an Integral Value Transformation(IVT) are two well established mathematical models which evolve in discrete time steps. Theoretically, studies on CA suggest that CA is capable of producing a great variety of evolution patterns. However computation of non-linear CA or higher dimensional CA maybe complex, whereas IVTs can be manipulated easily. The main purpose of this paper is to study the link between a transition function of a one-dimensional CA and that of IVTs. Mathematically, we have also established the algebraic structures of a set of transition functions of a one-dimensional CA as well as that of a set of IVTs using binary operations.

**Keywords** Cellular automaton · Integral value transformations · Transition functions

## 1 Introduction

Cellular Automaton(pl. cellular automata, abbrev. CA) is a discrete model which has applications in computer science, mathematics, physics, complexity science, theoretical biology and micro structure modeling. This model was introduced by J.von Neumann and S.Ulam in 1940 for designing self replicating systems [19, 22, 23].

S. Ghosh (✉)
Department of Applied Mathematics, University of Calcutta, Kolkata 700009, India

S. Ghosh · J. K. Das · P. P. Choudhury
Applied Statistics Unit, Indian Statistical Institute, Kolkata 700108, India

S. Sahoo
Institute of Mathematics and Applications, Bhubaneshwar 751029, India

Sk. S. Hassan
Department of Mathematics, Pingla Thana Mahavidyalaya, Paschim Medinipur 721140, India

J. K. Das
School of Medicine, Johns Hopkins University, Baltimore, MD 21287, USA



S. Das and G. J. Martinez (eds.), *Proceedings of First Asian Symposium on Cellular Automata Technology*, Advances in Intelligent Systems and Computing 1425,
https://doi.org/10.1007/978-981-19-0542-1_7

A CA consists of a finite/countably infinite number of finite-state semi-automata [11, 12] known as 'cells' arranged in an ordered $n$-dimensional grid. Each cell receives input from the neighbouring cells and changes according to the transition function. The transitions at each of the cells together induces a change of the grid pattern. The simplest CA is a CA where the grid is a one-dimensional line. Stephen Wolfram's work in the 1980s contributed to a systematic study of one-dimensional CA, providing the first qualitative classification of their behaviour [24]. CA has been studied for solving many interesting problems on utilizing mathematical bases such as polynomial, matrix algebra, Boolean derivative etc. [4, 6]. Further, complex and dynamic behaviours of CA can also be studied using various mathematical tools [25, 26] that help to understand the crucial properties and modeling of various classes of discrete dynamical systems [1, 2, 9, 18].

An Integral Value Transformation (abbrev. IVT), a class of discrete dynamical system, were first introduced during 2009–10 ([15]). An IVT of $k$-dimension is a function defined using $p$-adic numbers over $\mathbb{N}_0^k$ where $\mathbb{N}_0 = \mathbb{N} \cup \{0\}$, $p \in \mathbb{N}$. The IVTs have been studied in different viewpoints including the understanding of the evolution of integer sequences, behavioral patterns of integers in discrete time points as well as number preserving CA rules [8, 14, 21]. Unlike CA, in case of IVTs the updating of cells are independent of overlapping neighbourhood interaction or any boundary conditions [21].

In this paper, Wolfram code of an elementary CA and global transition function of a one-dimensional CA has been represented through IVTs. The algebraic structure of a set of transition functions of a one-dimensional CA as well as the algebraic structure of a set of IVTs under some binary operations have been studied.

## 2 Mathematical Preliminaries

**Definition 2.1** Let $Q$ be a finite set of memory elements also called the **state set**. A **global configuration** is a mapping from the group of integers $\mathbb{Z}$ to the set $Q$ given by $C : \mathbb{Z} \to Q$. The set $Q^{\mathbb{Z}}$ is the set of all global configurations where $Q^{\mathbb{Z}} = \{C | C : \mathbb{Z} \to Q\}$.

**Definition 2.2** A mapping $\tau : Q^{\mathbb{Z}} \to Q^{\mathbb{Z}}$ is called a **global transition function**. A **CA**(denoted by $\mathcal{C}_{\tau}^{Q}$)(reported in [10, 13, 17]) is a triplet $(Q, Q^{\mathbb{Z}}, \tau)$, where, $Q$ is the finite state set, $Q^{\mathbb{Z}}$ is the set of all configurations, $\tau$ is the global transition function.

**Definition 2.3** The set $\mathcal{Q}^{\mathbb{Z}} = \{\tau | \tau : Q^{\mathbb{Z}} \to Q^{\mathbb{Z}}\}$ is the set of all possible global transition functions of CA having state set $Q$. A mapping $\tau$ is invertible if $\forall C_i, C_j \in Q^{\mathbb{Z}}$, $\exists \tau^{-1}$ such that $\tau(C_i) = C_j \Leftrightarrow \tau^{-1}(C_j) = C_i$.

**Definition 2.4** For $i \in \mathbb{Z}, r \in \mathbb{N}$, let $S_i = \{i - r, ..., i - 1, i, i + 1, ..., i + r\} \subseteq \mathbb{Z}$. $S_i$ is the neighbourhood of the $i$th cell, $r$ is the radius of the neighbourhood of a cell. It follows that $\mathbb{Z} = \bigcup_i S_i$.

A restriction from $\mathbb{Z}$ to $S_i$ induces a restriction of $C$ to $\overline{c_i}$ given by $\overline{c_i} : S_i \rightarrow Q$; where $\overline{c_i}$ may be called **local configuration** of the $i$th cell.

The mapping $\mu_i : Q^{S_i} \rightarrow Q$ is known as a **local transition function** for the $i$th cell having radius $r$. Thus $\forall i \in \mathbb{Z}$, $\mu_i(\overline{c_i}) \in Q$ and it follows that,

$$\tau(C) = \tau(..., c_{i-1}, c_i, c_{i+1}, ...) = ...\mu_{i-1}(\overline{c_{i-1}}).\mu_i(\overline{c_i}).\mu_{i+1}(\overline{c_{i+1}})...$$

**Definition 2.5** The set $M = \{\mu_i | \mu_i : Q^{S_i} \rightarrow Q, i \in \mathbb{Z}\}$ is the set of all possible local transition functions of CA having state set $Q$.

**Definition 2.6** If for a particular CA, $|Q| = 2$ so that we can write $Q = \{0, 1\}$, then the CA is said to be a **binary CA** or a **Boolean CA**. A Boolean CA having radius 1 is known as an **Elementary CA(ECA).**

**Definition 2.7** **Wolfram code** is a naming system often used for a one-dimensional CA introduced by Stephen Wolfram (see [24]).

For a one-dimensional CA with $Q$ states, the local rule $\mu_i$ for some $i$th cell, $i \in \mathbb{Z}$ of radius $r$(neighbourhood $2r + 1$) can be specified by an $Q^{2r+1}$-bit sequence. The decimal equivalent form of this sequence is known as the Wolfram code.

Thus, the Wolfram code for a particular rule is a number in the range from 0 to $Q^{Q^{2r+1}} - 1$, converted from $Q$-ary to decimal notation.

*Example 2.1* Let the local rule of an ECA for some $i$th cell, $i \in \mathbb{Z}$ be,

$$\mu_i(\overline{c_i}) = \mu_i(c_{i-1}, c_i, c_{i+1}) = (c_{i-1} \vee c_{i+1}) \wedge c_i$$

where $'\vee'$ stands for OR operation, $'\wedge'$ stands for AND operation, $c_j$ is the $j$th cell configuration for $j = i - 1, i, i + 1$. Then we get the $2^3$-bit sequence as,

$$\mu(111)\mu(110)\mu(101)\mu(100)\mu(011)\mu(010)\mu(001)\mu(000) = 1\,1\,0\,0\,1\,0\,0\,0$$

The decimal equivalent number for 11001000 is 200 and so the Wolfram code is RULE **200**.

**Definition 2.8** A $p$-adic $k-$dimensional **Integral Value Transformation(IVT)** denoted by $IVT_j^{p,k}$ for $p \in \mathbb{N}$, $k \in \mathbb{N}$ is a function of $p$-base numbers from $\mathbb{N}_0^k$ to $\mathbb{N}_0$ defined(in [16]) as

$$IVT_j^{p,k}(n_1, n_2, ..., n_k) = (f_j(a_0^{n_1}, ..., a_0^{n_k})f_j(a_1^{n_1}, ..., a_1^{n_k})...f_j(a_{l-1}^{n_1}, ..., a_{l-1}^{n_k}))_p = m$$

where $\mathbb{N}_0 = \mathbb{N} \cup \{0\}$, $p \in \mathbb{N}$, $n_s = (a_0^{n_s} a_1^{n_s} ... a_{l-1}^{n_s})_p$ for $s = 1, 2, ..., k$.
$f_j$ is a function from $\{0, 1, ..., p - 1\}^k$ to $\{0, 1, ..., p - 1\}$ for $j = 0, 1, ..., p^{p^k} - 1$, $m$ is the decimal conversion of the $p$-base number.

*Remark 2.1* In particular [3, 5, 7, 20], if $\forall i = 0, 1, ..., l - 1$, the function $f_j$ be defined as

1. $f_j(a_i^{n_1}, ..., a_i^{n_k}) = \lfloor (a_i^{n_1} + ... + a_i^{n_k})/p \rfloor$, then $IVT_j^{p,k}$ is known as a **Modified Carry Value Transformation(MCVT)**.
   Again, $MCVT$ with a 0 padding at the right end is known as a **Carry Value Transformation(CVT)**.
2. $f_j(a_i^{n_1}, ..., a_i^{n_k}) = (a_i^{n_1} + ... + a_i^{n_k}) \ mod \ p$, then $IVT_j^{p,k}$ is known as an **Exclusive OR Transformation(XORT)**.
3. $f_j(a_i^{n_1}, ..., a_i^{n_k}) = max(a_i^{n_1}, ..., a_i^{n_k})$, then $IVT_j^{p,k}$ is known as an **Extreme Value Transformation(EVT)**.

## 3 Wolfram Code of CA and IVT

For an ECA, Wolfram code for a local rule $\mu$ is the decimal number $j \in \{0, 1, 2, ..., 255\}$ obtained from the 8-bit sequence $\mu(111)\mu(110)\mu(101)\mu(100)\mu(011)\mu(010)\mu(001)\mu(000) = j$. Therefore, Wolfram code for each local transition function of a 3-neighbourhood Boolean CA can be represented by a 2 base 3-dimensional IVT. The function $\mu$ in the above 8-bit sequence can be represented by a function $f_j : \{0, 1\}^3 \rightarrow \{0, 1\}$ which gives

$$(f_j(111) f_j(110) f_j(101) f_j(100) f_j(011) f_j(010) f_j(001) f_j(000))_2$$

$$= IVT_j^{2,3}(11110000_2, 11001100_2, 10101010_2) = IVT_j^{2,3}(240_{10}, 204_{10}, 170_{10})$$

Hence it follows that for a 3−neighbourhood Boolean CA, any Wolfram code $j \in \{0, 1, 2, ...255\}$ can be equivalently represented by $IVT_j^{2,3}(240_{10}, 204_{10}, 170_{10})$.

*Example 3.1* Wolfram code 200 can be equivalently represented as
$200_{10} = (11001000)_2 = IVT_{200}^{2,3}(240_{10}, 204_{10}, 170_{10})$

However the following example shows that for $j \in \{0, 1, 2, ...255\}$, any $IVT_j^{2,3}$ may not correspond to a Wolfram code.

*Example 3.2* $IVT_j^{2,3}(240_{10}, 204_{10}, 171_{10}) = IVT_j^{2,3}(11110000_2, 11001100_2, 10101011_2) = (f_j(111) f_j(110) f_j(101) f_j(100) f_j(011) f_j(010) f_j(001) f_j(001))_2$

In this 8-bit sequence, $f_j(000)$ is missing and so this cannot correspond to any Wolfram code.

For a 3-state 3 neighbourhood CA, the local rule can be represented by a function $f_j : \{0, 1, 2\}^3 \rightarrow \{0, 1, 2\}$

*Example 3.3* CA transition rule 200 can be equivalently represented in 3-base, 3-neighbourhood system as $200_{10} = (21102)_3 = IVT_{200}^{3,3}(222222222111111111000000000_3, 222111000222111000222111000_3, 210210210210210210210210210_3) = IVT_{200}^{3,3}(7625403764901_{10}, 7479532539765_{10}, 6159136430181_{10})$

For a 2-state 5 neighbourhood CA, the local rule can be represented by a function $f_j : \{0, 1\}^5 \rightarrow \{0, 1\}$.

*Example 3.4* CA transition rule 69800 can be equivalently represented in 2-base, 5-neighbourhood system as $69800_{10} = (10001000010101000)_2 = IVT^{2,5}_{69800}$ $(11111111111111110000000000000000_2, 11111111000000001111111100000000_2, 11110000111100001111000011110000_2, 11001100110011001100110011001100_2, 10101010101010101010101010101010_2) = IVT^{2,5}_{69800}(4294901760_{10}, 4278255360_{10}, 4042322160_{10}, 3435973836_{10}, 2863311530_{10})$

## 3.1 Some Particular Wolfram Codes and Particular IVTs

We know that transformations such as MCVT, XORT, EVT are particular cases of IVTs. Again, any Wolfram code can be represented by an IVT. Therefore some particular Wolfram codes which can be represented by an MCVT, XORT or EVT are as follows.

1. Let the local rule of an ECA for some $i$th cell, $i \in \mathbb{Z}$ be, $\mu_i(\overline{c_i}) = \mu_i(c_{i-1}, c_i, c_{i+1}) = (c_{i-1} \wedge c_i) \vee ((c_{i-1} \vee c_i) \wedge c_{i+1})$.
   Then we get the $2^3$-bit sequence as 11101000 and Wolfram code 232.
   Here, $\mu_i$ can be represented by function $f_j : \{0, 1\}^3 \rightarrow \{0, 1\}$ given by

   $$f_j(c_{i-1}, c_i, c_{i+1}) = (c_{i-1} \wedge c_i) \vee ((c_{i-1} \vee c_i) \wedge c_{i+1}) = \lfloor (c_{i-1} + c_i + c_{i+1})/2 \rfloor$$

   Thus, Wolfram code 232 can be represented as

   $$(f_{232}(111) f_{232}(110) f_{232}(101) f_{232}(100) f_{232}(011) f_{232}(010) f_{232}(001) f_{232}(000))_2$$

   $$= (\lfloor 3/2 \rfloor \lfloor 2/2 \rfloor \lfloor 2/2 \rfloor \lfloor 1/2 \rfloor \lfloor 2/2 \rfloor \lfloor 1/2 \rfloor \lfloor 1/2 \rfloor \lfloor 0/2 \rfloor)_2$$

   $$= MCVT^{2,3}_{232}(11110000_2, 11001100_2, 10101010_2) = MCVT^{2,3}_{232}(240_{10}, 204_{10}, 170_{10})$$

2. Let the local rule of an ECA for some $i$th cell, $i \in \mathbb{Z}$ be, $\mu_i(\overline{c_i}) = \mu_i(c_{i-1}, c_i, c_{i+1}) = (c_{i-1} \underline{\vee} c_i \underline{\vee} c_{i+1})$.
   Then we get the $2^3$-bit sequence as 10010110 and Wolfram code 150.
   Here, $\mu_i$ can be represented by function $f_j : \{0, 1\}^3 \rightarrow \{0, 1\}$ given by

   $$f_j(c_{i-1}, c_i, c_{i+1}) = (c_{i-1} \underline{\vee} c_i \underline{\vee} c_{i+1}) = (c_{i-1} + c_i + c_{i+1}) \ mod\ 2$$

   Thus, Wolfram code 150 can be represented as

   $$(f_{150}(111) f_{150}(110) f_{150}(101) f_{150}(100) f_{150}(011) f_{150}(010) f_{150}(001) f_{150}(000))_2$$

   $$= ((3 mod 2)(2 mod 2)(2 mod 2)(1 mod 2)(2 mod 2)(1 mod 2)(1 mod 2)(0 mod 2))_2$$

   $$= XORT^{2,3}_{150}(11110000_2, 11001100_2, 10101010_2) = XORT^{2,3}_{150}(240_{10}, 204_{10}, 170_{10})$$

3. Let the local rule of an ECA for some $i$th cell, $i \in \mathbb{Z}$ be, $\mu_i(\overline{c_i}) = \mu_i(c_{i-1}, c_i, c_{i+1}) = (c_{i-1} \vee c_i \vee c_{i+1})$.
Then we get the $2^3$-bit sequence as 11111110 and Wolfram code 254.
Here, $\mu_i$ can be represented by function $f_j : \{0, 1\}^3 \rightarrow \{0, 1\}$ given by

$$f_j(c_{i-1}, c_i, c_{i+1}) = (c_{i-1} \vee c_i \vee c_{i+1}) = max\{c_{i-1}, c_i, c_{i+1}\}$$

Thus, Wolfram code 254 can be represented as
$(f_{254}(111)f_{254}(110)f_{254}(101)f_{254}(100)f_{254}(011)f_{254}(010)f_{254}(001)f_{254}(000))_2 = (11111110)_2$

$$= EVT_{254}^{2,3}(11110000_2, 11001100_2, 10101010_2) = EVT_{254}^{2,3}(240_{10}, 204_{10}, 170_{10})$$

## 4 Transition Function of a CA and IVT

**Definition 4.1** The set $T^{p,k} = \{IVT_j^{p,k} | IVT_j^{p,k} : \mathbb{N}_0^k \rightarrow \mathbb{N}_0\}$ is the set of all $p$-base $k$-dimensional IVTs for $p \in \mathbb{N}, k \in \mathbb{N}$.

**Definition 4.2** The $v$-fold Cartesian product $\underbrace{T^{p,k} \times ... \times T^{p,k}}_{v\ times}$ for $v \in \mathbb{N}$, denoted by $\mathscr{T}^v$ is given by

$$\mathscr{T}^v = \{(IVT_{j_1}^{p,k}, IVT_{j_2}^{p,k}, ..., IVT_{j_v}^{p,k})\}$$

where $IVT_{j_s}^{p,k} : \mathbb{N}_0^k \rightarrow \mathbb{N}_0 \in T^{p,k}$, for $s = 1, 2, ..., v(\in \mathbb{N})$

**Definition 4.3** Let a restriction on $IVT_j^{p,k}$ from $\mathbb{N}_0^k$ to $Q^k = \{0, 1, ..., p-1\}^k$ for $p \in \mathbb{N}$ be denoted by $\underline{IVT}_j^{p,k}$. The set $T_{|Q}^{p,k} = \{\underline{IVT}_j^{p,k} | \underline{IVT}_j^{p,k} : Q^k \rightarrow Q\}$ is the set of all $p$-base $k$-dimensional IVTs when $\mathbb{N}_0^k$ is restricted to the subset $Q^k$. Therefore for $(\eta_1, \eta_2, ..., \eta_k) \in Q^k$ we get,

$$\underline{IVT}_j^{p,k}(\eta_1, \eta_2, ..., \eta_k) = f_j(\eta_1, \eta_2, ..., \eta_k)_p = m \in Q$$

**Definition 4.4** The set $\mathscr{T}^v_{|Q} = \{(\underline{IVT}_{j_1}^{p,k}, \underline{IVT}_{j_2}^{p,k}, ..., \underline{IVT}_{j_v}^{p,k})\}$ is the $v$-fold Cartesian product $\underbrace{T_{|Q}^{p,k} \times ... \times T_{|Q}^{p,k}}_{v\ times}$ when $\mathbb{N}_0^k$ is restricted to subset $Q^k$ where

$$\underline{IVT}_{j_s}^{p,k} : Q^k \rightarrow Q \in T_{|Q}^{p,k} \text{ for } s = 1, 2, ..., v(\in \mathbb{N})$$

Now, a local transition function of any $i$th cell of a one-dimensional CA having $p(\in \mathbb{N})$ states and radius $r(\in \mathbb{N})$ will be of the form

$$\mu_i(\overline{c_i}) = \mu_i(c_{i-r}, ..., c_i, ..., c_{i+r}) = c_i^*$$

where $c_{i-r}, ..., c_i, ..., c_{i+r}, c_i^* \in \{0, 1, ..., p-1\}$.
This can be represented by some $\underline{IVT}_{j_i}^{p,k}$, where $p \in \mathbb{N}, k = 2r+1$ and $j_i \in \{0, 1, 2, ..., (p^{p^k} - 1)\}$ is the underlying Wolfram code(which is in turn equal to $\underline{IVT}_{j_i}^{p,k}(240_{10}, 204_{10}, 170_{10})$).
Thus $\forall i \in \mathbb{Z}$,

$$\mu_i(\overline{c_i}) \equiv \underline{IVT}_{j_i}^{p,k}(\overline{c_i})$$

Now if $\tau$ be the global transition function of any $v(\in \mathbb{N} \geq k)$-celled CA, then for a global configuration $C = (c_1, c_2, ..., c_v)$, we get

$$\tau(C) = \tau(c_1, c_2, ..., c_v) = \mu_1(\overline{c_1}).\mu_2(\overline{c_2})....\mu_v(\overline{c_v})$$

Therefore it follows that,

$$\tau(c_1, c_2, ..., c_v) \equiv (\underline{IVT}_{j_1}^{p,k}(\overline{c_1}), \underline{IVT}_{j_2}^{p,k}(\overline{c_2}), ..., \underline{IVT}_{j_v}^{p,k}(\overline{c_v}))$$

Hence for any $v$-celled ECA it follows that,

$$\tau(C) \equiv (\underline{IVT}_{j_1}^{2,3}(\overline{c_1}), \underline{IVT}_{j_2}^{2,3}(\overline{c_2}), ..., \underline{IVT}_{j_v}^{2,3}(\overline{c_v}))$$

where $j_1, j_2, ..., j_v \in \{0, 1, ..., 255\}$.
Now if $\tau$ be the global transition function of a countably infinite celled CA with cells having $k$ neighbourhood, then for any global configuration $C = (...c_{i-1}, c_i, c_{i+1}, ...)$, we get

$$\tau(C) = ...\mu_{i-1}(\overline{c_{i-1}}).\mu_i(\overline{c_i}).\mu_{i+1}(\overline{c_{i+1}})...$$

Therefore it follows that,

$$\tau(C) \equiv (..., \underline{IVT}_{j_{i-1}}^{p,k}(\overline{c_{i-1}}), \underline{IVT}_{j_i}^{p,k}(\overline{c_i}), \underline{IVT}_{j_{i+1}}^{p,k}(\overline{c_{i+1}}), ...)$$

where $...j_{i-1}, j_i, j_{i+1}, ... \in \{0, 1, 2, ..., (p^{p^k} - 1)\}$.

*Example 4.1* Let the initial configuration of a CA having state set $\{0, 1, 2\}$ be $C_0 = (01201)_3$ and the transition function be given by

$$\tau(01201)_3 = \mu_1(101)\mu_2(012)\mu_3(120)\mu_4(201)\mu_5(010) = (02001)_3$$

where $\mu_1, \mu_3, \mu_5$ follow Wolfram code 377 and $\mu_2, \mu_4$ follow Wolfram code 588 for 3-state CA.
A local transition function can be equivalently represented by some $\underline{IVT}_j^{3,3}$ and thus it follows that $\tau(01201)_3$ is equivalent to

$$\underline{IVT}_{377}^{3,3}(1,0,1)\underline{IVT}_{588}^{3,3}(0,1,2)\underline{IVT}_{377}^{3,3}(1,2,0)\underline{IVT}_{588}^{3,3}(2,0,1)\underline{IVT}_{377}^{3,3}(0,1,0)$$

Conversely, for any $\underline{IVT}_j^{p,k}$ if $k$ be odd, i.e. if $k = 2r + 1$ for some $r < k(\in \mathbb{N}_0)$, then $\underline{IVT}_j^{p,k}(\eta_1, \eta_2, ..., \eta_{2r+1})$ will be equivalent to a local transition function of the $(r+1)$th cell in a CA with $p(\in \mathbb{N})$ states and $2r+1$ neighbourhood whose underlying Wolfram code is $j \in \{0, 1, ..., p^{p^k} - 1\}$, given by

$$\underline{IVT}_j^{q,l}(\eta_1, \eta_2, ..., \eta_{2r+1}) = \underline{IVT}_j^{p,k}(\overline{\eta}_{r+1}) \equiv \mu_{r+1}(\overline{\eta}_{r+1})$$

*Remark 4.1* For a $p(\in \mathbb{N})$-state $k(= 2r+1 \in \mathbb{N})$-neighbourhood CA, clearly $\forall \overline{c_i} \in Q^{S_i} \subseteq Q^{\mathbb{Z}}$ $\mu_j(\overline{c_i}) \equiv \underline{IVT}_j^{p,k}(\overline{c_i})$ where $j \in \{0, 1, ..., p^{p^k} - 1\}$ is the underlying Wolfram Code. It follows that the set of local transition functions $M$ is equivalent to the set $T^{p,k}_{|Q}$ and vice-versa.

Hence a function $\phi : M \to T^{p,k}_{|Q}$ defined by $\phi(\mu_j)(\overline{c_i}) = \underline{IVT}_j^{p,k}(\overline{c_i})$ is an isomorphism.

Moreover, for any $v(\in \mathbb{N} \geq k)$-celled CA, having a global configuration $C = (c_1, c_2, ..., c_v)$, if $\mathscr{Q}^v = \{\tau | \tau : Q^v \to Q^v\}$, then $\forall\ \overline{c_i} \in Q^k \subseteq Q^v, i = 1, 2, ..., v$, it follows that a function $\phi : \mathscr{Q}^v \to \mathscr{T}^v_{|Q}$ defined by

$$\phi(\tau)(c_1, c_2, ..., c_v) \equiv (\underline{IVT}_{j_1}^{p,k}(\overline{c_1}), \underline{IVT}_{j_2}^{p,k}(\overline{c_2}), ..., \underline{IVT}_{j_v}^{p,k}(\overline{c_v}))$$

is an isomorphism.

## 5 Some Algebraic Results on CA and IVT

**Theorem 5.1** $(\mathscr{Q}^{\mathbb{Z}}, \circ)$ *forms a monoid w.r.t. composition of global transition functions.*

***Proof*** Clearly $\mathscr{Q}^{\mathbb{Z}}$ is closed and associative under composition of global transition functions.

The transition function $\mu_e$ such that $\forall c_i \in Q$, $\mu_e(\overline{c_i}) = c_i$ is the local identity and it follows that $\tau_e \in \mathscr{Q}^{\mathbb{Z}}$ is the global identity such that $\forall C \in Q^{\mathbb{Z}}, i \in \mathbb{Z}$, $\tau_e(C) = ...\mu_e(\overline{c_{i-1}})\mu_e(\overline{c_i})\mu_e(\overline{c_{i+1}})... = C$

Hence the theorem. □

**Corollary 5.1** $(\overline{\mathscr{Q}^{\mathbb{Z}}}, \circ)$ *forms a group w.r.t. composition of global transition functions where* $\overline{\mathscr{Q}^{\mathbb{Z}}} \subseteq \mathscr{Q}^{\mathbb{Z}}$ *is the set of all invertible global transition functions of CA having state set* $Q$.

***Proof*** Since any $\tau \in \overline{\mathscr{Q}^{\mathbb{Z}}} \subseteq \mathscr{Q}^{\mathbb{Z}}$ is invertible, the corollary holds true. □

**Theorem 5.2** $(\mathscr{T}^v, \circ)$ *forms a monoid w.r.t. composition of IVTs.*

***Proof*** The set $\mathscr{T}^v$, for $v \in \mathbb{N}$, is closed under composition $'\circ'$ since for any $(IVT_{i_1}^{p,k}, IVT_{i_2}^{p,k}, ..., IVT_{i_v}^{p,k}), (IVT_{j_1}^{p,k}, IVT_{j_2}^{p,k}, ..., IVT_{j_v}^{p,k}) \in \mathscr{T}^v$ and for any $(n_1, n_2..., n_v) \in \mathbb{N}_0^v \ \exists \ (m_1, m_2, ..., m_v) \in \mathbb{N}_0^v$ such that,

$$\begin{aligned}&\Big((IVT_{i_1}^{p,k}, IVT_{i_2}^{p,k}, ..., IVT_{i_v}^{p,k}) \circ (IVT_{j_1}^{p,k}, IVT_{j_2}^{p,k}, ..., IVT_{j_v}^{p,k})\Big)(n_1, n_2, ..., n_v)\\ &= \Big((IVT_{i_1}^{p,k} \circ IVT_{j_1}^{p,k}), (IVT_{i_2}^{p,k} \circ IVT_{j_2}^{p,k}), ..., (IVT_{i_v}^{p,k} \circ IVT_{j_v}^{p,k})\Big)(n_1, n_2, ...n_v)\\ &= \Big((IVT_{i_1}^{p,k} \circ IVT_{j_1}^{p,k})(\overline{n_1}), (IVT_{i_2}^{p,k} \circ IVT_{j_2}^{p,k})(\overline{n_2}), ..., (IVT_{i_v}^{p,k} \circ IVT_{j_v}^{p,k})(\overline{n_v})\Big)\\ &= \Big(IVT_{i_1}^{p,k}(IVT_{j_1}^{p,k}(\overline{n_1})), IVT_{i_2}^{p,k}(IVT_{j_2}^{p,k}(\overline{n_2})), ..., IVT_{i_v}^{p,k}(IVT_{j_v}^{p,k}(\overline{n_v}))\Big)\\ &= IVT_{i_1}^{p,k}(\overline{n_1^*})IVT_{i_2}^{p,k}(\overline{n_2^*})...IVT_{i_v}^{p,k}(\overline{n_v^*}) = (m_1, m_2, ..., m_v)_p\end{aligned}$$

where $n_s^* = IVT_{j_s}^{p,k}(\overline{n_s}) = IVT_{j_s}^{p,k}(\eta_1^s, \eta_2^s, ..., \eta_k^s)$ for $s = 1, 2, ..., v$.
$\mathscr{T}^v$ is associative under $'\circ'$ since composition of functions are associative.
Again, for $(n_1, n_2..., n_v) \in \mathbb{N}_0^v \ \exists \ (IVT_{e_1}^{p,k}, IVT_{e_2}^{p,k}, ..., IVT_{e_v}^{p,k}) \in \mathscr{T}^v$ such that

$$(IVT_{e_1}^{p,k}, IVT_{e_2}^{p,k}, ..., IVT_{e_v}^{p,k})(n_1, n_2, ..., n_v)$$

$$= IVT_{e_1}^{p,k}(\overline{n_1})IVT_{e_2}^{p,k}(\overline{n_2})...IVT_{e_v}^{p,k}(\overline{n_v}) = (n_1, n_2, ..., n_v)$$

Since $IVT_{e_s}^{p,k}(\overline{n_s}) = n_s \ \forall s = 1, ..., v$, it follows that $e_1 = e_2 = ... = e_v = e(say)$
Thus $(IVT_e^{p,k}, IVT_e^{p,k}, ..., IVT_e^{p,k})$ is the identity element of $\mathscr{T}^v$
Hence the theorem. □

**Definition 5.1** Modular addition and multiplication of two local transition functions for a $p(\in \mathbb{N})$-state CA are defined $\forall i \in \mathbb{Z}$ as
$(\mu_i^1 \oplus_p \mu_i^2)(\overline{c_i}) = \mu_i^1(\overline{c_i}) \oplus_p \mu_i^2(\overline{c_i})$ *and* $(\mu_i^1 \otimes_p \mu_i^2)(\overline{c_i}) = \mu_i^1(\overline{c_i}) \otimes_p \mu_i^2(\overline{c_i})$

**Definition 5.2** Modular addition and multiplication of two $p(\in \mathbb{N})$-base $k(\in \mathbb{N})$-dimensional IVTs (see [16]) are defined $\forall (n_1, n_2, ..., n_k) \in \mathbb{N}_0^k$ and $j_1, j_2 \in \{0, 1, ..., p^{p^k} - 1\}$, as
$(IVT_{j_1}^{p,k} \oplus_p IVT_{j_2}^{p,k})(n_1, n_2, ..., n_k) = IVT_{j_1}^{p,k}(n_1, n_2, ..., n_k) \oplus_p IVT_{j_2}^{p,k}$
$(n_1, n_2, ..., n_k)$
$= \big(f_{j_1}(a_0^{n_1}, ..., a_0^{n_k}) \oplus_p f_{j_2}(a_0^{n_1}, ..., a_0^{n_k})...f_{j_1}(a_{l-1}^{n_1}, ..., a_{l-1}^{n_k}) \oplus_p f_{j_2}(a_{l-1}^{n_1}, ..., a_{l-1}^{n_k})\big)_p$
*and*,
$(IVT_{j_1}^{p,k} \otimes_p IVT_{j_2}^{p,k})(n_1, n_2, ..., n_k) = IVT_{j_1}^{p,k}(n_1, n_2, ..., n_k) \otimes_p IVT_{j_2}^{p,k}(n_1,$
$n_2, ..., n_k)$
$= \big(f_{j_1}(a_0^{n_1}, ..., a_0^{n_k}) \otimes_p f_{j_2}(a_0^{n_1}, ..., a_0^{n_k})...f_{j_1}(a_{l-1}^{n_1}, ..., a_{l-1}^{n_k}) \otimes_p f_{j_2}(a_{l-1}^{n_1}, ..., a_{l-1}^{n_k})\big)_p$
where, $n_s = (a_0^{n_s}, a_1^{n_s} ..., a_{l-1}^{n_s})_p$ for $s = 1, 2, ..., k$.

**Theorem 5.3** $(\mathscr{Q}^{\mathbb{Z}}, \oplus_p, \otimes_p)$ *forms a commutative ring with identity under the operations* $\oplus_p$ *and* $\otimes_p$ *defined as*

$$(\tau_1 \oplus_p \tau_2)(C) = ...(\mu_{i-1}^1 \oplus_p \mu_{i-1}^2)(\overline{c_{i-1}}).(\mu_i^1 \oplus_p \mu_i^2)(\overline{c_i}).(\mu_{i+1}^1 \oplus_p \mu_{i+1}^2)(\overline{c_{i+1}})... \ and$$

$$(\tau_1 \otimes_p \tau_2)(C) = ...(\mu_{i-1}^1 \otimes_p \mu_{i-1}^2)(\overline{c_{i-1}}).(\mu_i^1 \otimes_p \mu_i^2)(\overline{c_i}).(\mu_{i+1}^1 \otimes_p \mu_{i+1}^2)(\overline{c_{i+1}})...$$

*where* $\tau_1, \tau_2 \in \mathscr{Q}^{\mathbb{Z}}$, $C \in Q^{\mathbb{Z}}$, $\oplus_p$ *denotes addition modulo* $p$ *and* $\otimes_p$ *denotes multiplication modulo* $p$ *for* $|Q| = p(\in \mathbb{N})$.

***Proof*** In a $p(\in \mathbb{N})$-state CA, for any $\tau_1, \tau_2 \in \mathscr{Q}^{\mathbb{Z}}$,
$(\tau_1 \oplus_p \tau_2)(C) \in Q^{\mathbb{Z}}$ *and* $(\tau_1 \otimes_p \tau_2)(C) \in Q^{\mathbb{Z}}$ since $\forall i \in \mathbb{Z}$,

$$(\mu_i^1 \oplus_p \mu_i^2)(\overline{c_i}) \in Q \ \ and \ \ (\mu_i^1 \otimes_p \mu_i^2)(\overline{c_i}) \in Q$$

Therefore $\mathscr{Q}^{\mathbb{Z}}$ is closed w.r.t. $\oplus_p$ and $\otimes_p$.
Associativity follows from associativity of $\oplus_p$ and $\otimes_p$.
The local transition function $\mu_0$ such that $\forall c_i \in Q$, $\mu_0(\overline{c_i}) = 0$ is the additive identity since $\forall i \in \mathbb{Z}$ $(\mu_0 \oplus_p \mu_i)(\overline{c_i}) = (\mu_i \oplus_p \mu_0)(\overline{c_i}) = \mu_i(\overline{c_i})$
It follows that $\tau_0 \in \mathscr{Q}^{\mathbb{Z}}$ is the additive identity such that $\forall C \in Q^{\mathbb{Z}}$,

$$\tau_0(C) = ...\mu_0(\overline{c_{i-1}})\mu_0(\overline{c_i})\mu_0(\overline{c_{i+1}})...$$

The local transition function $\mu_{id}$ such that $\forall c_i \in Q$, $\mu_{id}(\overline{c_i}) = 1$ is the multiplicative identity since $\forall i \in \mathbb{Z}$, $(\mu_{id} \otimes_p \mu_i)(\overline{c_i}) = (\mu_i \otimes_p \mu_{id})(\overline{c_i}) = \mu_i(\overline{c_i})$
It follows that $\tau_{id} \in \mathscr{Q}^{\mathbb{Z}}$ is the multiplicative identity such that $\forall C \in Q^{\mathbb{Z}}$,

$$\tau_{id}(C) = ...\mu_{id}(\overline{c_{i-1}})\mu_{id}(\overline{c_i})\mu_{id}(\overline{c_{i+1}})...$$

Now any local transition function $\mu_i$, is a $p^k$-bit sequence $b_1 b_2 ... b_{p^k}$(say), for a $p(\in \mathbb{N})$-state $k(\in \mathbb{N})$-neighbourhood CA where $b_1, ..., b_{p^k} \in Q = \{0, 1, ..., p\}$.
Clearly for any $b_s \in Q$, $\exists b_s^{-1} \in Q$ such that $\forall s = 1, 2, ..., p^k$, $b_s \oplus_p b_s^{-1} = 0$.
Thus for any $\mu_i (\equiv b_1 ... b_{p^k})$, $\exists \mu_i^{-1} (\equiv b_1^{-1} ... b_{p^k}^{-1})$ such that $\forall c_i \in Q$,

$$(\mu_i^{-1} \oplus_p \mu_i)(\overline{c_i}) = (\mu_i \oplus_p \mu_i^{-1})(\overline{c_i}) = \mu_0(\overline{c_i}) = 0$$

It follows that for any $\tau \in \mathscr{Q}^{\mathbb{Z}}$, $\exists \tau^{-1} \in \mathscr{Q}^{\mathbb{Z}}$ such that $\forall C \in Q$,

$$(\tau \oplus_p \tau^{-1})(C) = ...(\mu_{i-1} \oplus_p \mu_{i-1}^{-1})(\overline{c_{i-1}})(\mu_i \oplus_p \mu_i^{-1})(\overline{c_i})(\mu_{i+1} \oplus_p \mu_{i+1}^{-1})(\overline{c_{i+1}})... = \tau_0(C),$$

$$(\tau^{-1} \oplus_p \tau)(C) = ...(\mu_{i-1}^{-1} \oplus_p \mu_{i-1})(\overline{c_{i-1}})(\mu_i^{-1} \oplus_p \mu_i)(\overline{c_i})(\mu_{i+1}^{-1} \oplus_p \mu_{i+1})(\overline{c_{i+1}})... = \tau_0(C)$$

Thus for any global transition function $\tau$, its additive inverse exists in $\mathscr{Q}^{\mathbb{Z}}$.
Commutativity follows from commutativity of $\oplus_p$ and $\otimes_p$.
Now for any local transitions $\mu_i^1, \mu_i^2, \mu_i^3$, $\forall c_i \in Q, i \in \mathbb{Z}$ we get,

$$(\mu_i^1 \otimes_p (\mu_i^2 \oplus_p \mu_i^3))(\overline{c_i}) = (\mu_i^1 \otimes_p \mu_i^2)(\overline{c_i}) \oplus_p (\mu_i^1 \otimes_p \mu_i^3)(\overline{c_i}) \ \ and,$$

$$((\mu_i^2 \oplus_p \mu_i^3) \otimes_p \mu_i^1)(\overline{c_i}) = (\mu_i^2 \otimes_p \mu_i^1)(\overline{c_i}) \oplus_p (\mu_i^3 \otimes_p \mu_i^1)(\overline{c_i})$$

Therefore for $\tau_1, \tau_2, \tau_3 \in (Q)^{\mathbb{Z}}$, $\forall C \in Q^{\mathbb{Z}}$ we get,

$$(\tau_1 \otimes_p (\tau_2 \oplus_p \tau_3))(C) = (\tau_1 \otimes_p \tau_2)(C) \oplus_p (\tau_1 \otimes_p \tau_3)(C) \ \ and,$$

$$((\tau_2 \oplus_p \tau_3) \otimes_p \tau_1)(C) = (\tau_2 \otimes_p \tau_1)(C) \oplus_p (\tau_3 \otimes_p \tau_1)(C)$$

Hence the theorem. □

*Remark 5.1* The following example shows that $(\mathscr{Q}^{\mathbb{Z}}, \oplus_p, \otimes_p)$ will not form a field under the operations $\oplus_p$ and $\otimes_p$ even if $|Q| = p$ is prime.

*Example 5.1* For initial configuration $C = (10001)_2$, let $\tau(10001)_2 = (\mu(110)\ \mu(100)\ \mu(000)\mu(001)\mu(011))_2$
Let $\mu$ follow Wolfram code $3(\neg(c_{i-1} \vee c_i))$.
Then $\tau(10001)_2 = (\mu_3(110)\mu_3(100)\mu_3(000)\mu_3(001)\mu_3(011))_2 = (00110)_2$.
Now, $\tau_{id}(10001)_2 = (\mu_{id}(110)\mu_{id}(100)\mu_{id}(000)\mu_{id}(001)\mu_{id}(011))_2 = (11111)_2$.
But $\nexists\ \tau^{-1}$ such that $(\tau \otimes_2 \tau^{-1})(10001)_2 = (11111)_2$ since $\nexists\ \mu^{-1}$ such that for all local configurations $\mu \otimes_2 \mu^{-1} = 1$.

**Theorem 5.4** *$(\mathscr{T}^v, \oplus_p, \otimes_p)$ forms a commutative ring under the operations $\oplus_p$ and $\otimes_p$ defined as*

$$\Big((IVT_{i_1}^{p,k}, IVT_{i_2}^{p,k}, ..., IVT_{i_v}^{p,k}) \oplus_p (IVT_{j_1}^{p,k}, IVT_{j_2}^{p,k}, ..., IVT_{j_v}^{p,k})\Big)\,(n_1, n_2, ..., n_v)$$

$$= \Big((IVT_{i_1}^{p,k} \oplus_p IVT_{j_1}^{p,k}), (IVT_{i_2}^{p,k} \oplus_p IVT_{j_2}^{p,k}), ......, (IVT_{i_v}^{p,k} \oplus_p IVT_{j_v}^{p,k})\Big)$$
$$(n_1, n_2, ..., n_v)$$

$$= \Big((IVT_{i_1}^{p,k} \oplus_p IVT_{j_1}^{p,k})(\overline{n_1}), (IVT_{i_2}^{p,k} \oplus_p IVT_{j_2}^{p,k})(\overline{n_2}), ....., (IVT_{i_v}^{p,k} \oplus_p IVT_{j_v}^{p,k})(\overline{n_v})\Big)$$

$$and, \Big((IVT_{i_1}^{p,k}, IVT_{i_2}^{p,k}, ..., IVT_{i_v}^{p,k}) \otimes_p (IVT_{j_1}^{p,k}, IVT_{j_2}^{p,k}, ..., IVT_{j_v}^{p,k})\Big)$$
$$(n_1, n_2, ..., n_v)$$

$$= \Big((IVT_{i_1}^{p,k} \otimes_p IVT_{j_1}^{p,k}), (IVT_{i_2}^{p,k} \otimes_p IVT_{j_2}^{p,k}), ....., (IVT_{i_v}^{p,k} \otimes_p IVT_{j_v}^{p,k})\Big)$$
$$(n_1, n_2, ..., n_v)$$

$$= \Big((IVT_{i_1}^{p,k} \otimes_p IVT_{j_1}^{p,k})(\overline{n_1}), (IVT_{i_2}^{p,k} \otimes_p IVT_{j_2}^{p,k})(\overline{n_2}), ....., (IVT_{i_v}^{p,k} \otimes_p IVT_{j_v}^{p,k})(\overline{n_v})\Big)$$

*where* $i_s, j_s \in \{0, 1, ..., p^{p^k} - 1\}$, $p, k \in \mathbb{N}$, $\overline{n_s} = (\eta_1^s, \eta_2^s, ..., \eta_k^s)$ *for* $s = 1, 2, ..., v$

***Proof*** For any $(IVT^{p,k}_{i_1}, IVT^{p,k}_{i_2}, ..., IVT^{p,k}_{i_v}), (IVT^{p,k}_{j_1}, IVT^{p,k}_{j_2}, ..., IVT^{p,k}_{j_v}) \in \mathscr{T}^v$, and $(n_1, n_2, ..., n_v) \in \mathbb{N}_0^v$,

$$\left((IVT^{p,k}_{i_1}, IVT^{p,k}_{i_2}, ..., IVT^{p,k}_{i_v}) \oplus_p (IVT^{p,k}_{j_1}, IVT^{p,k}_{j_2}, ..., IVT^{p,k}_{j_v})\right)(n_1, n_2, ..., n_v) \in \mathbb{N}^v$$

$$and, \left((IVT^{p,k}_{i_1}, IVT^{p,k}_{i_2}, ..., IVT^{p,k}_{i_v}) \otimes_p (IVT^{p,k}_{j_1}, IVT^{p,k}_{j_2}, ..., IVT^{p,k}_{j_v})\right)$$
$$(n_1, n_2, ..., n_v) \in \mathbb{N}^v$$

since for every $s = 1, 2, ..., v$,

$$(IVT^{p,k}_{i_s} \oplus_p IVT^{p,k}_{j_s})(\overline{n_s}) \in \mathbb{N}_0 \ and \ (IVT^{p,k}_{i_s} \otimes_p IVT^{p,k}_{j_s})(\overline{n_s}) \in \mathbb{N}_0$$

where $\overline{n_s} = (\eta_1^s, \eta_2^s, ..., \eta_k^s)$
Therefore $\mathscr{T}^v$ is closed w.r.t. $\oplus_p$ and $\otimes_p$ for $p \in \mathbb{N}$.
Associativity follows from associativity of $\oplus_p$ and $\otimes_p$.
$IVT^{p,k}_0 \in T^{p,k}$ is such that $\forall \overline{n_s} \in \mathbb{N}_0^k$, $IVT^{p,k}_0(\overline{n_s}) = 0$ is the additive identity of $T^{p,k}$ since $\forall s = 1, 2, ..., v$

$$(IVT^{p,k}_0 \oplus_p IVT^{p,k}_s)(\overline{n_s}) = (IVT^{p,k}_s \oplus_p IVT^{p,k}_0)(\overline{n_s}) = IVT^{p,k}_s(\overline{n_s})$$

where, for each $(\overline{n_s}) = (\eta_1, \eta_2, ..., \eta_k) \in \mathbb{N}_0^k$,

$$IVT^{p,k}_0(\eta_1, \eta_2, ..., \eta_k) = (f_0(a_0^{\eta_1}, ..., a_0^{\eta_k}), ..., f_0(a_{l-1}^{\eta_1}, ..., a_{l-1}^{\eta_k})) = 0$$

where, $\eta_u = (a_0^{\eta_u} a_1^{\eta_u} ... a_{l-1}^{\eta_u})_p$ for $u = 1, 2, ..., k$.
It follows that $(IVT^{p,k}_0, IVT^{p,k}_0, ..., IVT^{p,k}_0)$ is the additive identity of $\mathscr{T}^v$ such that for any $(n_1, n_2, ..., n_v) \in \mathbb{N}_0^v$

$$(IVT^{p,k}_0, IVT^{p,k}_0, ..., IVT^{p,k}_0)(n_1, n_2, ..., n_v) = \left(IVT^{p,k}_0(\overline{n_1}), IVT^{p,k}_0(\overline{n_2}), ..., IVT^{p,k}_0(\overline{n_v})\right)$$

For any $IVT^{p,k}_s \in T^{p,k}$ $\exists$ $IVT^{p,k}_{-s} \in T^{p,k}$ such that $\forall \overline{n_s} \in \mathbb{N}_0^k$,

$$(IVT^{p,k}_s \oplus_p IVT^{p,k}_{-s})(\overline{n_s}) = (IVT^{p,k}_{-s} \oplus_p IVT^{p,k}_s)(\overline{n_s}) = IVT^{p,k}_0(\overline{n_s}) = 0$$

It follows that for any $(IVT^{p,k}_{i_1}, IVT^{p,k}_{i_2}, ..., IVT^{p,k}_{i_v}) \in \mathscr{T}^v$, $\exists$ $(IVT^{p,k}_{-i_1}, IVT^{p,k}_{-i_2}, ..., IVT^{p,k}_{-i_v}) \in \mathscr{T}^v$ such that $\forall (n_1, n_2, ..., n_v) \in \mathbb{N}_0^v$,

$$\left((IVT^{p,k}_{i_1}, IVT^{p,k}_{i_2}, ..., IVT^{p,k}_{i_v}) \oplus_p (IVT^{p,k}_{-i_1}, IVT^{p,k}_{-i_2}, ..., IVT^{p,k}_{-i_v})\right)(n_1, n_2, ..., n_v)$$
$$= (IVT^{p,k}_0, IVT^{p,k}_0, ..., IVT^{p,k}_0)(n_1, n_2, ..., n_v)$$

$$\left((IVT^{p,k}_{-i_1}, IVT^{p,k}_{-i_2}, ..., IVT^{p,k}_{-i_v}) \oplus_p (IVT^{p,k}_{i_1}, IVT^{p,k}_{i_2}, ..., IVT^{p,k}_{i_v})\right)(n_1, n_2, ..., n_v)$$

Thus for any element of $\mathscr{T}^v$ its additive inverse exists in $\mathscr{T}^v$.
Commutativity follows from commutativity of $\oplus_p$ and $\otimes_p$.
Now for any $IVT_h^{p,k}, IVT_i^{p,k}, IVT_j^{p,k} \in T^{p,k}, \forall \overline{n_s} \in \mathbb{N}_0^k$, we get,

$$(IVT_h^{p,k} \otimes_p (IVT_i^{p,k} \oplus_p IVT_j^{p,k}))(\overline{n_s}) = (IVT_h^{p,k} \otimes_p IVT_i^{p,k})(\overline{n_s}) \oplus_p (IVT_h^{p,k} \oplus_p IVT_j^{p,k})(\overline{n_s}),$$

$$((IVT_i^{p,k} \oplus_p IVT_j^{p,k}) \otimes_p IVT_h^{p,k})(\overline{n_s}) = (IVT_i^{p,k} \otimes_p IVT_h^{p,k})(\overline{n_s}) \oplus_p (IVT_j^{p,k} \oplus_p IVT_h^{p,k})(\overline{n_s})$$

Thus for any $(IVT_{h_1}^{p,k}, ..., IVT_{h_v}^{p,k}), (IVT_{i_1}^{p,k}, ..., IVT_{i_v}^{p,k}), (IVT_{j_1}^{p,k}, ..., IVT_{j_v}^{p,k}) \in \mathscr{T}^v, \ \forall (n_1, ..., n_v) \in \mathbb{N}_0^v$,

$$\left(IVT_{h_1}^{p,k}, ..., IVT_{h_v}^{p,k}\right) \otimes_p \left((IVT_{i_1}^{p,k}, ..., IVT_{i_v}^{p,k}) \oplus_p (IVT_{j_1}^{p,k}, ..., IVT_{j_v}^{p,k})\right)\Big) (n_1, ..., n_v)$$

$$= \left((IVT_{h_1}^{p,k}, ..., IVT_{h_v}^{p,k}) \otimes_p (IVT_{i_1}^{p,k}, ..., IVT_{i_v}^{p,k})\right) (n_1, ..., n_v)$$

$$\oplus_p \left((IVT_{h_1}^{p,k}, ..., IVT_{h_v}^{p,k}) \otimes_p (IVT_{j_1}^{p,k}, ..., IVT_{j_v}^{p,k})\right) (n_1, ..., n_v), \textit{and}$$

$$\left(\left((IVT_{i_1}^{p,k}, ..., IVT_{i_v}^{p,k}) \oplus_p (IVT_{j_1}^{p,k}, ..., IVT_{j_p}^{p,k})\right) \otimes_p (IVT_{h_1}^{p,k}, ..., IVT_{h_v}^{p,k})\right) (n_1, ..., n_v)$$

$$= \left((IVT_{i_1}^{p,k}, ..., IVT_{i_v}^{p,k}) \otimes_p (IVT_{h_1}^{p,k}, ..., IVT_{h_v}^{p,k})\right) (n_1, ..., n_v)$$

$$\oplus_p \left((IVT_{j_1}^{p,k}, ..., IVT_{j_v}^{p,k}) \otimes_p (IVT_{h_1}^{p,k}, ..., IVT_{h_v}^{p,k})\right) (n_1, ..., n_v)$$

Hence the theorem. □

**Theorem 5.5** *$(\mathscr{T}^v_{|Q}, \oplus_p, \otimes_p)$ forms a commutative ring with identity under the operations $\oplus_p$ and $\otimes_p$.*

***Proof*** Clearly $(\mathscr{T}^v_{|Q}, \oplus_p, \otimes_p)$ for $p, k, v \in \mathbb{N}$ is a commutative ring since $\mathscr{T}^v_{|Q} \subseteq \mathscr{T}^v$ when $\mathbb{N}_0^v$ is restricted to $Q^v$. Now, for any $(\eta_1, \eta_2, ..., \eta_k) \in Q^k$, $\exists$ multiplicative identity $\underline{IVT}_{id}^{p,k} \in T_{|Q}^{p,k}$ such that

$$\underline{IVT}_{id}^{p,k}(\eta_1, \eta_2, ..., \eta_k) = (f_{id}(\eta_1, \eta_2, ..., \eta_k))_p = 1$$

$$\textit{Since } (\underline{IVT}_j^{p,k} \otimes_p \underline{IVT}_{id}^{p,k})(\eta_1, \eta_2, ..., \eta_k) = \underline{IVT}_j^{p,k}(\eta_1, \eta_2, ..., \eta_k) = \left(f_j(\eta_1, \eta_2, ..., \eta_k)\right)_p,$$

$$\textit{And } (\underline{IVT}_{id}^{p,k} \otimes_p \underline{IVT}_j^{p,k})(\eta_1, \eta_2, ..., \eta_k) = \underline{IVT}_j^{p,k}(\eta_1, \eta_2, ..., \eta_k) = \left(f_j(\eta_1, \eta_2, ..., \eta_k)\right)_p$$

It follows that $(\underline{IVT}_{id}^{p,k}, \underline{IVT}_{id}^{p,k}, ..., \underline{IVT}_{id}^{p,k}) \in \mathscr{T}^{v}{}_{|Q}$ is the multiplicative identity of $\mathscr{T}^{v}{}_{|Q}$ such that $\forall(n_1, n_2, ..., n_v) \in Q^v$,

$$(\underline{IVT}_{id}^{p,k}, \underline{IVT}_{id}^{p,k}, ..., \underline{IVT}_{id}^{p,k})(n_1, n_2, ..., n_v) = \underline{IVT}_{id}^{p,k}(\overline{n_1})\underline{IVT}_{id}^{p,k}(\overline{n_2})...\underline{IVT}_{id}^{p,k}(\overline{n_v})$$

where $\underline{IVT}_{id}^{p,k}(\overline{n_s}) = \underline{IVT}_{id}^{p,k}(\eta_1^s, \eta_2^s, ..., \eta_k^s) = 1$ for each $s = 1, 2, ..., v$.
Hence the theorem. □

## 6 Conclusion

In this paper Wolfram code and transition function of a one-dimensional CA has been represented in terms of an IVT. An IVT is a convenient mathematical tool for representation of any Wolfram code. Moreover, if the domain of IVTs is restricted from positive integers to $p(\in \mathbb{N})$-base numbers, then an odd dimensional IVT is equivalent to a local transition function of a CA. It could be proved that the set of all transition functions of a $p(\in \mathbb{N})$-state CA, forms a commutative ring with identity, whereas the set of $p$-base IVTs forms only a commutative ring under modular addition and multiplication. However, for the set of $p$-base IVTs multiplicative identity under modular addition and multiplication exists, when the domain of IVTs is restricted from positive integers to $p(\in \mathbb{N})$-base numbers.

## References

1. Adamatzky, A.: Designing Beauty: The Art of Cellular Automata, vol. 20, Chap. Excitable Automata, pp. 173–179. Springer (2016)
2. Basu, S., Ghosh, S.: Fuzzy cellular automata model for discrete dynamical system representing spread of mers and covid-19 virus. In: Internet of Medical Things for Smart Healthcare, pp. 267–304. Springer (2020)
3. Choudhury, P.P., Hassan, S.S., Sahoo, S., Nayak, B.K.: Act of cvt and evt in the formation of number theoretic fractals. Int. J. Comput. Cogn. **9**(1), 1–8 (2011)
4. Choudhury, P.P., Sahoo, S., Chakraborty, M., Bhandari, S.K., Pal, A.: Investigation of the global dynamics of cellular automata using boolean derivatives. Comput. Math. Appl. **57**(8), 1337–1351 (2009)
5. Choudhury, P.P., Sahoo, S., Nayak, B.K., Hassan, S.: Theory of carry value transformation (cvt) and its application in fractal formation (2009). arXiv:0904.0534
6. Das, A.K., Sanyal, A., Palchaudhuri, P.: On characterization of cellular automata with matrix algebra. Inf. Sci. **61**(3), 251–277 (1992)
7. Das, J.K., Choudhury, P.P., Sahoo, S.: Multi-number cvt-xor arithmetic operations in any base system and its significant properties. In: 2016 IEEE 6th International Conference on Advanced Computing (IACC), pp. 769–773. IEEE (2016)
8. Das, J.K., Sahoo, S., Hassan, S., Choudhury, P.P.: Two dimensional discrete dynamics of integral value transformations (2020). arXiv:1709.05205
9. Edwards, R., Maignan, A.: A class of discrete dynamical systems with properties of both cellular automata and l-systems. Nat. Comput. 1–33 (2019)

10. Ghosh, S.: Evolutions of some one-dimensional homogeneous cellular automata. Complex Syst. **30**, 75–92 (2021)
11. Ghosh, S., Basu, S.: Evolution patterns of finite celled synchronous cellular automata having atmost one active cell. In: Proceedings of The 10th International Conference MSAST 2016, pp. 154–164 (2016)
12. Ghosh, S., Basu, S.: Evolution patterns of some boolean cellular automata having atmost one active cell to model simple dynamical systems. Bull. Cal. Math. Soc. **108**(6), 449–464 (2016)
13. Ghosh, S., Basu, S.: Some algebraic properties of linear synchronous cellular automata (2017). arXiv:1708.09751
14. Hassan, S.S., Choudhury, P.P., Nayak, B.K., Ghosh, A., Banerjee, J.: Integral value transformations: a class of affine discrete dynamical systems and an application (2011). arXiv:1110.0724
15. Hassan, S.S., Choudhury, P.P., Sahoo, S., Singh, R.B., Nayak, B.K.: Collatz function like integral value transformations. Alex. J. Math. **1**(2), 30–35 (2012)
16. Hassan, S.S., Nayak, B.K., Choudhury, P.P.: One-dimensional p-adic integral value transformation. J. Orissa Math. Soc. **31**(2), 116–124 (2012)
17. Kari, J.: Theory of cellular automata: a survey. Theor. Comput. Sci. **334**(1–3), 3–33 (2005)
18. Martinez, G.J.: Designing Beauty: The Art of Cellular Automata, vol. 20, Chap. Gliders in One-Dimensional Cellular Automata, pp. 167–172. Springer (2016)
19. von Neumann, J., Burks, A.W., et al.: Theory of self-reproducing automata. IEEE Trans. Neural Netw. **5**(1), 3–14 (1966)
20. Pal, S., Sahoo, S., Nayak, B.K.: Properties of carry value transformation. Int. J. Math. Math. Sci. **2012** (2012)
21. Pal, S., Sahoo, S., Nayak, B.K.: Study of number conserving cellular automata rules through integral value transformations. Math. Forum **28(1)** (2020). ISSN : 0972-9852
22. Schiff, J.L.: Cellular Automata: A Discrete View of the World, vol. 45. Wiley (2011)
23. Ulam, S.: On some mathematical problems connected with patterns of growth of figures. In: Proceedings of Symposia in Applied Mathematics, vol. 14, pp. 215–224. Providence (1962). ( Am. Math. Soc. Vol. 14)
24. Wolfram, S.: A New Kind of Science. Wolfram Media Inc, Champaign, IL, USA (2002)
25. Wuensche, A.: Discrete dynamics lab: Tools for investigating cellular automata and discrete dynamical networks. Kybernetes (2003)
26. Xu, X., Song, Y., Banks, S.P.: On the dynamical behavior of cellular automata. Int. J. Bifurc. Chaos **19**(04), 1147–1156 (2009)

# Emulating Mersenne Twister with Cellular Automata

**Kamalika Bhattacharjee, Nitin More, Shobhit Kumar Singh, and Nikhil Verma**

**Abstract** Mersenne Twister is by far the most widely used general-purpose pseudo-random number generator (PRNG) but it has been 30 years since it was introduced. This paper aims to use the advancement in technology and knowledge of a cellular automaton (CA) and its properties to create a PRNG that can use resources more effectively while giving an equivalent or a better result than `MT19937`. The research involves creation of suitable generator by finalizing an appropriate length of the CA that fits perfectly with space-time trade-off, finding the seed, running the PRNG on various testbeds like Dieharder, TestU01 and progressively improving it. In the end, it is shown that our PRNG is at least at par with `MT19937` and also a good challenger to today's more advanced pseudo-random number generators like SFMT and TinyMT.

**Keywords** Pseudo-random number generator (PRNG) · Mersenne Twister · Sophie Germain prime · Cellular automaton (CA) · Maximal length cellular automata · CA(150′)

## 1 Introduction

Pseudo-random numbers play an important role in many areas of communications and computing. Pseudo-random number generators (PRNGs) have a long history of development—its modern journey starting in the late 19th century to the early 20th century and evolving and getting more powerful ever since. Several papers target to survey this development of random number generators, see for example [2]. Usually, the latest PRNG claims to be superior than the previous ones. This claim is based

---

K. Bhattacharjee (✉) · N. More · S. K. Singh · N. Verma
Department of Computer Science and Engineering, National Institute of Technology, Tiruchirappalli 620015, Tamilnadu, India



S. Das and G. J. Martinez (eds.), *Proceedings of First Asian Symposium on Cellular Automata Technology*, Advances in Intelligent Systems and Computing 1425,
https://doi.org/10.1007/978-981-19-0542-1_8

on the PRNG's performance in some statistical tests, like Dieharder [3], BigCrush of TestU01 [4] etc. battery of tests, which empirically detect non-randomness in the generated numbers.

One of the most important of these PRNGs is Mersenne Twister (MT) developed by Makoto Matsumoto and Takuji Nishimura in 1997 [5]. This PRNG has several interesting properties. In particular, building MT-based generators with a very large state space is easy and the largest version (`MT19937`) with 19937 bits of the state became very popular. The fact that the sequence generated would not repeat before $2^{19937} - 1$ 32-bit integers has been met with enthusiasm, and quickly the Mersenne Twister was adopted as the standard generator in many environments. Also, `MT19937` can be easily implemented and hence a plethora of language implementers decided to use it. For example, the stock PRNG of the GCC compiler and Python, as well as of the Maple mathematical computing framework, is some version of the Mersenne Twister.

However, Mersenne Twister `MT19937` has several shortcomings. It needs to compute and store all 623 numbers together even when the application requires a fairly lower number of these random numbers. Also it fails statistical tests designed to find bias in linear generators, such as Marsaglia's binary-rank test and the linear-complexity test. Recently, much stronger PRNGs based on MTs were also proposed, like SFMT (SIMD-oriented Fast Mersenne Twister) [6] and TinyMT (Tiny Mersenne Twister) [7]. But, unfortunately these two stronger PRNGs also can not fool BigCrush testbed and fail a couple of linearity tests.

Hence, primary objective of this paper is to create a new pseudo-random generator emulating Mersenne Twister that could give a better performance than `MT19937` on different testbeds. More specifically, our PRNG has to follow the same style of working like Mersenne Twister but has to pass the statistical tests designed to find bias in linear generators. Moreover, to compete with `MT19937` we need to make sure that our PRNG is easy to use and understand, and easily implementable but still strong enough to beat `MT19937` as well as its variants in the major statistical tests. The PRNG also needs to balance space-time trade-off to ensure that it outperforms all the above PRNGs while not taking considerably more amount of space and time than them.

We propose a pseudo-random number generator that uses cellular automata (CAs) as the base component to produce pseudo-random numbers. Reason for choosing CAs is, like LFSR, CA is also a popular well-known option to develop PRNGs. Further, CAs can also be easily implemented in hardware—each cell consists of a memory element to store its state and a combinational logic circuit to find the next state of the cell. For this work, however, we employ a CA named CA(150′) which uses Rule 90 on the first bit and Rule 150 on all others following null-boundary condition [1]. This CA has a special property—it has the potential to have maximal cycle length of $2^n - 1$ for some CA size $n$. That is, for those $n$, the sequence does not repeat beforehand just like the Mersenne Twister. Further, we shall use twisting and tempering function like `MT19937` on this simple CA. Our target is to bring down the storage requirements to make it lightweight while keeping the period length sufficiently large. Moreover, we also need to find a good seed to avoid *zero-excess initial state* problem where, as a result of a fair number of bits being 0, the PRNG might produce some non-chaotic patterns. Finally, we want our PRNG to be comparable to MT family and easily implementable in any programming language.

## 2 Background

In this section, we briefly discuss about Mersenne Twister (MT19937) and cellular automata, especially CA(150′), along with the related terminologies.

### *2.1 Mersenne Twister*

The Mersenne Twister (MT) is one of the most widely used general-purpose PRNG. It is a variant of Twisted Generalized Feedback Shift Register (TGFSR) which takes an incomplete array to realize a Mersenne prime $2^{19937} - 1$ as its period. It is considered to be fast, because it avoids multiplications and divisions and also uses the advantages of caches, pipelines to efficiently use memory as only 624 words are needed for the working area.
**Algorithm**: The working principle of Mersenne Twister is as follows: Here, a 32-bit integer is taken as seed which is used to initialize the starting state of the 19937 length TGFSR. Every time *Twist* function is used to generate the next state of the PRNG (Note that, $32 \times 624 = 19936$). But, instead of directly producing each state as output, each of the internal states of Mersenne Twister is divided into 624 32-bit words where each word (integer) is tempered using the *Temper* function before getting produced as output. So, after each application of *Twist*, the current state of the TGFSR is used to generate 624 32-bit integers consecutively. Once these 624 numbers are exhausted, then only again the state of the TGFSR is updated using the *Twist* function.

Tempering is done by the following transformations in succession:

$$\mathbf{y} = \mathbf{x} \oplus (\mathbf{x} >> u) \tag{1}$$

$$\mathbf{y} = \mathbf{y} \oplus ((\mathbf{y} << s) \text{ AND } \mathbf{b}) \tag{2}$$

$$\mathbf{y} = \mathbf{y} \oplus ((\mathbf{y} << t) \text{ AND } \mathbf{c}) \tag{3}$$

$$\mathbf{z} = \mathbf{y} \oplus (\mathbf{y} >> l) \tag{4}$$

where $l, s, t, u$ are integers called tempering parameters, **b** and **c** are suitable bitmasks of size $w$ and **z** is the returned vector. Note that, all the operations are shifts, ANDs, ORs, and XORs. The associated parameters for MT19937 32-bit and 64-bit are

**Table 1** Parameters of 32-Bit and 64-bit MT19937

| Parameter | Value | |
|---|---|---|
| | 32-bit | 64-bit |
| n | 624 | 312 |
| w | 32 | 64 |
| r | 31 | 31 |
| m | 397 | 156 |
| a | 0x9908B0DF | 0xB5026F5AA96619E916 |
| u | 11 | 29 |
| s | 7 | 17 |
| t | 15 | 37 |
| l | 18 | 43 |
| b | 0x9D2C5680 | 0x71D67FFFEDA6000016 |
| c | 0xEFC60000 | 0xFFF7EEE00000000016 |

shown in Table 1. These parameters ensure that the period is the Mersenne prime $2^{19937} - 1$ which is longer than any other random number generator proposed before and is one of the reasons for the popularity of Mersenne Twister. However, as *Twist* function is not a one-way function, MT is not cryptographically secure.

**Limitation and Improvement**: One of the limitations of MT is the initialization. If the initial state has too many zeros then the generated sequence may also contain many zeros for more than 10000 generations and the output sequences will be correlated. Another limitation of it is MT is not preferred for cryptographic purposes. It is easy to predict the next state given the present outputs in MT because of its linearity and the large output of 19937 bits. We target to overcome these limitations with our PRNG. Two other variants are also proposed as part of the improvement.

**SFMT**: It stands for single instruction multiple data (SIMD)- oriented Fast Mersenne Twister. It is an LFSR PRNG that uses all features of MT along with multi-stage pipelines and Single Instruction Multiple Data (SIMD) (like 128-bit integer) operations of today's computer system. Its period length is the same as `MT19937`, we call it `SFMT19937`. It can generate both 32-bit and 64-bit unsigned integer numbers, as well as double-precision floating-point numbers. However, in our empirical study, we have not used the floating-point numbers for testing. It is roughly twice as fast as Mersenne Twister.

**TinyMT**: TinyMT has a far shorter period ($2^{127} - 1$) than MT. The merit of TinyMT is in the small size of the 127-bit internal state, far smaller than the 19937 bits of MT. The outputs of TinyMT satisfy several statistical tests for non-cryptographic randomness, including BigCrush in TestU01. The `TinyMT32` PRNG must be initialized with a parameter set that needs to be well chosen. In this specification, the following parameter set has been used:

∘ mat1 = Ox8F7011EE = 2406486510

o mat2 = 0xFC78FF1F = 4235788063
o tmat = 0x3793FDFF = 932445695

## 2.2 Cellular Automata

In this work we consider 2-state 3-neighborhood non-uniform $n$-cell cellular automata under null boundary condition where cells of the CA may take different rules. In that case, the CA is represented in terms of a *rule-vector* $\mathcal{R} = \langle R_0, R_1, \cdots, R_{n-1}\rangle$ which means for each $i$, $0 \le i \le n-1$, cell $i$ uses rule $R_i$ to update its next state. We consider only two rules:

$$\text{Rule 90: } R_{90}(x, y, z) = x \oplus z$$

$$\text{Rule 150: } R_{150}(x, y, z) = x \oplus y \oplus z$$

Hence, these rules are *linear* rules. A non-uniform CA with these rules can be a *maximal length* CA.

**Theorem 1** ([1]) *A linear CA is maximal length iff the rule vector of the CA consists of rule 90 and rule 150, and its characteristic polynomial is primitive over GF(2).*

Therefore, a maximal length CA can be formed using only rule 90 and rule 150, but an arbitrary sequence of 90 and 150 does not form a maximal length CA. In this paper, we are interested in one special CA formed with these two rules which has the potentiality to become a maximal length CA.

**Definition 1** A $n$-cell rule vector $\mathcal{R} = \langle R_0, R_1, \cdots, R_{n-1}\rangle$ is called CA(150′) if $R_0 = 90$ and $R_i = 150$ where $1 \le i \le n-1$.

This CA was introduced in the paper–Adak and Das [1]. The paper describes three greedy strategies for finding primitive polynomials of a large degree over GF(2). The proposed strategies greedily synthesize (linear) CAs of this type, which are almost always maximal length CAs. Since the characteristic polynomial of a maximal length CA is primitive, the characteristic polynomials of the CAs which are synthesized from the strategies are claimed as primitive. They have given a list of CA sizes for which CA(150′) is a maximal length CA. For the sake of completeness, we reproduce the table here as Table 2 which lists those CA sizes. Here, [a] denotes false positives. Note that, this CA has all cells following rule 150 except the first one having rule 90 whose presence helps this CA to achieve its maximal length property for some $n$. To get the idea of such $n$, we reprint the following conjecture from the paper–Adak and Das [1]:

**Conjecture 1** ([1]) Characteristic polynomials of CA(150′) of size $n$ are primitive over $GF(2)$ if $n$ and $2n+1$ both are primes (i.e. $n$ is a Sophie Germain prime).

By looking at the above table we can observe that, for a prime $p$, if $2p+1$ is prime then CA(150′) of size $p$ shows maximality. That is, for *Sophie Germain prime*,

**Table 2** Sizes where CA(150′) is maximal length

| | | | | | | | | | | | | | | |
|---|---|---|---|---|---|---|---|---|---|---|---|---|---|---|
| 2 | 3 | 5 | 9 | 11 | 14 | 23 | 26 | 29 | 35 | 39 | 41 | 50[a] | 53 | 65 |
| 69 | 74 | 81 | 83 | 86 | 89 | 95 | 98[a] | 99[a] | 105 | 113 | 119 | 131 | 134[a] | 146 |
| 155 | 158 | 173 | 179 | 189 | 191 | 194[a] | 209 | 221 | 230 | 231 | 233 | 239 | 243 | 251 |
| 254 | 278[a] | 281 | 293 | 299 | 303 | 323 | 326 | 329 | 338[a] | 350[a] | 359 | 371 | 375 | 386 |
| 398 | 410[a] | 411 | 413 | 419 | 429 | 431 | 443 | 453 | 470 | 473 | 491 | 509 | 515 | 519 |
| 530 | 531 | 543 | 554 | 561 | 575 | 593 | 614[a] | 615 | 629 | 638 | 639 | 641 | 645[a] | 650[a] |
| 653 | 659 | 683 | 686[a] | 713 | 719 | 723 | 725 | 741[a] | 743 | 746 | 749 | 761 | 779 | 783[a] |
| 785 | 803[a] | 809 | 818 | 831 | 833 | 866 | 873 | 893 | 911 | 923 | 950[a] | 953 | 965 | 974[a] |
| 975 | 986[a] | 989 | 993 | 998 | 1013 | 1019 | 1031 | 1034[a] | 1041 | 1043 | 1049 | 1070 | 1103 | 1106[a] |
| 1118 | 1119 | 1121 | 1133 | 1154 | 1155 | 1166 | 1169 | 1178[a] | 1185 | | | | | |

[a]denotes false positives

CA(150′) is a maximal length CA. There are 42 Sophie Germain primes up to 1200 and for all of them, CA(150′) shows maximality. Therefore, in our work, we target to use CA(150′) with $n$ as a Sophie Germain prime.

## 3 Phase I: Making Our Own PRNG

To develop our own PRNG, we choose the size of CA to be a Sophie Germain prime. Henceforth all further work is done using CA(150′) where $n$ is a Sophie Germain prime. The next task is to decide which of such $n$ is to be used.

### 3.1 *Choosing the Size of CA(150′)*

Before choosing $n$, we have to verify the maximality claim since the success rate of the strategies mentioned in the paper was not 100%. What we need to do for this is, we have to run the CA and check if all but 1 configurations are part of the same cycle. However, there is a problem with this approach. For a large $n$ it is practically impossible to compute the complete cycle length. In this situation, we take help of the property of CA(150′):

*Property 1* ([1]) The CA(150′) with size $n$ is reversible for any $n$ except $n \in 3\mathbb{N} + 1$.

Hence, if we choose the length $n$ as a Sophie Germain prime such that $n \notin 3\mathbb{N} + 1$, then, it is guaranteed that the initial configuration will eventually repeat and all configurations are part of some cycles. Using this information, we develop the following strategy to test maximality: *We declare the CA(150′) with size n where n is a Sophie Germain prime and $n \notin 3\mathbb{N} + 1$, a maximal length CA if either the complete cycle length test was possible, or, the initial configuration has not returned even after* $10^9$

**Table 3** Observed cycle lengths of CA(150′)

| n | #Configurations | Observed Cycle | n | #Configurations | Observed Cycle |
|---|---|---|---|---|---|
| **3** | **8** | **7** | 20 | 1048576 | 1023 |
| 4 | 16 | | 21 | 2097152 | 127 |
| **5** | **32** | **31** | 22 | 4194304 | |
| 6 | 64 | 21 | **23** | **8388608** | **8388607** |
| 7 | 128 | | 24 | 16777216 | 2097151 |
| 8 | 256 | 15 | 25 | 33554432 | |
| **9** | **512** | **511** | **26** | **67108864** | **67108863** |
| 10 | 1024 | | 27 | 134217728 | 1048575 |
| **11** | **2048** | **2047** | 28 | 268435456 | |
| 12 | 4096 | 1023 | **29** | **536870912** | **536870911** |
| 13 | 8192 | | 30 | 1073741824 | 17043521 |
| **14** | **16384** | **16383** | 31 | 2147483648 | |
| 15 | 32768 | 31 | 32 | 4294967296 | 63 |
| 16 | 65536 | | 33 | 8589934592 | 1227133513 |
| 17 | 131072 | 4095 | 34 | 17179869184 | |
| 18 | 262144 | 29127 | **35** | **34359738368** | **34359738367** |
| 19 | 524288 | | 36 | 68719476736 | 511 |

*steps, that is, the cycle length is larger than* $10^9$. Therefore, our maximality testing method is:

1 For any $n$, check if CA(150′) has configuration $000\cdots00$ as a single length cycle (this configuration repeats itself). If CA is maximal length all other configurations must be in a single cycle.
2 Take any other configuration as seed and run CA(150′) on it till the seed is obtained again as a next configuration. A CA is a maximal length for size $n$, if the seed is obtained again after exactly $2^n - 1$ steps.
3 For large $n$, check if the seed is repeated within $10^9$ steps. If not, declare the CA as a maximal length CA.

Table 3 shows the results for $3 \leq n \leq 36$ where the seed is $00\cdots001$. The rows highlighted in bold indicate the sizes $n$ where CA(150′) is a maximal length CA. Empty observed length values indicate that the CA is not reversible. Note that, irreversibility is observed for $n \in 3N + 1$, which verifies the property mentioned above. It can be seen that for all Sophie Germain primes till 36, CA(150′) is a maximal length CA. Further, similar test on large values of $n$ (up to 2000) suggested that all Sophie Germain primes in that range give a period length more than $10^9$.

For finalizing the size of CA for our PRNG, we took inspiration from `MT19937` that keeps the length in the form of $32 \times n + 1$. Since the PRNG produces 32-bit output, the size of form $32 \times n + 1$ will generate x outputs from a single configuration such that only one bit is left unused. So, as a trade-off to make size of each configuration not too big whereas the cycle length is not too small, we choose the size $n$ for the PRNG as 1409, a Sophie Germain prime of form $32 \times x + 1$. Here, every single configuration produces 44 32-bit random numbers.

### 3.2 Need of Temper Function

Figure 1 shows the space-time diagram of CA(150′) when $n = 89$. It can be clearly seen that there are some visible triangular patterns present there. These patterns are always present in CA(150′) irrespective of the chosen seed. But it is a property of a good PRNG that there are no visible patterns in the space-time diagram. The presence of these patterns also makes sure that if the 32-bit output is generated from CA(150′), it will not perform well on various testbeds. To eliminate these patterns, we take help of a mechanism called *tempering* to our PRNG, which is similar to the tempering used by `MT19937`. For our PRNG, the *Temper* Function is defined same as that of the `MT19937` (see Eqs. 1–4) where y is the 32-bit number generated by the consecutive 32 bits of a configuration of the CA. Other coefficients are:

- (u, d) = (11, FFFFFFFF16)
- (s, b) = (7, 9D2C568016)
- (t, c) = (15, EFC6000016)
- l = 18

We have observed that, use of Tempering eliminates the visible triangular patterns. It also makes sure that the state of the PRNG cannot be identified from a random output.

### 3.3 Finding The Best Seed

A pseudo-random number generator's number sequence is completely determined by the seed: if the generator is reinitialized with the same seed, it will produce the same sequence of numbers. Hence, the choice of a good random seed is crucial in the field of computer security – when a secret encryption key is pseudo-randomly generated, having the seed will allow one to obtain the key.

To obtain the best performing seed for our generator, we do the following:

1. Generate 1000 random seeds using `rand` where each seed has an equal distribution of 0s and 1s for every 32 bit. This is to remove any bias of 0s and 1s. A seed with equal distribution of 0s and 1s may ensure even distribution throughout a cycle.
2. For each seed, generate 12MB of binary file of the 32-bit random outputs of the PRNG.

**Fig. 1** Space Time diagram for CA(150′) ($n = 89$)

3. Test the binary file on the *Dieharder* [3] test suite which contains 114 tests.
4. For a small file size of 12MB, it will fail many tests on Dieharder. Once the seed fails more than 6 tests, testing stops.
5. Out of all the seeds, those passing more than 40 tests before stopping are considered.

After complete analysis, seed number 187 and 367 are found to be the best performing seeds. In the next section, we use only these two seeds.

## 3.4 Results and Comparison with MT

We test our generator using Dieharder and BigCrush of TestU01 test-suites. Dieharder test-suite contains 114 tests where BigCrush of TestU01 contains 106 most stringent tests. For comparison, we have taken `MT19937`, `SFMT19937` and `TinyMT` for 32-bit with their default seeds. Following are the results.

**Table 4** Comparison using Dieharder

| File Size | CA(150′)+tempering | | | | Competitors | | | | | |
|---|---|---|---|---|---|---|---|---|---|---|
| | Seed 187 | | Seed 367 | | MT | | TinyMT | | SFMT | |
| | Weak | Failed | Weak | Failed | Weak | Failed | Weak | Failed | Weak | Failed |
| 1.2 GB | 7 | 2 | 6 | 1 | 5 | 3 | 5 | 2 | 7 | 5 |
| 12 GB | 4 | 0 | 2 | 0 | 2 | 0 | 3 | 0 | 2 | 0 |

### 3.4.1 Dieharder

Dieharder test gives reliable output when tested on a file size >10 GB. However, due to time constraint, for finding the best seed, we used only 12 MB binary data. For testing our PRNG (on seed number 187 and 367) and to get a meaningful comparison, we are now using two file sizes of 1.2 and 12 GB. 1.2 GB file size is used because all the generators compared here do not fail a single test on a 12 GB input file.

Table 4 shows that our PRNG is performing on par with MT, SFMT, and TinyMT in terms of randomness. As seed number 367 is the better performing seed on Dieharder, we choose this seed as our default seed to perform other tests.

### 3.4.2 TestU01 BigCrush

Table 5 shows the failed tests for our PRNG and its competitors using the default seed on BigCrush of TestU01 library (for our PRNG, the default seed is seed no 367). Here, for each of the failed tests, the $p$-value is either *eps* meaning a value $< e^{-300}$ or *1-eps1* which means a value $< e^{-15}$. These results show that even though Dieharder results of our PRNG were comparable to MT and its variants, it is still not up to the levels of MT and other variants in terms of randomness. It fails 4 tests on BigCrush. Note that, the TinyMT (32-bit) tested here also fails many tests on BigCrush, but its 64-bit version passes all the tests on BigCrush.

As our PRNG fails the Matrix Rank and LinearComp test, let us look into more details on what these tests signify. The description of the Matrix Rank and LinearComp test are given below.

**MatrixRank**: [4] A sequence of uniform bits is generated and $m = k \times l$ bits are taken from each. Then a matrix of size $k \times l$ is filled one row at a time. The test then computes the *rank* of the matrix. This is repeated $n$ times generating $n$ matrices. Finally, it compares the empirical distribution of these $n$ matrices with the theoretical distribution of the rank of a random matrix via a chi-square test after merging classes, if needed.

**LinearComp**: [4] Here the *jump complexity* test and the *jump size* test, two tests based on the linear complexity of a sequence of bits, are applied simultaneously. A sequence of $n$ bits is constructed by taking $s$ bits from each random number where both $s$, $n$ are input arguments. Then, using *Berlekamp-Massey algorithm*, the linear

**Table 5** Comparison using BigCrush of TestU01. Here, only failed tests are shown

| **Our PRNG** | | **MT and SFMT** | | **TinyMT** | | | |
|---|---|---|---|---|---|---|---|
| # | Test Name | # | Test Name | # | Test Name | # | Test Name |
| 70 | MatrixRank, r = 15, s = 15 | 80 | LinearComp, r=0 | 8 | CollisionOver, t=7 | 69 | MatrixRank, L=1000, r=26 |
| 71 | MatrixRank, r = 0, s = 30 | 81 | LinearComp, r=29 | 10 | CollisionOver, t=14 | 70 | MatrixRank, r = 15, s = 15 |
| 80 | LinearComp, r=0 | | | 12 | CollisionOver, t=21 | 71 | MatrixRank, r = 0, s = 30 |
| 81 | LinearComp, r=29 | | | 19 | BirthdaySpacings, t=8 | 81 | LinearComp, r=29 |
| | | | | 21 | BirthdaySpacings, t=8 | 87 | LongestHeadRun, r=27 |
| | | | | 27 | SimpPoker, r=27 | 102 | Run of bits, r=27 |
| | | | | 58 | AppearanceSpacings, r=27 | | |

complexity of the subsequence formed by the first $l$ bits is computed for each l, $1 \leq l \leq n$.

Therefore, failure of our PRNG in these two tests suggests that the element of linearity in it is the cause of this failure. In the next phase, our target is to add non-linearity to our generator to further improve its randomness quality.

## 4 Phase II: Improving our PRNG by Non-linearity

The previous section indicates that our PRNG has an element of linearity, which is considered to be an undesirable property. In this section, we explore some ways to add nonlinearity to our generator and see if we can produce comparable results to MT and its variants. We mainly look at two approaches -

1. Using procedure *Twist*, which is taken from `MT19937`. We have further two ways to incorporate this into our generator–
   (a) Twist on Output
   (b) Twist on CA Configuration
2. Using basic non-linear component, like using *XOR* followed by *Multiply*, performed after *Twist* on the output.

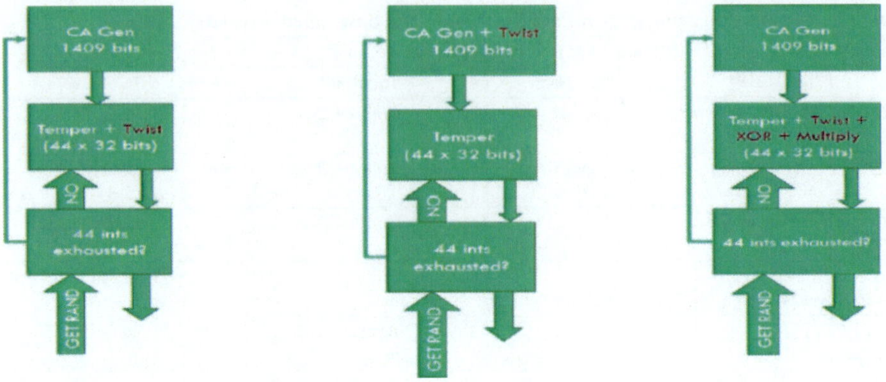


(a) Twist on Output (b) Twist on Configuration (c) Using XOR + Multiply

**Fig. 2** Applying Non-linearity to our PRNG

## *4.1 Using Twist*

The *Twist* function is utilized in `MT19937` for generating the 623-word state. After the state is generated using Twist, a word is extracted and *Temper* is applied on the word to produce the output. Even though *Twist* produces the 623-word state in a single iteration, it does so by manipulating the individual words one at a time. We can use this fact to draw out two approaches of incorporating Twist in our generator.

- **On output**—we were producing output by applying *Temper*, but here, we will apply *Temper* followed by Twist to generate the output.
- **On CA configuration**—this is similar in principle to what `MT19937` does. We first generate the next configuration of CA and then apply Twist to generate the final form of PRNG state. Then we continue the usual process to extract random numbers, i.e, apply *Temper*.

The flowchart of the above two approaches are given in Fig. 2a and b respectively. We test these two approaches on our testbeds again considering seed no. 367. Table 6 shows the results. We observe that our generator with *Temper+Twist on Output* produces the overall best result in Dieharder, beating others at both 1.2 GB as well as 12 GB input size. Also, careful observation reveals that SFMT, which is supposed to be better than MT, actually produces worse results in Dieharder (see Table 4 for comparison). However, while testing on BigCrush, our generator again fails the same tests of the previous phase. Therefore, the *Twist* function is not sufficient to remove the element of linearity from our PRNG. Next, we explore another method to add nonlinearity.

**Table 6** Results after applying *Twist* on Diehard and BigCrush of TestU01 library

| Diehard | | | | | | | | BigCrush on Both | |
|---|---|---|---|---|---|---|---|---|---|
| Twist on output | | | | Twist on configuration | | | | # | Test name |
| 1.2 GB | | 12 GB | | 1.2 GB | | 12 GB | | | |
| Weak | Failed | Weak | Failed | Weak | Failed | Weak | Failed | | |
| 4 | 1 | 1 | 0 | 9 | 3 | 2 | 0 | 70 | MatrixRank, r = 15, s = 15 |
| | | | | | | | | 71 | MatrixRank, r = 0, s = 30 |
| | | | | | | | | 80 | LinearComp, r=0 |
| | | | | | | | | 80 | LinearComp, r=29 |

### 4.2 *XOR + MULTIPLY*

In this approach, we use the *Temper+Twist on Output* method and apply *XOR* as well as *Multiply* operation on the output. The idea is—as, multiplication is a nonlinear operation, so, when we apply multiplication with numbers close to Sophie Germain primes, nonlinearity is added. The XOR part is included just to add more randomness. The flowchart of this approach is given in Fig. 2c.

However, unfortunately, there no improvement in performance in BigCrush. The same tests are failed again (likewise Table 6). At this point, our PRNG is also burdened with many functions, to the extent that when the speed of our generator to test is nearly 5 times that of `MT19937`. This is because, even though the rules of CA use only bitwise operations which are efficient in terms of CPU cycles, multiplication is quite costly and cannot be directly reduced to bitwise operations. Since we are applying so many manipulations on the output, the number of function calls to these procedures also increases, thereby significantly increasing the execution time. This, combined with operations included in procedures *Temper* and *Twist* further increase the overhead.

Therefore, we propose not to use this unnecessary *XOR + Multiply* operation and choose the version of Fig. 2a, that is, *Temper+Twist on Output* on CA(150′) with seed no. 367 as the default seed as our final PRNG. As per Conjecture 1 of Ref. [1], this PRNG has a period length of $2^{1409} - 1$.

## 5 Conclusion and Next Steps

We have developed a pseudo-random number generator that uses a cellular automaton as the base component to produce pseudo-random numbers. We have employed CA size as 1409 (that is $32 \times 44 + 1$) and used a CA named CA(150′) which constitutes of Rule 90 on the first bit and Rule 150 on all others following null-boundary condition. In each iteration, our PRNG can produce 44 numbers which when compared to its competitors is a fairly balanced number because it does not produce unnecessarily huge numbers in each iteration. This reduces the storage requirements and makes it lightweight. The PRNG requires a specific seed to initiate the process and we have selected a good seed for this purpose. We have named this seed as *seed no. 367*. We have used the same technology of `MT19937` with CA(150′) to compete with `MT19937` and `SFMT19937`. Our PRNG contains *Temper* and *Twist* function like `MT19937` and can pass all tests of Dieharder which is even better than its peers `MT19937` and `SFMT19937`.

However, it fails to pass the LinearComp and Matrix Rank tests of BigCrush. In fact, we are still far away (as indicated by p-values of failed tests) from achieving our desired result, that is, passing all tests of BigCrush. Further addition of any function is impractical as it will hamper the efficiency of the generator. Therefore, we need to take another approach altogether. As the problem may be due to the linearity in our CA(150′), the obvious suggestion is to move on to a nonlinear CA. The idea is instead of using a linear CA and adding nonlinearity in it externally, to somehow incorporate nonlinearity in our CA itself. But the challenge lies here is to find a suitable non-linear CA which has maximal length.

As a parallel to this, there are other linear CAs like CA(150′) that have similar characteristics, like CA(90′) which uses rule 90 at all cells except the first cell which follows rule 150. This CA is also introduced in Ref. [1] and is a *potential* maximal length CA. Work may be directed to see it can also be exploited to develop a stronger PRNG that can fool the randomness testbeds. Further, identifying other linear CAs that has potential to be maximal length as well as are good source of randomness might be another research direction.

**Acknowledgements** The authors are grateful to Prof. Sukanta Das for his support, comments, guidance and encouragement to do this work.

## References

1. Adak, S., Das, S.: imperfect) strategies to generate primitive polynomials over gf(2. Theoretical Computer Science **872**, 79–96 (2021)
2. Bhattacharjee, K., Maity, K., Das, S.: A search for good pseudo-random number generators: survey and empirical studies (2018). CoRR, arXiv:abs/1811.04035
3. Brown, R.G., Eddelbuettel, D., Bauer, D.: Dieharder: a random number test suite. https://webhome.phy.duke.edu/~rgb/General/dieharder.php
4. L'Ecuyer, P., Simard, R.: TestU01: A C library for empirical testing of random number generators. ACM Trans. Math. Softw. **33**(4), 22:1–22:40 (2007)
5. Matsumoto, M., Nishimura, T.: Mersenne Twister: A 623-dimensionally equidistributed uniform pseudo-random number generator. ACM Trans. Model. Comput. Simul. **8**(1), 3–30 (1998)
6. Saito, M., Matsumoto, M.: SIMD-oriented fast Mersenne Twister: a 128-bit pseudorandom number generator. In: 7th International Conference on Monte Carlo and Quasi-Monte Carlo Methods in Scientific Computing, pp. 607–622. Springer, Berlin, Heidelberg (2008)
7. Saito, M., Matsumoto, M.: A high quality pseudo random number generator with small internal state. IPSJ SIG Notes **3**, 1–6 (2011)

# Defining Reachability Tree Under Adiabatic and Reflexive Boundary Conditions

**Nazma Naskar and B. K. Sivaraj**

**Abstract** In this paper the elementary cellular automata is explored under adiabatic boundary condition and reflexive boundary condition, using the reachability tree. To construct the reachability tree under the mentioned conditions, structures of those boundary conditions are analysed. The valid RMTs of root level and leaf level are identified. It was found that only at root level and leaf level, the structures and arrangements differ slightly with the reachability tree of null boundary condition.

**Keywords** Cellular automata · Boundary condition · RMT · Rule mean term · Reachability tree · Adiabatic boundary condition · Reflexive boundary condition

## 1 Introduction

In the world of cellular automata most of the work considered either null or periodic boundary conditions [5], and only a very few number of works were done under other boundary conditions like Adiabatic, Reflexive and Intermediate boundary conditions. Both the null boundary and periodic boundary conditions for elementary cellular automata is being widely used in domains like pattern classification, pattern clustering, cryptography [6], simulation of markets [3], image processing [4], generating pseudo random number etc. In some these fields, the boundary condition affects results significantly, like, in pattern classification, for some dataset the accuracy is much higher in null boundary than the periodic boundary condition and vice-versa is also true for some other datasets. In this scenario, the usage of other not widely used boundary conditions might lead to better results, and provides us more conditions to experiment with. This is the motive of the paper, to define less known boundary

N. Naskar (✉)
School of Computer Science, KIIT Deemed to be University,
Bhubaneswar 751024, Odisha, India
e-mail: nazma.naskarfcs@kiit.ac.in

B. K. Sivaraj
Department of Metallurgical and Materials Engineering, National Institute
of Technology Tiruchirapalli (NIT-T), Tiruchirapalli 620015, Tamil Nadu, India

 
S. Das and G. J. Martinez (eds.), *Proceedings of First Asian Symposium on Cellular Automata Technology*, Advances in Intelligent Systems and Computing 1425,
https://doi.org/10.1007/978-981-19-0542-1_9

conditions, and describe their behaviour. The first demand of this comparative study is a uniform characterization tool.

The *Reachability Tree* [2] proved its efficiency as an analysis tool of elementary cellular automata under null boundary condition and as well as periodic boundary conditions. For a given CA, tasks like identifying the reachable configuration or non-reachable configuration, counting number of non-reachable (and reachable states), deciding whether a CA is reversible or not, counting and identifying point attractors are easily performed by utilizing reachability tree. Along with *Linked Tree*, which is an extension of reachability tree, additionally we are able to identify and count the number of cyclic states (and acyclic states) [7], determine whether a given CA has only point attractor or not, predict cycle structures etc. Moreover, in the vast area of cellular automata, using reachability tree is not restricted to the elementary cellular Automata, and all the above mentioned tasks can also be performed for any $d$-state, one dimensional, 3 neighborhood cellular automata.

Therefore, to harness the characteristics of adiabatic boundary condition (ABC) and reflexive boundary condition (RBC), their logic and structure has to be understood, to construct reachability tree, which is discussed in this paper. The applications of this paper includes identifying CAs with point state attractors, pattern classification, along with wider adoption of ABC and RBC.

## 2 Basics of Cellular Automata

A cellular automaton is a regular arrangement of cells. Each cell has integer value, referred as *State*. The state of cells evolve with discrete time and space, and transit to possible states using a localised *Transition Function*. These transition functions are generally are based on a neighbourhood condition, utilising the state values of the neighborhood. The neighbourhood of a given cell consists of the considered cell itself and the cells adjacent to it. Conventional cellular automata have infinite lattice size. However, for any practical application, finite CA, with finite number of cells is utilised, for obvious reasons. Here we consider finite cell CA with two states (0 and 1), with 1 dimensional three neighborhood dependency.

Consider the $i$th cell ( $1 \leq i \leq n$, n is lattice size) at time t, with state $S_i^t$. Hence, the left and the right neighbours are denoted $S_{i-1}^t and S_{i+1}^t$, respectively. The radius of the neighbourhood is defined as the number of neighbours depended by a cell in a direction. Radius of 1 implies that each cell considers one neighbour on each side. Hence, the next state of the $i$th cell at time $(t+1)$ is $S_i^{t+1}$, is obtained by applying the*Transition Function* (or Local Rule), denoted by F, such that $S_i^{t+1}$ = F $(S_{i-1}^t, S_i^t, S_{i+1}^t)$. Due to 2-state three neighborhood condition, we have 256 $(2^{2^3})$ transition functions, traditionally called as *Rules* $(\mathcal{R})$. A cellular automaton is represented with the rule that each cell follows, as $\langle \mathcal{R}_0, \mathcal{R}_1, \cdots, \mathcal{R}_i, \cdots, \mathcal{R}_{n-1} \rangle$ which is called as the *Rule Vector* of the CA. If each cell of CA follows same transition function, then the CA is said to be *Uniform CA*, otherwise, the CA is referred as *Non-uniform CA*.

## 2.1 Rule Min Term (RMT)

Rule Min Term (RMT) is the possible combinations of input to the local rule of a cell, i.e., possible combinations of present states of the cell and its neighbours. It is generally represented as a string of state values of the neighbourhood of a cell, from left to right. This means that an RMT consists of $(2 * radius) + 1$ number of state values, and total number of RMTs possible will be $states^{(2*radius)+1}$. In our case, the RMTs will consist of 3 values, say xyz, and there are 8 possible RMTs. This representation is Min term of three variables, according to switching function.

We represent RMTs by their decimal equivalents, for our convenience. This means that, in our case, $000 = 0, 001 = 1, 010 = 2, \cdots, 111 = 7$, are the possible combinations of input possible for local rule. The first row of Tables 1 and 2 represents the possible RMTs. Now, for each RMT, the next RMT in the sequence can be calculated by suffixing the two possible states (0 and 1), and disregarding the first cell of the RMT. Thus, each RMT has two possible next RMTs, called as *Sibling RMTs*.

For example, RMT 011 (3) can be suffixed with 0 and 1, which yields the new RMTs 110 (6) and 111 (7), which are the sibling RMTs of 011. Similarly, RMT 001 yields the RMTs 010 and 011. Some RMTs have same sibling RMTs, and those RMTs are called as *Equivalent RMTs*. For instance, like the RMT 011 (3), RMT 111 (7) also yields the same sibling RMTs 110 (6) and 111 (7), hence RMTs 011 (3) and 111 (7) are equivalent RMTs.

The relationship between $i$th and $(i + 1)$th RMTs, along with equivalent RMTs and their sibling RMTs are shown in Table 3. The meaning of $d$ is don't care state, i.e. the states of those RMTs does not effect the nature of considered automaton. The reason behind marking '$d$' for some specific RMTs are discussed in the construction procedures of Reachability Tree (Sects. 3.2 and 3.3) .

**Table 1** CA ⟨162, 136, 130, 202⟩ under Adiabatic boundary condition

| PS (RMT) | 111 (7) | 110 (6) | 101 (5) | 100 (4) | 011 (3) | 010 (2) | 001 (1) | 000 (0) | Rule |
|---|---|---|---|---|---|---|---|---|---|
| NS | 1 | 0 | d | d | d | d | 1 | 0 | 162 |
| NS | 1 | 0 | 0 | 0 | 1 | 0 | 0 | 0 | 136 |
| NS | 1 | 0 | 0 | 0 | 0 | 0 | 1 | 0 | 130 |
| NS | 1 | d | d | 0 | 1 | d | d | 0 | 202 |

**Table 2** CA ⟨162, 136, 130, 202⟩ under Reflexive boundary condition

| PS (RMT) | 111 (7) | 110 (6) | 101 (5) | 100 (4) | 011 (3) | 010 (2) | 001 (1) | 000 (0) | Rule |
|---|---|---|---|---|---|---|---|---|---|
| NS | 1 | d | 1 | d | d | 0 | d | 0 | 162 |
| NS | 1 | 0 | 0 | 0 | 1 | 0 | 0 | 0 | 136 |
| NS | 1 | 0 | 0 | 0 | 0 | 0 | 1 | 0 | 130 |
| NS | 1 | d | 0 | d | d | 0 | d | 0 | 202 |

**Table 3** Relation between $i$th and $(i+1)$th RMT

| $i$th (Equivalent) RMTs | $(i+1)$th Sibling RMTs |
|---|---|
| 0,4 | 0,1 |
| 1,5 | 2,3 |
| 2,6 | 4,5 |
| 3,7 | 6,7 |

## 2.2 *State Transition Diagram*

The next state of each cell is obtained by simultaneously applying the local rule to each cell, i.e., parallelly to all the cells of the CA. At time $t$, collectively, all the states together is called *Configuration of CA*, $S^t$ $(S_1^t, S_2^t, \cdots, S_{n-1}^t)$. It is equivalent to a Global Transition Function being applied on a configuration to obtain the next configuration. The graphical representation of evolution of configurations is called *State Transition Diagram*. It helps us to understand dynamic behaviour of a given cellular automaton. Figures 1 and 2 show the state transition diagram of (162, 136, 130, 202) CA under adiabatic boundary condition and under reflexive boundary condition, respectively. The configurations can be grouped into two categories, namely Cyclic and Acyclic configurations.

A state (here a configuration is a *state of the diagram* ) is called cyclic, if it is a part of some cyclic transition. For example, configurations ⟨0001⟩, ⟨1111⟩, ⟨0011⟩, ⟨0000⟩ (Fig. 1) and ⟨0000⟩ (Fig. 2) are cyclic states. The contrast are the acylic states. Further, the acyclic configurations are of two types, reachable and non-reachable. For instance, in Fig. 1, ⟨1001⟩, ⟨1000⟩, ⟨1100⟩ are reachable but acyclic states, while ⟨1101⟩, ⟨1011⟩, ⟨0110⟩ are non reachable acyclic states.

## 2.3 *Boundary of a Finite CA*

Conventionally, cellular automata are considered with infinite lattice size. The infinite lattice size is computationally difficult to implement CA as modelling tool for technical fields [1]. Hence, the concept of a finite cellular automaton evolved. It has a

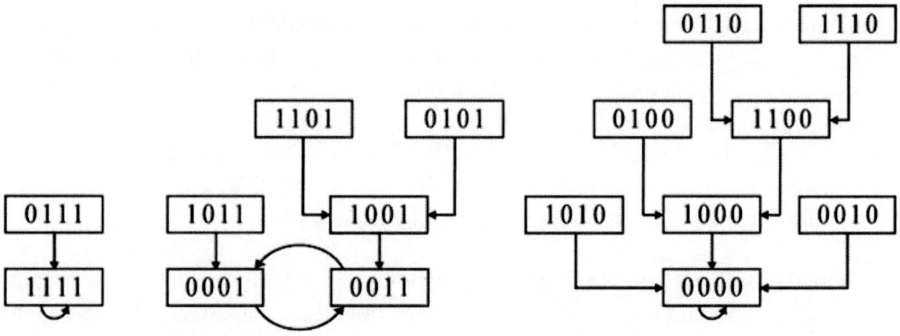


**Fig. 1** State transition diagram of (162, 136, 130, 202) under ABC

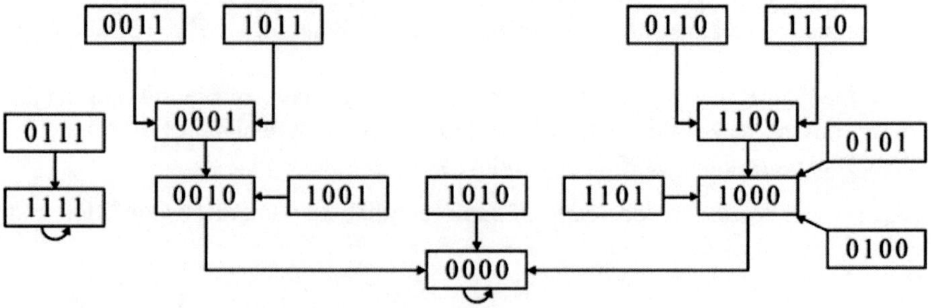


**Fig. 2** State transition diagram of (162, 136, 130, 202) under RBC

finite number of cells, with $n$ denoting the number of cells. They possess an additional interesting feature, which is the boundary of the lattice of cells. Boundary essentially means that there are no cells further across the end of the lattice. This means that cells in the interior of the lattice, away from boundary, have enough number of neighbours to compute their next state, while the boundary cells lack neighbours. In order to obtain the next state, the missing neighbour is compensated by providing another state in order to apply the local rule to the boundary cell, which is given according to the type of boundary condition we define.

Consider the following one dimensional finite elementary CA of radius 1, with $n$ number of cells. The neighbourhood of $i$th cell consists of its left neighbour, itself and right neighbour. Let the initial time be $t$, at which the configuration is the set $S^t$ $(S_0^t, S_1^t, \cdots, S_{n-2}^t, S_{n-1}^t)$. Note that the cell index starts from 0 and index of last cell is $n-1$.

The missing neighbours, at the right and left boundaries are denoted by the indices $n$th and $-1$th respectively, as shown above. Their arbitrary states, which is yet to be fixed, are $S_n^t$ and $S_{-1}^t$ . Now, we have several boundary conditions, according to which their states are given. These are:

1. Null Boundary Condition: As the name suggests, the states of the missing neighbours are given as 0 ( i.e null). It means that $S_n^t$ = $S_{-1}^t$= 0. It is the most simplest boundary condition, and most widely applied.
2. Periodic Boundary Condition: Here, the states of the missing neighbours are given states as if the both boundaries meet together. It is as if the $n$ cell CA is periodically repeated at the front end and back end of the original $n$ cell lattice. Hence last cell becomes the missing neighbour of first cell and vice versa, i.e $S_{-1}^t$ = $S_{n-1}^t$and $S_n^t$ = $S_0^t$.
3. Adiabatic Boundary Condition : In this boundary condition, the states of the missing neighbours are same as the states of boundary cells itself. Therefore, it can written as $S_{-1}^t$ = $S_0^t$ and $S_n^t$ = $S_{n-1}^t$ .
4. Reflexive Boundary Condition: Here, the states of the missing neighbours are the reflection with respect to the boundary cell. It means that missing neighbour of the boundary cell has same state of the present neighbour, i.e $S_{-1}^t$ = $S_1^t$ and $S_n^t$ = $S_{n-2}^t$.
5. Intermediate Boundary Condition: In this case, the states of the missing neighbours are the states of cell which is next to the present neighbour of the boundary cell. It is represented as $S_{-1}^t$ = $S_2^t$ and $S_n^t$ = $S_{n-3}^t$.

In our present work, we consider and explore Adiabatic Boundary Condition and Reflexive Boundary Condition.

## 3 Reachability Tree

Even though state transition diagrams are very useful for understanding the behaviour of a CA, it can be tedious to obtain such a diagram. The dynamics of a given cellular automata, like identifying reachable and non reachable configurations, verifying reversibility property etc, are efficiently revealed by the tool called *Reachability Tree*.

The Reachability Tree is a labelled binary tree that visualises the reachable states of a CA. For a $n$ cell CA, there are $n + 1$ levels, starting from level 0 to level $n$. Level 0 is called as Root, and level $n$ is called as Leaf level, like the concept of tree. Mainly, it has two components : Nodes and Edges. The nodes of the tree are a subset of set of all RMTs, which are representative of next reached RMT from the previous RMT. Every edge is associated with set of RMTs called label $l_{i.j}$. The nodes are represented by $N_{i.j}$ ,where $i$ is the level index $(0 \leq i \leq n)$, and j is the node number at $i$th level $(0 \leq j \leq 2^{i-1})$.

The nodes are the subset of RMTs of rules $N_{i.j} \subseteq Z_8^i$. The numbering of nodes in each level starts from left side and towards the right. Since $Z_8^n = \phi$ for an n-cell CA, the leaf nodes are empty. The edges are denoted by $E_{i.2j} = (N_{i.j}, N_{[i+1].2j}, l_{i.2j})$ and $E_{i.[2j+1]} = (N_{i.j}, N_{[i+1].[2j+1]}, l_{i.[2j+1]})$ in which $(0 \leq i \leq n-1)$, and $(0 \leq j \leq 2^{i+1} - 1)$, where $l_{i.2j} \subseteq N_{i.j}$ and $l_{i.[2j+1]} \subseteq N_{i.j}$ are the labels of the edges, such that $l_{i.2j} \cup l_{i.[2j+1]} = N_{i.j}$ and $l_{i.2j} \cap l_{i.[2j+1]} = \phi$. The existence of any edge $E_{i.j}$ depend on $R_i[r]$, where $R_i$ is the rule for the $i$th cell, and $r \in N_{[i\_1].[j/2]}$. If $R_i[r] = 0$, then RMT $r$ appears on label $l_{i.2j}$; else if $R_i[r] = 1$, then RMT $r$ appears on label $l_{i.[2j+1]}$.

This essentially means that edges represent the application of local rule for each cell, and segregates the nodes based on next state reached. All the leftward edges represent reaching the state 0, while rightward edges denote reaching state of 1. The leaves of last level represent the state of last cell of the finite CA. Traversing from root towards the leaf node along the edges will determine the reachable configurations of the given CA.

Apart from this basic mechanism, the reachable states are influenced by the different set of procedures followed for each boundary condition.

## 3.1 Construction of Reachability tree

1. The RMT table is first determined, based on logical interpretation of boundary condition. This involves plotting the rule vector and states involved (i.e binary form of the rule number for each cell), and selecting the possible RMTs for the root node and possible RMTs for the n-1 node, by constraints from the boundary condition followed.
2. The root node is fixed as the set of possible RMTs that begins the RMT sequence.
3. Based on next state of first cell as given by the local rule, the RMTs from root node are classified by two edges, one for state 0 (left edge), another for state 1 (right edge). RMT table is referred for this step.
4. The next node is obtained by determining the sibling RMTs of the RMTs present on the edges.
5. Again, based on the next next state of second cell as given by the local rule, the RMTs from nodes of level 1 are classified by two edges, one for state 0 (left edge), another for state 1 (right edge).
6. Steps 4 and 5 are repeated until we reach $n - 1$th level.
7. After reaching $n - 1$th level, the non reachable RMTs for last cell are discarded from the nodes of $n - 1$th level, based on the logic of boundary condition. Then, based rule for last cell, the possible RMTs are segregated into two edges each, giving rise to leaf nodes.
8. Finally, traverse the tree from root towards the leaf nodes along the edges, to determine the reachable configurations of the given CA.

We can formally define the reachability tree as:

**Definition 1** The reachability tree for a n-cell cellular automaton is a rooted and edge-labeled binary tree with n+1 levels, with edges $E_{i.2j} = (N_{i.j}, N_{[i+1].2j}, l_{i.2j})$ *and* $E_{i.[2j+1]} = (N_{i.j}, N_{[i+1].[2j+1]}, l_{i.[2j+1]})$ where $(0 \leq i \leq n-1)$ and $(0 \leq j \leq 2^{i+1} - 1)$. The nodes are represented by $N_{i.j}$ such that $N_{i.j} \subseteq Z_8^i$. The edge $E_{i.2j}$ is between the nodes $N_{i.j}$, $N_{[i+1].2j}$ with label $l_{i.2j}$, and the edge $E_{i.[2j+1]}$ is between the nodes $N_{i.j}$, $N_{[i+1].[2j+1]}$ with label $l_{i.[2j+1]}$. Here, $l_{i.2j} \subseteq N_{i.j}$ and $l_{i.[2j+1]} \subseteq N_{i.j}$ and $l_{i.2j} \cup$

$l_{i.[2j+1]} = N_{i.j}$. The existence of any edge $E_{i.j}$ depend on $R_i[r]$, where $R_i$ is the rule for the $i$th cell, and $r \in N_{[i\_1].[j/2]}$. If $R_i[r] = 0$, then RMT $r$ appears on label $l_{i.2j}$; else if $R_i[r] = 1$, then RMT $r$ appears on label $l_{i.[2j+1]}$.

Depending upon the boundary condition used, the reachability tree has small variations at the root level and leaf level. The reachability tree for Null boundary condition and Periodic boundary conditions has been already framed. We shall now see the construction of reachability tree for Adiabatic boundary condition and Reflexive boundary condition.

## 3.2 *Reachability Tree for ABC*

**Valid RMTs at root level and last level under ABC**: Adiabatic Boundary Condition is represented as $S^t_{-1} = S^t_0$ and $S^t_n = S^t_{n-1}$, as discussed in Sect. 2.3. This condition narrows down the possible RMTs at root level and last level.

To satisfy the boundary condition, two constraints must be followed:

**(a) The first and second bits of the first RMT must be same**

At the left boundary under ABC, as we saw in Sect. 2.3, $S^t_{-1} = S^t_0$. It means that the first and second bits of the first RMT are same, and can be either both 0 or both 1. If both the bits are 0, then third bit can be either 0 or 1, implying 000 and 001 are the possible RMTs to begin the RMT sequence under ABC. If both the bits are 1, then third bit can be either 0 or 1, implying 110 and 111 are also the possible RMTs to begin the RMT sequence. **Hence the valid RMTs to begin the sequence are : 0, 1, 6, and 7**, and all other RMTs are discarded.

**(b) The second and third bits of the last RMT must be same**

At the right boundary under ABC, as we saw in Sect. 2.3, $S^t_n = S^t_{n-1}$. It means that the second and third bits of the last RMT are same, and can be either both 0 or both 1. If both the bits are 0, then first bit can be either 0 or 1, implying 000 and 100 are the possible RMTs to end the RMT sequence under ABC. If both the bits are 1, then first bit can be either 0 or 1, implying 011 and 111 are also the possible RMTs to end the sequence. **Thus, RMTs 0, 3, 4 and 7 are the only possible RMTs of the nodes of $n - 1$th level**, and all other RMTs are discarded (Fig. 3).

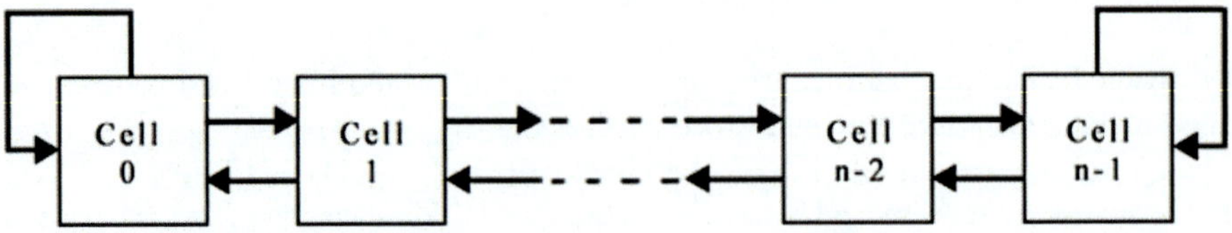


**Fig. 3** Structure of adiabatic boundary condition

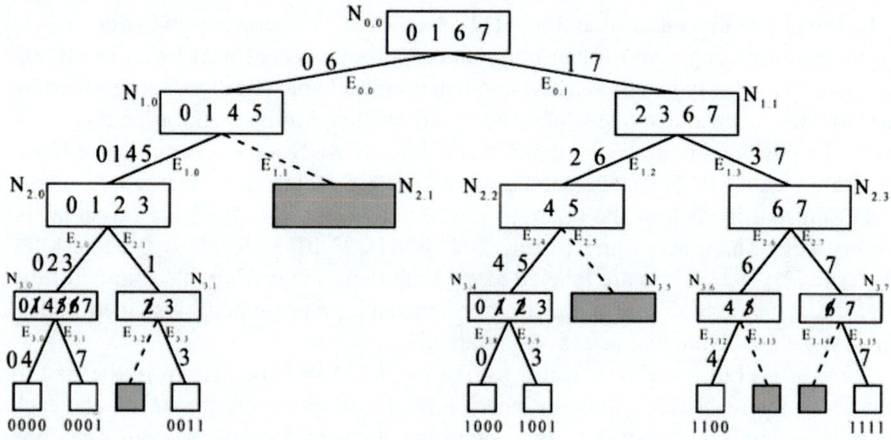


**Fig. 4** Reachability Tree for the CA under ABC

The discarded RMTs are represented as d in RMT table. After applying these constraints, the remaining construction of the reachability tree is same as for any reachability tree, as per Sect. 3.1. It can be described as follows:

**Definition 2**

1. $\bigcup_{0\leq j\leq d^i-1} N_{i.j} = Z_8^i$, where $(0 \leq i \leq n-1)$. Note that $\{r \in Z_8^0 \mid r = xxy\}$ and $\{r \in Z_8^{n-1} \mid r = xyy\}$, where $x, y = \{0, 1\}$ in adiabatic boundary condition.
2. $\bigcup_{0\leq j\leq 7}\{l_{i.2j}, l_{i.[2j+1]}\} = N_{i-1.j}$.
3. $\forall r \in l_{i.[2j+x]}, R_i[r] = x$, and RMTs $2r + x \pmod{2}^3$ of $R_{i+1}$ are in $N_{i+1.2j+x}$, where $x = \{0, 1\}$.

**Example 1** Consider a non uniform 4 cell elementary CA, with the rule vector (162, 136, 130, 202). Under Adiabatic Boundary Condition, the RMT table for this CA is given in Table 1, Sect. 2.1. Figure 4 shows the reachability tree for the considered CA. Here, due to the boundary constraint, only four RMTs are valid for root level (left most cell) i.e. $Z_0^4 = (0, 1, 6, 7)$ and last level (right most cell) i.e. $Z_{n-1}^4 = (0, 3, 4, 7)$, as discussed above. Beginning from the root, we have $N_{0.0} = \{0, 1, 6, 7\}$. From the RMT table (Table 1), the edges are decided by values of the possible RMTs according to the rule of the cell. For cell 0, according to rule 162, RMTs 0 and 6 have the value 0, represented by edge $E_{0.0}$, while RMTs 1 and 7 have the value 1, represented by edge $E_{0.1}$. Then, the next nodes are determined by finding the set of all sibling RMTs for all the RMTs present on the edges leading to a node. Thus, we get $N_{1.0} = \{0, 1, 4, 5\}$ and $N_{1.1} = \{2, 3, 6, 7\}$. Then, according to the rule of that cell, the RMTs on the node are split into two edges, and these two steps continue.

At level 1 we consider the rule 136 for cell 1, which gives $\mathcal{R}_1[0] = \mathcal{R}_1[1] = \mathcal{R}_1[4] = \mathcal{R}_1[5] = 0$, hence $l_{1.1} = \phi$. This means all the RMTs on the node $N_{1.0}$ (i.e

0, 1, 4, 5) have the value 0, and no RMT has value 1. Therefore this edge $E_{1.1}$ is a non-reachable edge, and it can not connect to any node of next level, rendering that node $N_{2.1}$ also to be non reachable. Such non reachable edges are represented by dashed lines, while the non reachable nodes are shaded. Similarly, the edge $E_{2.5} = \phi$, hence the edge and node $N_{3.5}$ are non reachable. As we reach the penultimate level, the "don't care" RMTs (represented as d in RMT table) are discarded from the RMTs present in each of the node, as only the RMTs {0, 3, 4, 7} are possible for the last cell. Then, according to rule 202, RMTs $\mathcal{R}_3[0] = \mathcal{R}_3[4] = 0$ and RMTs $\mathcal{R}_3[3] = \mathcal{R}_3[7] = 1$, and are labeled along with their edges. Here also, some of the edges, and hence their leaf node, are non reachable, represented by shaded nodes. Finally, the reachable leaf nodes are obtained.

Now, to find the reachable states for the given CA, we traverse from the root to leaf, along the reachable edges, with each left edge representing the state 0 and each right edge representing state 1, and obtain the states of each of the cell. Thus one route from the root to a reachable leaf node represents the reachable states of all the cells of the CA, and we get the collection of reachable states, which is a reachable configuration of the CA. The number of reachable leaf nodes represents the total number of reachable states of the CA. Thus in this example, configurations ⟨0000⟩, ⟨0001⟩, ⟨0011⟩, ⟨1000⟩, ⟨1001⟩, ⟨1100⟩ and ⟨1111⟩ are the reachable configurations of the considered CA, while other configurations are non reachable. This is also visible clearly from the state transition diagram of the CA, in Fig. 1, where the above 7 configurations are reachable, while the other 9 configurations are non reachable.

### 3.3 Reachability Tree for RBC

**Valid RMTs at root level and last level under RBC:** Reflexive Boundary Condition is represented as $S^t_{-1} = S^t_1$ and $S^t_n = S^t_{n-2}$, as discussed in Sect. 2.3. This condition narrows down the possible RMTs at root level and last level (Fig. 5).

To satisfy the boundary condition, two constraints must be followed:

**(a) The first and third bits of the first RMT must be same**

At the left boundary under RBC, as we saw in Sect. 2.3, $S^t_{-1} = S^t_1$. It means that the first and third bits of the first RMT are same, and can be either both 0 or both 1. If both the bits are 0, then second bit can be either 0 or 1, implying 000 and 010 are the

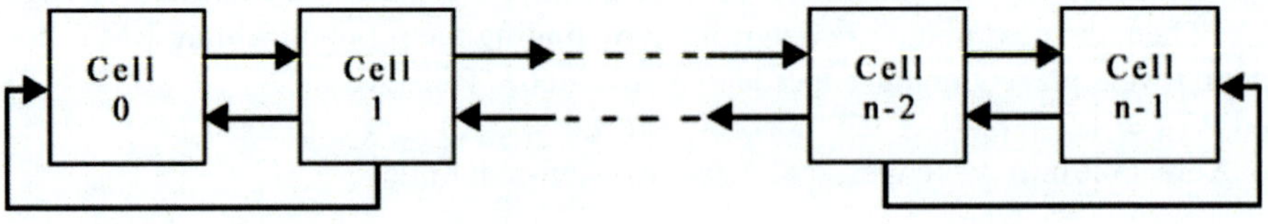


**Fig. 5** Structure of reflexive boundary condition

possible RMTs to begin the RMT sequence under RBC. If both the bits are 1, then second bit can be either 0 or 1, implying 101 and 111 are also the possible RMTs to begin the sequence. **Hence the valid RMTs at the root node are : 0, 2, 5, and 7**, and all other RMTs are discarded.

**(b) The first and third bits of the last RMT must be same**
At the right boundary under RBC, as we saw in Sect. 2.3, $S_n^t = S_{n-2}^t$. Here also, the conditions for possible RMTs are same as that of first RMT, hence the same set of RMTs are only possible. **Thus, the RMTs 0, 2, 5 and 7 are the only possible RMTs of the nodes of $n-1$th level, and all other RMTs are discarded.**

The discarded RMTs are represented as d in RMT table. After applying these constraints, the remaining construction of the reachability tree is same as for any reachability tree, as per Sect. 3.1. It can be described as follows:

**Definition 3**

1. $\bigcup_{0\le j\le d^i-1} N_{i.j} = Z_8^i$, where $(0 \le i \le n-1)$. Note that $\{r \in Z_8^0 \mid r = yxy\}$ and $\{r \in Z_8^{n-1} \mid r = xyx\}$, where $x, y = \{0, 1\}$ in reflexive boundary condition.
2. $\bigcup_{0\le j\le 7}\{l_{i.2j}, l_{i.[2j+1]}\} = N_{i-1.j}$.
3. $\forall r \in l_{i.[2j+x]}$, $R_i[r] = x$, and RMTs $2r + x \pmod 2)^3$ of $R_{i+1}$ are in $N_{i+1.2j+x}$, where $x = \{0, 1\}$.

**Example 2** Consider a non uniform 4 cell elementary CA, with the rule vector (162, 136, 130, 202).Under Reflexive Boundary Condition, the RMT table for this CA is given in Table 2, Sect. 2.1. Figure 6 shows the reachability tree for the considered CA. Here, due to the boundary constraint, only four RMTs are valid for left most and

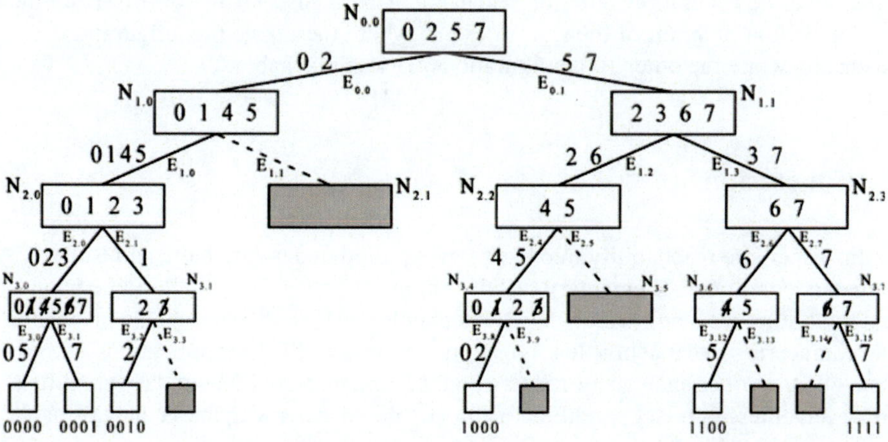


Fig. 6 Reachability Tree for the CA under RBC

right most cells, as discussed above, and $Z_0^8 = Z_{n-1}^8 = (0, 2, 5, 7)$. Beginning from the root, we have $N_{0.0} = Z_0^8$. From the RMT table (Table 2), the edges are decided by values of the possible RMTs according to the rule of the cell. For cell 0, according to rule 162, RMTs 0 and 2 have the value 0, represented by edge $E_{0.0}$, while RMTs 5 and 7 have the value 1, represented by edge $E_{0.1}$. Then, the next nodes are determined by finding the set of all sibling RMTs for all the RMTs present on the edges leading to a node. Thus, we get $N_{1.0} = \{0, 1, 4, 5\}$ and $N_{1.1} = \{2, 3, 6, 7\}$. Then, according to the rule of that cell, the RMTs on the node are split into two edges, and these two steps continue.

At level 1 we consider the rule 136 for cell 1, which gives $\mathcal{R}_1[0] = \mathcal{R}_1[1] = \mathcal{R}_1[4] = \mathcal{R}_1[5] = 0$, hence $l_{1.1} = \phi$. This means all the RMTs on the node $N_{1.0}$ (i.e $0, 1, 4, 5$) have the value 0, and no RMT has value 1. Therefore this edge $E_{1.1}$ is a non-reachable edge, and it can not connect to any node of next level, rendering that node $N_{2.1}$ also to be non reachable.Similarly, the edge $E_{2.5} = \phi$, hence the edge and node $N_{3.5}$ are non reachable. As we reach the penultimate level, the "don't care" RMTs (represented as d in RMT table) are discarded from the RMTs present in each of the node, as only the RMTs $\{0, 2, 5, 7\}$ are possible for the last cell. Then, according to rule 202, RMTs $\mathcal{R}_3[0] = \mathcal{R}_3[2] = \mathcal{R}_3[5] = 0$ and RMT $\mathcal{R}_3[7] = 1$, and are labeled along with their edges. Here also, some of the edges, and hence their leaf node, are non reachable, represented by shaded nodes. Finally, the reachable leaf nodes are obtained.

Now, to find the reachable states for the given CA, we traverse from the root to leaf, along the reachable edges, with each left edge representing the state 0 and each right edge representing state 1, and obtain the states of each of the cell. Thus one route from the root to a reachable leaf node represents the reachable states of all the cells of the CA, and we get the collection of reachable states, which is a reachable configuration of the CA. Thus in this example, configurations $\langle 0000 \rangle$, $\langle 0001 \rangle$, $\langle 0010 \rangle$, $\langle 1000 \rangle$, $\langle 1100 \rangle$ and $\langle 1111 \rangle$ are the reachable configurations of the considered CA, while other configurations are non reachable. This is also visible clearly from the state transition diagram of the CA, in Fig. 2, where the above 6 configurations are reachable, while the other 10 configurations are non reachable.

## 4 Conclusion

In this paper, the reachability tree has been explored under Adiabatic and Reflexive boundary conditions. Elementary cellular automata under Null and Periodic boundary conditions are being widely used as technology [1] in different fields of science and engineering like machine learning, cryptography etc. These applications are possible due to proper analysis of the CA under null and periodic boundary condition. However, other boundary conditions are not explored at par with these, and this paper serves as a base for exploring the adiabatic and reflexive boundary conditions for such applications.

### 4.1 Future scope

It will be interesting to explore the vast applications in various fields with other boundary conditions, such as:

1. Pattern classification domain, implementation (with high efficiency) requires only point state attractor Cellular Automata. Such CA can be easily implemented using reachability and linked tree, and leveraging the behaviour of these two boundary conditions.
2. Psuedo number pattern generator, demands reversible CA, which is easily identified and synthesized using reachability tree. Suitable CA can be framed using these relatively unexplored boundary conditions.

## References

1. Bhattacharjee, K., Naskar, N., Roy, S., Das, S.: A survey of cellular automata: types, dynamics, non-uniformity and applications. Nat. Comput. **19**(2), 433–461 (2020)
2. Das, S.: Theory and Applications of Nonlinear Cellular Automata in VLSI Design. Ph.D. thesis, BE College (2006)
3. Goldenberg, J., Efroni, S.: Using cellular automata modeling of the emergence of innovations. Technol. Forecast. Soc. Chang. **68**(3), 293–308 (2001)
4. Mohammed, J., Nayak, D.R.: An efficient edge detection technique by two dimensional rectangular cellular automata (2013)
5. Nandi, S., Chaudhuri, P.P.: Analysis of periodic and intermediate boundary 90/150 cellular automata. IEEE Trans. Comput. **45**(1), 1–12 (1996)
6. Nandi, S., Kar, B.K., Chaudhuri, P.P.: Theory and applications of cellular automata in cryptography. IEEE Trans. Comput. **43**(12), 1346–1357 (1994)
7. Naskar, N., Chakraborty, A., Maji, P., Das, S.: Analysis of reachability tree for identification of cyclic and acyclic ca states. In: International Conference on Cellular Automata, pp. 63–72. Springer (2012)

# First Degree Cellular Automata as Pseudo-Random Number Generators

**Kamalika Bhattacharjee**

**Abstract** This paper depicts simple cellular automata (CAs) as potential pseudo-random number generators (PRNGs). It introduces a technique to represent CAs as first degree equation identified by some tuples. Some 3-neighborhood 10-state CAs, identified by 6 tuples, are selected following greedy criteria. Concept of window is used to extract a number from each configuration. We observe that, our scheme of choosing the CAs has potentiality to beat all other existing PRNGs.

**Keywords** Psudo-random number generator (PRNG) · 10-state cellular automata (CAs) · First degree CAs · Diehard · TestU01 · NIST · Space-time diagram

## 1 Introduction

In late 1980s, cellular automaton (CA) was introduced as a source of pseudo-randomness. The complexity of its global behavior that arises from simple local interaction and computation as well as massive parallelism, along with ease of hardware implementation and scalability have popularized CAs as an substantial research destination [1, 4, 8]. However, in most of the cases, binary (2-state) CAs have been used. Recently, we have studied randomness of 3-neighborhood 10-state CAs [2]. (Here, we will use 'random' to mean 'pseudo-random' only.) We have observed that, randomness of 10-state CAs is extremely good and these CA based random number generators are comparable to the best PRNGs available today, like `SFMT-19937` [2]. Therefore, increasing number of states effectively increases the randomness quality. So, by changing number of states, we may get CA based PRNGs suitable for our application.

However, there are two problems with this approach. First, the total number of neighborhood combinations increases with number of states and number of neighbors. For instance, for any 3-neighborhood 10-state CA rule, total number of neigh-

K. Bhattacharjee (✉)
Department of Computer Science and Engineering, National Institute of Technology, Tiruchirappalli, Tamilnadu 620015, India



S. Das and G. J. Martinez (eds.), *Proceedings of First Asian Symposium on Cellular Automata Technology*, Advances in Intelligent Systems and Computing 1425,
https://doi.org/10.1007/978-981-19-0542-1_10

borhood combinations is $10^3 = 1000$. So, it is very difficult to represent a rule containing all these combinations in pen and paper. Even in the simplified form, we needed an algorithm to get the rule of the CA [2]. Further, as total number of possible rules is gigantic, exhaustive searching of the CA rule-space to find CAs which are good in terms of randomness yet easy to describe, is practically impossible. So, in this situation, we come with a greedy solution to find good and easy to describe CAs (Sect. 3). To make sure our greedy strategy gives desired rules, we enforce that the randomness properties (Sect. 2.2) are satisfied by them. Our scheme works for any $d$-state CAs, but here we have considered only decimal CAs. Therefore, if not explicitly mentioned, in this paper, by a CA, we mean a 3-neighborhood decimal CA. We propose PRNG with some rules following our scheme (Sect. 4) and test them in standard statistical testbeds for randomness quality. Finally, we can see that, these simple systems are also comparable to the best PRNGs of today.

## 2 Background

In this section we describe some terminologies associated with the CAs and then depict some properties of a CA which are necessary to become a good source of randomness.

### 2.1 Definitions

Here, we consider 1-dimensional 3-neighborhood $d$-state CAs under periodic boundary condition where each cell assumes any of the states $S = \{0, 1, \ldots, d-1\}$. For this work, we consider $d = 10$. Any rule of the CA is represented by $R : S^3 \rightarrow S$. Domain of $R$ contains triplets $\langle x, y, z\rangle$, $x, y, z \in S$, where $x$, $y$, and $z$ represent the left, self and right neighbors of a cell respectively. For our work, we refer this combination as *Rule Min Term (RMT)* and denote the value $R(x, y, z)$ by $R[r]$ (represented by its decimal equivalent $r = x \times d^2 + y \times d + z$). Therefore, a 10-state CA rule consists of $10^3 = 1000$ RMTs. A CA can also be represented by a de Bruijn graph, where any edge $(xy, yz)$ depicts the RMT $xyz$.

**Definition 1** "Let $\Sigma$ be a set of symbols, and $k \geq 1$ be a number. Then, the de Bruijn graph is $B(k, \Sigma) = (V, E)$, where $V = \Sigma^k$ is the set of vertices, and $E = \{(xy, yz) | x, z \in \Sigma, y \in \Sigma^{k-1}\}$ is the set of edges" [2].

Figure 1a is a de Bruijn graph for any 3-neighborhood 2-state CA, whereas Fig. 1b shows the graph for ECA 30. In this graph, the connecting edge between nodes 11 and 10 is RMT 110 which has next state value 0 according to ECA 30. Hence, this edge is labeled by 0 in Fig. 1b and so on.

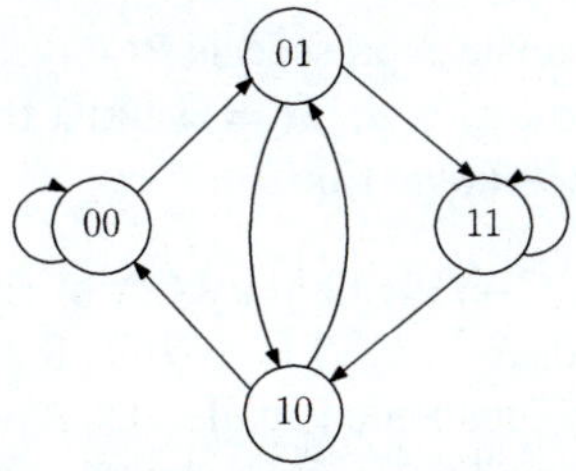


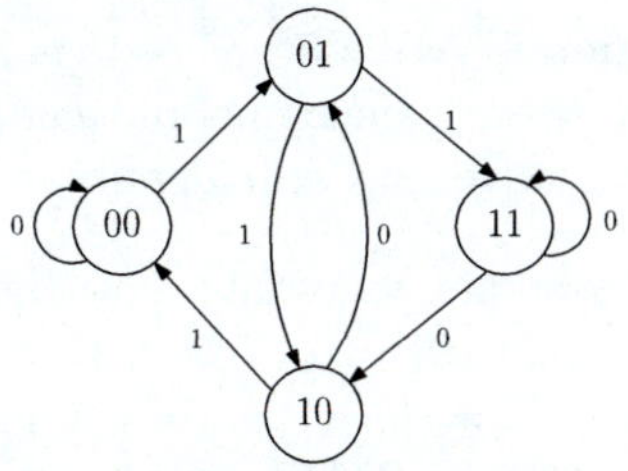


**Fig. 1** de Bruijn Graph representation of CA

The RMTs of a CA rule can be grouped together. In the de Bruijn graph, the incoming RMTs to each node are *equivalent* RMTs and outgoing RMTs from each node are *sibling* RMTs.

**Definition 2** "A set of $d$ RMTs $r_1, r_2, \ldots, r_d$ of a $d$-state CA rule are said to be equivalent to each other if for any $r_i, r_j$, where $r_i \neq r_j, r_i \equiv r_j \pmod{d^2}$" [2].

**Definition 3** "A set of $d$ RMTs $s_1, s_2, \ldots, s_d$ of a $d$-state CA rule are said to be sibling to each other if for any $s_i, s_j$, where $s_i \neq s_j$, $\lfloor \frac{s_i}{d} \rfloor = \lfloor \frac{s_j}{d} \rfloor$" [2].

For decimal CAs, there are 100 equivalent (resp, sibling) RMT sets. Each equivalent (resp. sibling) RMT set containing RMT $i$ and all of its equivalent (resp. sibling) RMTs, is termed as $Equi_i$ (resp., $Sibl_i$). Therefore, $Equi_i = \{i, 100 + i, \ldots, 900 + i\}$ and $Sibl_i = \{10i, 10i + 1, \ldots, 10i + 9\}$, where $0 \leq i \leq 100$. Hence, each configuration is a sequence of equivalent and sibling RMTs, where, if we choose an RMT from $Equi_i$, then the consecutive RMT of the sequence has to be chosen from $Sibl_i$.

**Definition 4** "Let $R : S^3 \rightarrow S$ be the local rule of a CA. Then, the CA is left-(resp. right-) permutive if and only if for any two equivalent (resp. sibling) RMTs $r$ and $s$, $R[r] \neq R[s]$" [2].

Therefore, if, for all $i$, all RMTs of $Sibl_i$ (resp. $Equi_i$) have distinct next state values, the CA is a right-permutive (resp. left-permutive) CA.

**Definition 5** "An RMT $r = x \times 10^2 + y \times 10 + z$ is said to be self-replicating, if $R(x, y, z) = y$, where $R$ is the rule of the CA" [2].

**Definition 6** "A fixed-point attractor is a configuration in CA, for which the next configuration is itself. That means, if a CA reaches a fixed-point attractor, then it remains at that particular configuration forever" [2].

**Definition 7** "A rule $R : S^3 \rightarrow S$ is called **balanced** if it contains 100 RMTs with each of the 10 possible states; otherwise it is an **unbalanced** rule" [2].

**Definition 8** A rule $R : S^3 \rightarrow S$ is called *linear* if $R$ can be expressed as $R(x, y, z) = c_1x + c_2y + c_3z$ under modulo $d$ addition, where $c_1, c_2, c_3 \in S$, $|S| = d$ and $S$ forms a commutative ring with identity; otherwise it is a *non-linear* rule.

For example, let us take a 3-neighborhood 2-state CA (Elementary CA or ECA) where $R(0, 0, 0) = 0$, $R(0, 0, 1) = 0$, $R(0, 1, 0) = 0$, $R(0, 1, 1) = 1$, $R(1, 0, 0) = 1$, $R(1, 0, 1) = 0$, $R(1, 1, 0) = 1$ and $R(1, 1, 1) = 0$. This Rule is called ECA 90 as the string of the RMTs, starting from 111 to 000 is 01011010 which is 90 in decimal. We can see that, for this CA, $R(x, y, z) = (1 \times x + 0 \times y + 1 \times z) \pmod 2$. Hence, it is a linear CA.

## 2.2 Desirable Properties of Cellular Automata

In a CA based random system, even a minute change of information in a cell has to eventually outspread to the whole system globally. To have this, it is expected that the CA has the following properties [2].

**Balancedness:** The rules of the CAs needs to be balanced such that there is no bias towards any particular state. However, there are $\frac{d^3!}{(d^2!)^d}$ number of balanced $d$-state 3-neighborhood CAs which is too large for exhaustive checking.

**Non-linearity:** There are several applications like cryptography where non-linearity is more applicable than the linearity. Therefore, we want our CAs to be nonlinear.

**Large Cycle Length:** To be a good PRNG, the period of the PRNG has to be very large. Because, larger the period, longer the sequence of unique numbers generated by the PRNG. However, in a periodic boundary CA, all configurations can not be part of the same cycle. So, in a CA-based PRNG, it is expected that, the configuration space of the CA is not partitioned into many sub-spaces resulting existence of large cycles. Also, the average cycle length of the CA has to be very large which grows exponentially with CA size $n$.

**Flow of Information:** In a CA-based PRNG, if the consecutive configurations are to be independent of each other, that means, the information of the previous configuration has been substantially changed while the cells of the CA update their states. Now, as update of cell is according to rule which takes into account the present states of the neighboring cells, this indirectly means the cells of the CA has to depend on their neighbors. That is, when the states of any neighboring cells are changed, the cells also update their states. In this case, even a tiny addition of information on any cell affects other cells and eventually propagates throughout the configuration. This situation makes the CAs unpredictable and desirable candidates as PRNGs.

Our CAs have 3-neighborhood dependency. So, change in state of the left neighbor (resp., right neighbor) can affect the next state of the cell. This can be estimated by observing the change of states in the sibling (resp. equivalent) RMT set. Therefore, if we consider any change in state of a cell as an information, then this information can flow in two directions—left and right. For any CA, the left (resp. right) directional information flow can be calculated as the cumulative sum of the change of states in each sibling (resp. equivalent) RMT set (except the self-replicating RMTs), divided by the total number of RMTs.

Flow of information in the CA plays a major role in its cyclic structure. Because, to complete a cycle, the CA has to return to its initial configuration that is part of the same cycle. Now, if the CA is *sensitive* or dependent on their neighbors, the information of the initial configuration will flow to the neighboring cells affecting them in the next step. And, this process will continue for some time until it stabilizes. Then the reverse process starts where the information flow helps the cells to come back to the starting configuration. Therefore, if the flow of information is more, more steps are needed for the CA to return to its initial configuration. That means, more number of configurations are part of the same cycle indicating the configuration space of the CA has less number of partitions.

Therefore, the CAs which satisfy the above properties are our desirable candidates. In the next sections, our target is to identify some simplified rules which have these properties.

## 3 The Selection and Rejection Criteria

Here we identify some filtering criteria to select rules which satisfy the necessary properties of randomness described in previous section.

### *3.1 Selection Criterion: Maximum Information Flow on One Side*

As information flow plays a significant role in randomness quality of the CAs, we develop our greedy strategy on the basis of information flow. Recall that, change of state in the sibling (resp. equivalent) RMTs implies information flow on left (resp. right) side. We, therefore, choose the CA rules such that RMTs of either $Sibl_i$ or $Equi_i$ have different next state values. This is to ensure that, there is maximum information flow on one direction. This condition is equivalent to the property of (right or left) permutivity. The concept of permutivity is well known and is also of significance in many applications of CA to cryptography which validates our strategy to be a good one.

**STRATEGY**: *Pick up the non-linear rules in which either the RMTs of each sibling set or each equivalent set have the different next state values, that is, no two RMTs of* $Sibl_i$ $(0 \leq i \leq d^2 - 1)$ *or* $Equi_i$ $(0 \leq i \leq d^2 - 1)$ *have same next state value, where* $d$ *is the number of states of the CA* [2].

Here, this 'or' is exclusive, meaning, we do not take those CA rules which have maximum information flow on both directions. This is because, maximum information flow on both sides create symmetry in the configuration space resulting in triangular patterns and fractals [2]. This is absolutely undesirable as the patterns make the numbers easy to detect and non-random.

Now, we identify the rejection criterion of the rules selected through our strategy which are unlikely to be random.

## 3.2 Rejection Criterion: Theoretical Filtering

In general, a CA has a long cycle length if it does not have an inclination towards converging to the fixed-point attractors(s). Moreover, if there are large number of fixed-point attractors, then also, the configurations of the CA are divided into large number of cycles resulting lesser cycle lengths. Therefore, to get large cycle length for the CAs, it is necessary that the CA has fewer fixed-point attractors. Further, as in the trivial configurations ($s^n, \forall s \in S$), there is no change of RMTs, we chose these trivial configurations to be separated from the other non-trivial configurations, whereas the non-trivial configurations do not form any fixed-point attractors. Hence, two filtering conditions are imposed—

1. No non-trivial fixed-point attractor(s)
2. Non-reachable trivial configurations.

### 3.2.1 No Non-trivial Fixed-Point Attractor(s)

Our target is to select those CAs which do not have any non-trivial fixed-point attractors for every cell length. These CAs may have, if exists, only trivial fixed-point attractors. Therefore, each CA has to be checked for every fixed-point attractors. This can be done by using the de Bruijn graph corresponding to the CA. We know that, any cycle in a de Bruijn graph represents an RMT sequence (see Sect. 2.1). Therefore, if the de Bruijn graph is modified to have only the edges with the self-replicating RMTs, then any cycle in this graph corresponds to a fixed-point attractor in the CA. In this way, we can detect all possible fixed-point attractors of the CA.

Note that, each of the CAs following our strategy has 100 self-replicating RMTs. Nevertheless, our target is to select those CAs following the strategy which do not have any cycle in the modified de Bruijn graph with only these self-replicating RMTs.

### 3.2.2 Non-reachable Trivial Configurations

The previous condition assures that, the CAs do not have non-trivial fixed-point attractors. But, this does not ensure that the non-trivial configurations are not converged to the trivial fixed-point attractors. However, to get large cycle length, the trivial configurations need to be disjoint from the non-trivial configurations. Therefore, we apply this second filtering condition.

Here also, to check whether the trivial configurations are reachable from the non-trivial configurations, we use de Bruijn graph. The process of checking reachability of a trivial configuration $s^n$, $\forall s \in S$ is as follows—Let us modify the de Bruijn graph such that it contains only the edges with next state value $s$. Now, any cycle in this de Bruijn graph represents an RMT sequence for which the next configuration is $s^n$. Therefore, if this graph has any cycle corresponding to a non-trivial configuration, then that implies, the trivial configuration is reachable from a non-trivial configuration. Hence, if for any trivial configurations, a cycle corresponding to a non-trivial configuration is detected, the CA is discarded.

Although huge number of rules can be discarded by applying these filtering conditions, still immense number of rules exist which follow all our greedy strategies. It is expected that, all these rules have very large cycle lengths and satisfy the desirable properties of a good PRNG. In the next section, we design some CA rules following our greedy strategy and test them as PRNGs.

## 4 Finding Rules for PRNG

As discussed in the previous section, we want to take rules which have maximum information flow on one direction and for which there is no non-trivial fixed point attractor as well as trivial configurations are not reachable from others. But, our primary condition is, such a CA has to be easy to represent and work with. For this reason we introduce *First Degree Rules* in cellular automata.

### 4.1 Rule of First Degree

We name a cellular automata rule $R : S^3 \rightarrow S$ is of first degree, if the rule can be represented in the following form:

$$R(x, y, z) = (c_0xyz + c_1xy + c_2xz + c_3yz + c_4x + c_5y + c_6z + c_7) \pmod{d} \tag{1}$$

Here, $d$ is the number of states of the CA and $x, y, z$ are the three neighbors of each cell where $x, y, z \in S$. For each rule, $c_i$ denotes a constant value fixed for the rule. As, we are using *modular arithmetic*, each $c_i \in \mathbb{Z}_d = \{0, 1, \dots, d-1\}$. Note that, in case of $m$ neighbors, this equation can be generalized accordingly. However, here we are concentrating over only 3-neighborhood CAs. As the Eq. 1 is of degree one, we named such CA rules as *rules of first degree* or simply, *first degree* CAs.

Now, not all rules following this equation will satisfy our selection criterion. For example, when $c_0 = c_1 = c_2 = c_3 = c_7 = 0$, the rules are linear. Taking $c_0 > 0$

ensures the CA is dependent on every neighbor, whereas, $c_1, c_2, c_3 > 0$ ensures that the CA depends on pairs of neighbors. Each of these components makes the CA non-linear. Further, $c_4$, $c_5$ and $c_6$ are to make the CA dependent on its left, self and right neighbor respectively. Hence, by choosing these constants appropriately, we can create CA rules which follows the randomness properties of Sect. 2.2. For example, Table 1 shows the value of different constants for ECAs when the CAs have maximum information flow in at least one direction. In this table, the entries marked in bold indicates that those ECAs do not satisfy our greedy strategy. They are rejected from the candidature of being a PRNG.

Observe that, $c_0$ is always 0, therefore, we can simply omit that. Hence, any rule following Eq. 1 can be depicted by giving the value of the other constants $\langle c_1, \ldots, c_7 \rangle$, which we call as the parameters for the first degree CA. If we take Eq. 1 over $d = 10$, then by varying these parameters, we can get 78400 rules which have maximum information flow in at least one side. Among these rules, 640 are linear and 1600 are having maximum information flow on both sides. Hence, they are discarded as they do not follow our strategy. Some example first degree CA rules that satisfy our strategy are shown in Table 2.

On these rules, we apply the theoretical filtering to reject those CAs which have non-trivial fixed points and reachable trivial configurations. The remaining CAs can be chosen as our PRNGs. Next subsection describes the scheme of PRNG and empirical results.

## *4.2 Development of PRNG*

One of the properties of a PRNG is, it generates all numbers in the range of the generator. However, for a periodic boundary CA, all possible configurations can never be part of an individual cycle. Therefore, taking whole configuration of a CA as the number will result into non-reachability of some numbers from a given seed. To address this issue, we take window based scheme [2]. A window of length $w < n$ is considered out of $n$ cells, where the numbers are extracted out of the window. To include all combinations of numbers in a cycle, the cell length $n$ needs to be very large. Size of $w$ and $n$ are variable; depending on the requirement, we can adjust the sizes.

To convert into 32-bit strings, the length of the decimal numbers have to be at least 10 digits. In this paper, we consider a window of size 12, from which last 32-bits of the corresponding binary string is taken as the generated number. The value of $n$ has to be large enough to accommodate all possible values of the window. For our experiment, $n$ is taken as 51. The seed is any arbitrary non-zero number to be supplied to the window, whereas, the other cells are initialized with $0^{n-w-1}k$, $k \in S$. This is to ensure that the initial configuration of the CA is not the trivial configuration $0^n$. Moreover, for any seed, the numbers are taken from $n + 1$st configuration onwards, as for a periodic boundary CA, first $n$ configurations may be used to spread the

**Table 1** First degree 3-neighborhood 2-state CA rules that follow our greedy strategy

| Constants | | | | | | | | ECA | | | | | | | | | Right side? | Left side? | Remarks |
|---|---|---|---|---|---|---|---|---|---|---|---|---|---|---|---|---|---|---|---|
| $c_0$ | $c_1$ | $c_2$ | $c_3$ | $c_4$ | $c_5$ | $c_6$ | $c_7$ | 111 | 110 | 101 | 100 | 011 | 010 | 001 | 000 | (Rule) | | | |
| 0 | 0 | 0 | 0 | 0 | 0 | 1 | 0 | 1 | 0 | 1 | 0 | 1 | 0 | 1 | 0 | (170) | False | True | **Linear** |
| 0 | 0 | 0 | 0 | 0 | 0 | 1 | 1 | 0 | 1 | 0 | 1 | 0 | 1 | 0 | 1 | (85) | False | True | Non-linear |
| 0 | 0 | 0 | 0 | 0 | 1 | 1 | 0 | 0 | 1 | 1 | 0 | 0 | 1 | 1 | 0 | (102) | False | True | **Linear** |
| 0 | 0 | 0 | 0 | 0 | 1 | 1 | 1 | 1 | 0 | 0 | 1 | 1 | 0 | 0 | 1 | (153) | False | True | Non-linear |
| 0 | 0 | 0 | 0 | 1 | 0 | 0 | 0 | 1 | 1 | 1 | 1 | 0 | 0 | 0 | 0 | (240) | True | False | **Linear** |
| 0 | 0 | 0 | 0 | 1 | 0 | 0 | 1 | 0 | 0 | 0 | 0 | 1 | 1 | 1 | 1 | (15) | True | False | Non-linear |
| 0 | 0 | 0 | 0 | 1 | 0 | 1 | 0 | 0 | 1 | 0 | 1 | 1 | 0 | 1 | 0 | (90) | **True** | **True** | **Linear** |
| 0 | 0 | 0 | 0 | 1 | 0 | 1 | 1 | 1 | 0 | 1 | 0 | 0 | 1 | 0 | 1 | (165) | **True** | **True** | Non-linear |
| 0 | 0 | 0 | 0 | 1 | 1 | 0 | 0 | 0 | 0 | 1 | 1 | 1 | 1 | 0 | 0 | (60) | True | False | **Linear** |
| 0 | 0 | 0 | 0 | 1 | 1 | 0 | 1 | 1 | 1 | 0 | 0 | 0 | 0 | 1 | 1 | (195) | True | False | Non-linear |
| 0 | 0 | 0 | 0 | 1 | 1 | 1 | 0 | 1 | 0 | 0 | 1 | 0 | 1 | 1 | 0 | (150) | **True** | **True** | **Linear** |
| 0 | 0 | 0 | 0 | 1 | 1 | 1 | 1 | 0 | 1 | 1 | 0 | 1 | 0 | 0 | 1 | (105) | **True** | **True** | Non-linear |
| 0 | 0 | 0 | 1 | 1 | 0 | 0 | 0 | 0 | 1 | 1 | 1 | 1 | 0 | 0 | 0 | (120) | True | False | Non-linear |
| 0 | 0 | 0 | 1 | 1 | 0 | 0 | 1 | 1 | 0 | 0 | 0 | 0 | 1 | 1 | 1 | (135) | True | False | Non-linear |
| 0 | 0 | 0 | 1 | 1 | 0 | 1 | 0 | 1 | 1 | 0 | 1 | 0 | 0 | 1 | 0 | (210) | True | False | Non-linear |
| 0 | 0 | 0 | 1 | 1 | 0 | 1 | 1 | 0 | 0 | 1 | 0 | 1 | 1 | 0 | 1 | (45) | True | False | Non-linear |
| 0 | 0 | 0 | 1 | 1 | 1 | 0 | 0 | 1 | 0 | 1 | 1 | 0 | 1 | 0 | 0 | (180) | True | False | Non-linear |
| 0 | 0 | 0 | 1 | 1 | 1 | 0 | 1 | 0 | 1 | 0 | 0 | 1 | 0 | 1 | 1 | (75) | True | False | Non-linear |
| 0 | 0 | 0 | 1 | 1 | 1 | 1 | 0 | 0 | 0 | 0 | 1 | 1 | 1 | 1 | 0 | (30) | True | False | Non-linear |
| 0 | 0 | 0 | 1 | 1 | 1 | 1 | 1 | 1 | 1 | 1 | 0 | 0 | 0 | 0 | 1 | (225) | True | False | Non-linear |
| 0 | 1 | 0 | 0 | 0 | 0 | 1 | 0 | 0 | 1 | 1 | 0 | 1 | 0 | 1 | 0 | (106) | False | True | Non-linear |
| 0 | 1 | 0 | 0 | 0 | 0 | 1 | 1 | 1 | 0 | 0 | 1 | 0 | 1 | 0 | 1 | (149) | False | True | Non-linear |
| 0 | 1 | 0 | 0 | 0 | 1 | 1 | 0 | 1 | 0 | 1 | 0 | 0 | 1 | 1 | 0 | (166) | False | True | Non-linear |
| 0 | 1 | 0 | 0 | 0 | 1 | 1 | 1 | 0 | 1 | 0 | 1 | 1 | 0 | 0 | 1 | (89) | False | True | Non-linear |
| 0 | 1 | 0 | 0 | 1 | 0 | 1 | 0 | 1 | 0 | 0 | 1 | 1 | 0 | 1 | 0 | (154) | False | True | Non-linear |
| 0 | 1 | 0 | 0 | 1 | 0 | 1 | 1 | 0 | 1 | 1 | 0 | 0 | 1 | 0 | 1 | (101) | False | True | Non-linear |
| 0 | 1 | 0 | 0 | 1 | 1 | 1 | 0 | 0 | 1 | 0 | 1 | 0 | 1 | 1 | 0 | (86) | False | True | Non-linear |
| 0 | 1 | 0 | 0 | 1 | 1 | 1 | 1 | 1 | 0 | 1 | 0 | 1 | 0 | 0 | 1 | (169) | False | True | Non-linear |

**Table 2** Example First degree 10-state CA rules with parameters

| Parameters | | | | | | | Rule Equation | Information Flow |
|---|---|---|---|---|---|---|---|---|
| $c_1$ | $c_2$ | $c_3$ | $c_4$ | $c_5$ | $c_6$ | $c_7$ | | |
| 1 | 0 | 0 | 0 | 0 | 1 | 2 | $xy + z + 2 \pmod{10}$ | Left |
| 1 | 0 | 0 | 0 | 0 | 1 | 6 | $xy + z + 6 \pmod{10}$ | Left |
| 1 | 0 | 0 | 7 | 7 | 1 | 0 | $xy + 7(x + y) + z \pmod{10}$ | Left |
| 0 | 3 | 0 | 7 | 0 | 1 | 6 | $3xz + 7x + z + 6 \pmod{10}$ | Left |
| 5 | 0 | 0 | 3 | 5 | 3 | 8 | $5xy + 3x + 5y + 3z + 8 \pmod{10}$ | Left |
| 0 | 0 | 3 | 7 | 0 | 5 | 4 | $3yz + 7x + 5z + 4 \pmod{10}$ | Right |
| 0 | 0 | 5 | 9 | 2 | 7 | 0 | $5yz + 9x + 2y + 7z \pmod{10}$ | Right |
| 3 | 0 | 0 | 1 | 5 | 3 | 1 | $3xy + x + 5y + 3z + 1 \pmod{10}$ | Left |
| 0 | 0 | 5 | 1 | 8 | 4 | 8 | $5yz + x + 8y + 4z + 8 \pmod{10}$ | Right |
| 0 | 0 | 6 | 9 | 3 | 3 | 8 | $6yz + 9x + 3y + 3z + 8 \pmod{10}$ | Right |
| 2 | 0 | 0 | 9 | 9 | 3 | 5 | $2xy + 9x + 9y + 3z + 5 \pmod{10}$ | Left |
| 0 | 0 | 8 | 7 | 2 | 5 | 1 | $8yz + 7x + 2y + 5z + 1 \pmod{10}$ | Right |
| 4 | 0 | 0 | 0 | 8 | 3 | 6 | $6xy + 9x + 3y + 3z + 8 \pmod{10}$ | Left |
| 6 | 0 | 0 | 4 | 0 | 9 | 3 | $6yz + 4x + 9z + 3 \pmod{10}$ | Right |
| 3 | 0 | 0 | 7 | 1 | 1 | 8 | $3xy + 7x + y + z + 8 \pmod{10}$ | Left |

change of information to all cells of the CA. These numbers are then tested with the empirical test-beds.

### 4.2.1 Empirical Test Results and Comparison

Target of empirical tests is to find the non-randomness in a pseudo-random number generator by using statistical and probabilistic measures. Some well-known battery of tests are Diehard tests [6], NIST test-suite [10] and TestU01 library [5]. Diehard battery test consists of 15 different tests and it is the most general purpose testbed. The NIST Statistical Test Suite tests a PRNG for cryptographic properties. This test-suite also contains 15 tests. However, TestU01 library contains the most versatile collection of tests. For our purpose, we have used battery *rabbit* from it which contains 25 tests.

Our CA-based PRNGs are tested with these empirical test-beds. To test a CA using Diehard and TestU01, a *.bin* file of size 11.5 MB is used, whereas, for NIST, a binary file of size 125MB is taken. A CA is considered to be a good source of randomness only when it consistently passes most of the tests for any seed. We test our PRNG with some random seeds. Table 3 shows some sample results for four arbitrary seeds—7, 1234, 12345 and 19650218. For each seed, the PRNG is tested on each testbed and the number of tests passed is recorded. It is observed that, our CA-based PRNGs pass 9–11 tests of Diehard battery of tests, all tests of battery *rabbit_file()* of TestU01 library and 14–15 tests of NIST statistical test-suite. That means, our PRNGs have good randomness quality.

We can now compare our proposed PRNGs with the existing PRNGs. For comparison, we have chosen the well-known PRNGs which are considered as *good* PRNGs [4], like `lrand48`, `rand`, `random` in GNU C Library, `MRG31k3p`, Knuth's `MMIX`, `PCG`, `WELL512a` and `WELL1024a`, `LFSR258`, `LFSR113`, `xorshift64*`, `xorshift128+` and `xorshift1024*M_8` generators, Mersenne Twisters `MT19937` (32-bit and 64-bit), `SMFT19937` (32-bit and 64-bit) and `dSFMT19937` (32-bit and 64-bit) along with the CA-based generators like non-linear 2-state CA, maximal-length CA, 3-state CA and decimal CA [2]. The codes used for this purpose are available at Ref. [3]. Table 3 depicts the result of this comparison for four arbitrary seeds.

From this table, we can observe that, our CA-based generators are comparable to the Mersenne Twister Family (both MT and SFMT) [7, 9] and decimal CA [2] for most of the seeds. Even, for some seeds, these PRNGs perform better than all other existing generators. Further, in terms of ease of implementation, portability and simplicity, we can claim that, these first degree CAs can beat every other PRNGs.

### 4.2.2 Verification with Space-Time Diagram

To verify the results of the statistical test-suites, we use a graphical test—space-time diagram. This diagram is a well-known theoretical tool in the field of CA, mostly used to observe and classify the evolution and behavior of a CA. It represents the

**Table 3** Comparison of First degree CAs of Table 2 with existing PRNGs using Diehard, TestU01 and NIST for 4 arbitrary seeds. The columns for Diehard, TestU01 and NIST notes the number of tests passed by the PRNG out of 15, 25 and 15 respectively for each seed

| Seeds ⟶ | | 7 | | | 1234 | | | 12345 | | | 19650218 | | |
|---|---|---|---|---|---|---|---|---|---|---|---|---|---|
| Name of the PRNGs | | Diehard | TestU01 | NIST | Diehard | TestU01 | NIST | Diehard | TestU01 | NIST | Diehard | TestU01 | NIST |
| LCGs | MMIX | 6 | 19 | 7 | 5 | 18 | 7 | 6 | 17 | 8 | 4 | 16 | 8 |
| | rand() | 1 | 1 | 2 | 1 | 1 | 2 | 1 | 3 | 2 | 1 | 2 | 2 |
| | lrand48() | 1 | 3 | 2 | 1 | 2 | 2 | 1 | 3 | 2 | 1 | 3 | 2 |
| | MRG31k3p | 1 | 2 | 1 | 1 | 1 | 1 | 1 | 2 | 1 | 1 | 1 | 1 |
| | PCG-32 | 9 | 25 | 15 | 9 | 25 | 14 | 11 | 25 | 14 | 10 | 24 | 15 |
| LFSRs | random() | 1 | 1 | 1 | 1 | 1 | 1 | 1 | 3 | 1 | 1 | 2 | 1 |
| | LFSR113 | 5 | 6 | 2 | 11 | 23 | 15 | 9 | 23 | 15 | 7 | 23 | 10 |
| | LFSR258 | 5 | 6 | 1 | 11 | 23 | 2 | 9 | 23 | 2 | 0 | 1 | 2 |
| | WELL512a | 9 | 23 | 15 | 10 | 23 | 14 | 10 | 23 | 15 | 8 | 23 | 15 |
| | WELL1024a | 9 | 25 | 15 | 10 | 24 | 15 | 9 | 24 | 14 | 9 | 25 | 15 |
| | MT19937-32 | 10 | 25 | 13 | 9 | 25 | 13 | 9 | 25 | 14 | 9 | 25 | 15 |
| | MT19937-64 | 10 | 25 | 15 | 10 | 24 | 15 | 8 | 24 | 15 | 11 | 25 | 15 |
| | SFMT-32 | 10 | 25 | 15 | 9 | 25 | 15 | 10 | 25 | 15 | 9 | 25 | 15 |
| | SFMT-64 | 11 | 25 | 15 | 10 | 25 | 15 | 10 | 25 | 15 | 9 | 25 | 15 |
| | dSFMT-32 | 7 | 25 | 15 | 8 | 25 | 15 | 11 | 24 | 13 | 11 | 25 | 15 |
| | dSFMT-64 | 0 | 5 | 1 | 0 | 5 | 1 | 0 | 4 | 1 | 0 | 5 | 1 |
| | XORShift64* | 0 | 25 | 15 | 8 | 25 | 15 | 10 | 25 | 15 | 10 | 25 | 15 |
| | XORShift1024*$M_8$ | 7 | 20 | 6 | 7 | 21 | 15 | 5 | 20 | 15 | 7 | 20 | 15 |
| | XORShift128+ | 9 | 25 | 14 | 9 | 25 | 14 | 10 | 24 | 15 | 10 | 25 | 15 |
| CAs | Maximal Length CA | 2 | 12 | 10 | 0 | 12 | 11 | 1 | 12 | 11 | 1 | 12 | 11 |
| | Non-linear 2-state CA | 6 | 11 | 4 | 8 | 10 | 2 | 5 | 12 | 3 | 5 | 12 | 4 |
| | 3-state CA | 3 | 12 | 0 | 3 | 12 | 0 | 3 | 11 | 6 | 2 | 11 | 5 |
| | Decimal CA | 9 | 25 | 15 | 10 | 25 | 15 | 11 | 25 | 15 | 11 | 25 | 15 |
| | First degree CAs | **11** | **25** | **15** | **10** | **25** | **15** | **11** | **25** | **15** | **11** | **25** | **15** |

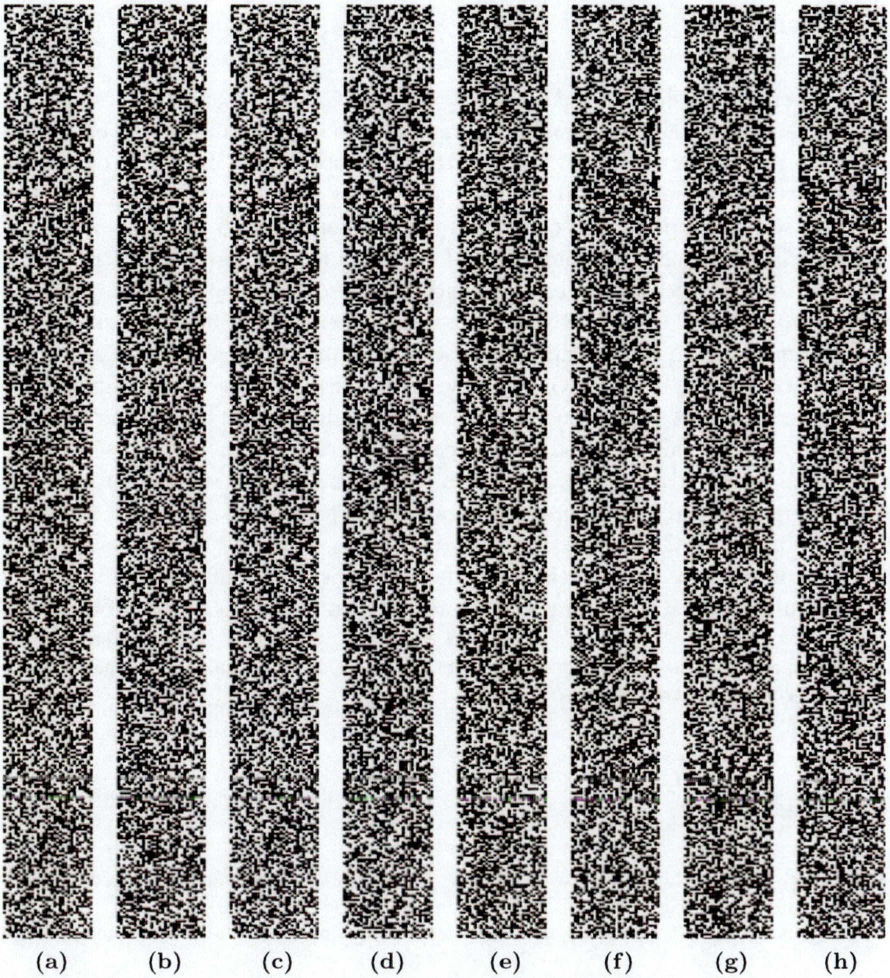


**Fig. 2** Space-time diagrams for one CA-based 32-bit PRNG (1st rule of Table 2). Here $n = 51$ and the seeds are taken arbitrarily. For each seed, 500 numbers are generated

configurations (on $x$-axis) at each time $t$ (on $y$-axis), where the states of each cell are represented by some color. But, it can also be used by PRNGs to visualize whether any pattern exists in the generated numbers [4].

Here, to test our PRNGs, we use the 32-bit output numbers for each seed and plot them against time. As numbers are represented in binary, only two colors—white for 0 and black for 1, are used. Some sample diagrams for one CA (1st rule of Table 2) are shown in Fig. 2. From this figure, it is evident that, there is no pattern in the generated numbers. Hence, the CAs selected through our scheme are good PRNGs.

## 5 Conclusion

In this paper, the possibility of using simple decimal CAs represented by first degree equation as pseudo-random number generators has been discussed. Here, each first degree CA rule can be effectively identified by a six-tuple which signify the constants for the first degree equation. To be used as PRNGs, these first degree CAs have to satisfy some criteria. These conditions are developed based on some desirable properties like balancedness, information flow, non-linearity, large cycle length etc. in the form of a greedy selection criterion and a theoretical filtering rejection criterion. While using selected CA as PRNG, we have considered the concept of window to extract numbers from configurations. It is observed that, randomness of these CAs are comparable to the best PRNGs available today. Further, considering simplicity of the scheme, these CA-based generators have potentiality to beat all existing PRNGs in terms of ease of implementation and speed. Nevertheless, though our scheme is shown with respect to only 3-neighborhood 10-state CAs, but, it will work for any $d$-state $m$-neighborhood CAs by properly choosing the parameters. However, studying the behavior of such CAs is part of future work. In the same way, our results are with respect to only degree one CAs. In future, it can be observed if increasing the degree of the cellular automata plays any role on improving randomness quality of the CAs.

**Acknowledgements** The author is grateful to Prof. Sukanta Das for his support, comments, guidance and encouragement to do this work.

## References

1. Bhattacharjee, K., Naskar, N., Roy, S., Das, S.: A survey of cellular automata: types, dynamics, non-uniformity and applications. Natl. Comput. **19**(2), 433–461 (2020)
2. Bhattacharjee, K., Das, S.: Random number generation using decimal cellular automata. Commun. Nonlinear Sci. Numer. Simul. **78**, 104878 (2019)
3. Bhattacharjee, K., Das, S.: PRNG library (2021). https://github.com/kamalikaB/. Accessed Nov 26, 2021
4. Bhattacharjee, K., Maity, K., Das, S.: A search for good pseudo-random number generators: survey and empirical studies (2018). abs/1811.04035
5. L'Ecuyer, P., Simard, R.: TestU01: A C library for empirical testing of random number generators. ACM Trans. Mathe. Softw. **33**(4), 22:1–22:40 (2007)
6. Marsaglia, G.: DIEHARD: a battery of tests of randomness (1996). http://stat.fsu.edu/~~geo/diehard.html
7. Matsumoto, Makoto, Nishimura, Takuji: Mersenne twister: a 623-dimensionally equidistributed uniform pseudo-random number generator. ACM Trans. Model. Comput. Simul. **8**(1), 3–30 (1998)
8. Pal Chaudhuri, P., Roy Chowdhury, D., Nandi, S., Chattopadhyay, S.: Additive Cellular Automata–Theory and Applications, vol. 1. IEEE Computer Society Press, USA (1997). ISBN 0-8186-7717-1

9. Saito, M., Matsumoto, M.: SIMD-oriented Fast Mersenne Twister (SFMT): twice faster than Mersenne Twister (2017). http://www.math.sci.hiroshima-u.ac.jp/~m-mat/MT/SFMT/#dSFMT
10. Soto, J.: Statistical testing of random number generators. In: Proceedings of the 22nd National Information Systems Security Conference, vol. 10, p. 12. NIST Gaithersburg, MD (1999)

# Asynchronous Cellular Automata as Randomness Enhancer

**Souvik Roy and Sumit Adak**

**Abstract** This work explores randomness enhancing capability of asynchronous cellular automata. This paper first reports theoretical results to identify elementary cellular automata under fully asynchronous updating scheme with large cycle of length $2^n$ or $2^n - 1$. To predict the pseudo-random number generation capability, we classify these large cycle asynchronous cellular automata into fully and partially exposed systems. Finally, it is observed that asynchronous cellular automata are able to improve randomness quality of a 'poor' random number generator.

**Keywords** Asynchronous cellular automata (ACAs) · Reversibility · Recurrent rules · Pseudo-Random Number Generator (PRNG)

## 1 Introduction

A cellular automaton (CA) is defined over a regular grid, each cell of which consists of a finite automaton that interacts with its neighbours to go to its next state [1, 2]. The signature feature of a CA is its complex global behaviour arising from local interaction. During the last three decades, due to the inherent simplicity, cellular automata (CAs) have been widely explored as technology, such as, VLSI design and test, pattern recognition, pseudo-random number generator, CAs-based cryptosystems etc. [3, 4]. In particular, in this article, we consider the well-known problem of generating pseudo-random number generator (PRNG) by exploiting the dynamical behaviour of CA. Traditionally, the popular two PRNG categories are linear congruential generators (LCGs) and linear feedback shift registers (LFSRs). In the literature of CAs, PRNG based on chaotic elementary CA 30 for cryptographic applications was first introduced by Wolfram [5]. Further, Damgård [6] introduced a more concrete construction of iterated hash function based on Wolfram's chaotic rule PRNG. For chaotic CA, slightly different initial configurations of the CA produce

S. Roy (✉) · S. Adak
Department of Information Technology, Indian Institute of Engineering Science and Technology, Shibpur Howrah 711103, West Bengal, India

 
S. Das and G. J. Martinez (eds.), *Proceedings of First Asian Symposium on Cellular Automata Technology*, Advances in Intelligent Systems and Computing 1425,
https://doi.org/10.1007/978-981-19-0542-1_11

different sequences of configuration having little or no correlation. Therefore, such chaotic CAs are unpredictable which is essential for a good pseudo-random number generation. Moreover, these CAs are also surjective [7] which implies that, all configurations are reachable from some initial configurations. In this direction, design of good PRNG demand CAs which are reversible having large length cycles. In fact, in the domain of PRNG and CA, research based on maximal length CA have been very popular where CA generates a cycle of length $2^n - 1$ (CA size is $n$). Some notable works using maximal length CAs are reported in Refs. [8–10]. Here, in this study, we do not try to find new CA that works well as PRNG. Instead, in a completely different direction from other studies, we study randomness enhancing capability of asynchronous cellular automata.

In this article, we consider fully asynchronous updating scheme where only a single cell is selected uniformly at random for update. In the literature of asynchronous CAs (ACAs), many authors consider that the fully asynchronous updating is the most "natural" updating method due to their *atomicity property* [11, 12]. Recently, fully asynchronous CAs have been explored as technology, such as, pattern classification [13]; cryptosystem design [14], pseudo-random number generator [15] etc. Note that, in fully asynchronous CAs, single cell is selected randomly in order to generate the update sequence. However, this randomness is not internal to the system but is itself generated by the use of PRNG. Therefore, ACA alone cannot be a source of randomness, rather can be seen as randomness enhancer.

In this direction, this study explores the randomness enhancing capability of ACAs. In this work, we first identify ACAs which can lead to large length cycle of length $2^n$ and $2^n - 1$ in their configuration space. However, large cycle length of a system sometimes induces some regularity, as an example, see maximal-length linear feedback shift register. To avoid these circumstances and predict the randomness enhancing capability, we classify these ACAs into fully and partially exposed systems. Our hypothesis is, fully exposed ACAs, where every configuration of a cycle is equally exposed, are expected to be 'good' randomness enhancer. Next, we move on to our second focus, here, we use Diehard [16], a basic yet well-known test bed to see how our fully exposed ACAs act as randomness enhancer.

## 2 Preliminaries

### 2.1 Cellular Automata

In this paper, we consider one-dimensional three-neighbourhood two-state (binary) cellular automata under periodic boundary condition, which is traditionally known as *elementary cellular automata* (ECAs). Here, the cells are arranged as a ring. The set of indices that represent each cell is denoted by $\mathcal{L} = \mathbb{Z}/n\mathbb{Z}$, where $n$ is the number of cells. A cell is assigned a state in $\{0,1\}$ at each time step $t \in \mathbb{N}$. The collection of all states at given time is called a *configuration* and the set of configurations is

**Table 1** Look-up table for ECA 43, 51 and 60

| (x,y,z) | 111 | 110 | 101 | 100 | 011 | 010 | 001 | 000 | Rule |
|---|---|---|---|---|---|---|---|---|---|
| (RMT) | (7) | (6) | (5) | (4) | (3) | (2) | (1) | (0) | |
| | 0 | 0 | 1 | 0 | 1 | 0 | 1 | 1 | 43 |
| $f(x,y,z)$ | 0 | 0 | 1 | 1 | 0 | 0 | 1 | 1 | 51 |
| | 0 | 0 | 1 | 1 | 1 | 1 | 0 | 0 | 60 |

denoted by $\varepsilon_n = \{0,1\}^{\mathcal{L}}$. For example, we will write $x = 001101$ to denote a particular configuration, where the ring size is $n = 6$. Here, $x_i$ depicts the state of a particular cell $i \in \mathcal{L}$. An ECA is defined by a local transition function $f\colon \{0,1\}^3 \mapsto \{0,1\}$ which can be expressed as a look-up table (see Table 1). We call each argument of $f$ as Rule Min Term (RMT). An RMT $(x, y, z)$ is generally represented by its decimal from $r = 4 * x + 2 * y + z$ (row 2 of Table 1). However, the decimal equivalent of the eight outputs is called "rule". There are $2^8 = 256$ ECA rules, out of which 88 are minimal representative rules and the rests are their equivalent [17]. Table 1 depicts three such rules 43, 51 and 60. An RMT $r$ of a rule $R$ is active if a CA cell flips its state (1 to 0 and 0 to 1) on $r$; otherwise, the RMT is passive. As an example, RMT 0 and RMT 3 are respectively active and passive for rule 43 (see Table 1).

**Definition 1** ([18]) A set of RMTs $P = \{r_1, r_2, \cdots, r_k\}$ is said to be a primary RMT set if both of the following conditions are satisfied.

1. if $r \in P$ then $s \in P$ where $s = 2r$ or $2r + 1$ (mod8).
2. No proper subset of $P$ can form an RMT sequence.

Note that, it is not possible to form an RMT sequence by an arbitrary arrangement of RMTs. The nature of primary RMT sets is that, in any RMT sequence, either all or none of the RMTs of such a set is used. We identify six such primary RMT sets for ECAs - {0}, {7}, {2,5} {4,1,2}, {5,3,6} and {4,1,3,6}. It is quite obvious that a proper subset of a primary RMT set $P$ cannot be included in an RMT sequence: if $r \in P$ is included $k$ ($k > 0$) times, then all the members of $P$ are included $k$ times in an RMT sequence [18].

**Definition 2** ([18]) A configuration, RMT sequence of which is formed with the RMTs of a primary RMT set only, is called homogeneous configuration.

We identify six *homogeneous* configuration for ECAs - **0** uses $\{0\}$, **1** uses $\{7\}$, **01** uses $\{2, 5\}$, **001** uses $\{4, 1, 2\}$, **011** uses $\{5, 3, 6\}$ and **0011** uses $\{4, 1, 3, 6\}$. In **0** and **1**, all the cells are in 0-state and 1-state respectively. Similarly, the cells are in alternate 0 and 1 states for **01**. And so on.

## 2.2 Asynchronous Cellular Automata & Reversibility

In this work, the CAs are updated asynchronously, i.e. a single cell is selected uniformly at random for update. Such CAs are known as fully asynchronous CA (ACA) [13, 19]. Therefore, a sequence $U = (u_t)_{t\in\mathbb{N}}$ of cells can be observed during the evolution of the system, where the cell updated at time $t$ is denoted by $u_t$. This sequence is known as *update pattern*. For an initial condition $x$ and an update pattern $U$, the evolution of the system is given by the sequence of states $(x^t)$ obtained by successive applications of the updates of $U$. Formally, we have: $x^{t+1} = F(x^t, u_t)$ and $x^0 = x$, with:

$$x_i^{t+1} = \begin{cases} f(x_{i-1}^t, x_i^t, x_{i+1}^t) & \text{if } i = u_t \\ x_i^t & \text{otherwise.} \end{cases}$$

We call a configuration $x$ as reachable if it is associated with at least one predecessor. Specifically, if there exists a configuration $y \in \varepsilon_n$ and an update position $u \in \mathcal{L}$ such that $F(y,u) = x$, then $x$ is reachable; otherwise the configuration is called non-reachable or a garden-of-Eden state. In an early work, Das et al. [20, 21] have defined asynchronous reversible CA where each configuration has to be returned back almost surely during the evolution of the system.

**Definition 3** ([21]) For a couple of configurations $x, y \in \{0,1\}^{\mathcal{L}}$, we say that $y$ is reachable from $x$ if there is a sequence of updates that leads from $x$ to $y$, that is:

$$\exists k \in \mathbb{N}^*, U = (u_0, \cdots, u_{k-1}), x^0 = x, x^k = y$$

and

$$x^{i+1} = F(x^i, u_i), \forall i \in \{0, \cdots, k-1\}$$

Next, we can define the main recurrence property:

**Definition 4** ([21]) A configuration $x \in \{0,1\}^{\mathcal{L}}$ is recurrent if for every configuration $y$ that is reachable from $x$, $x$ is also reachable from $y$. A configuration that is not recurrent is transient.

For transient configuration, there exists a particular sequence of updates which brings the system to a configuration from which it will never be possible to return back to the initial configuration. More generally, if $y$ is reachable from $x$ and $x$ is reachable from $y$, we say that $x$ and $y$ *communicate*. By convention, all configurations communicate with themselves. Clearly, the relationship "communicate" is an equivalence relation; this relation partitions the set of configurations into communicating classes.

To sum up, two major behaviours exist: for the transient configurations, the system remains a finite time in its communication class, then "escapes" this class and never returns back to it. In contrast, when the system starts from a recurrent configuration, it remains in the communication class of this configuration forever. Finally, an ACA is recurrent if each configuration of $\varepsilon_n$ is recurrent; otherwise it is irreversible.

## 3 ACAs with Large Cycles

Our first hypothesis is, ACAs with large cycle length, are expected to be 'good' randomness enhancer. Note that, the word 'large' indicates cycle length of $2^n$ or $2^n - 1$. To identify ACAs with large cycle, we explore following two possibilities.

1. Identify recurrent ACAs with size of communication class $2^n$ or $2^n - 1$.
2. Identify ACAs with multi-length attractor of attractor length $2^n - 1$.

### *3.1 Recurrent ACAs*

In a previous work, Sethi et al. [21] have classified the elementary CA rules into three classes—*recurrent*, *irreversible* and *strongly irreversible* rules. According to work of [21], 46 (18 minimal representative, out of 88) ECAs are recurrent in the sense that would always return to their initial condition an infinite number of times. Theorem 1 identifies the asynchronous recurrent ECAs.

**Theorem 1** ([21]) *Rule R ACA is recurrent if both the RMTs* 0 *and* 2 *(resp.* 5 *and* 7*) are either active or passive, and at least one of following conditions is satisfied:*

*(a) both the RMTs of at least one of the sets - {2,5}, {1,6}, {3,4}, {1,3}, {4,6} are active.*
*(b) RMTs 1,3,4, and 6 are passive.*

Following Theorem 1, 18 minimal representative recurrent ACAs are listed in Table 2. Fatès et al. [22] have also identified number of communication classes for these recurrent ACAs. According to [22], 6 recurrent ACAs (ECA 33, 35, 41, 43, 51, 57, see Table 2, in red) are associated with one single communication class, where the length of the cycle is $2^n$. On the other hand, 5 recurrent ACAs (ECA 38, 46, 54, 60, 62, see Table 2, in blue) are associated with two communication classes where one communication class is associated with cycle of length $2^n - 1$ and other communication class is of size 1 with only all 0 configuration. Note that, the rest recurrent ACAs are associated with linear and exponential number of communication class which are not the topic of interest in this particular journey. Therefore, these 11 recurrent ACAs with large cycle length are the candidate rules for the proposed study.

**Remark 1** ACA 33, 35, 41, 43, 51, 57, 38, 46, 54, 60 and 62, which are associated with cycle of length $2^n$ or $2^n - 1$, are expected to be 'good' randomness enhancer.

**Table 2** List of 18 minimum representative ACAs that are recurrent

| | | | | | | | | | | | | | | | | | |
|---|---|---|---|---|---|---|---|---|---|---|---|---|---|---|---|---|---|
| 33 | 35 | 38 | 41 | 43 | 46 | 51 | 54 | 57 | 60 | 62 | 105 | 108 | 134 | 142 | 150 | 156 | 204 |

### 3.2 ACAs with Multi-length Attractor

Next, we explore ACAs with multi-length attractor of cycle length $2^n - 1$. Note that, the maximal cycle length of multi-length attractor is $2^n - 1$ because multi-length attractor of length $2^n$ actually shows recurrent behaviour. Following theorem identifies the corresponding ACAs.

**Theorem 2** *Rule R ACA with n cells converges to multi-length attractor of length* $2^n$-1 *with all possible configurations except all 1's, if*

- *RMT 5 is passive.*
- *RMTs 0, 2 and 7 are active and at least one RMT from both the RMT sets {1,4} and {3,6} are active.*

***Proof*** Firstly, the configuration with only primary RMT set {7} (i.e. all 1 configuration) is excluded from the multi-length attractor. Therefore, we have to identify the responsible RMT(s) (active/passive) which restricts the journey towards all 1's configuration. In this direction, active RMT 7 is responsible for the transition from primary RMT set {7} to primary RMT set {5,3,6}. On the other hand, active RMT 5 is responsible for the reverse journey. Therefore, passive RMT 5 is responsible for restrict the transition from the multi-length attractor towards all 1 configuration. Moreover, active RMT 7 is responsible for restrict the 'recurrent' behaviour of these ACAs.

Hereafter, we need to identify the responsible RMTs (active/passive) for communication between rest primary RMT sets, i.e. {0}, {2,5}, {5,3,6}, {4,1,2} and {4,1,3,6}. Let us denote transition from primary RMT set {a} to primary RMT set {b} for active RMT $x$ by {a} $\overset{x}{\rightarrow}$ {b}. Now, active RMTs 0 and 2 are responsible for the transition from primary RMT set {0} to primary RMT set {4,1,2} and vice-versa. Therefore, {0} $\overset{0}{\rightarrow}$ {4,1,2} and {4,1,2} $\overset{2}{\rightarrow}$ {0}, i.e. {0} $\overset{[}{0}]2\leftrightarrows$ {4,1,2}. Similarly, active RMTs 0 and 2 are responsible for the transition from primary RMT set {0}{4,1,2} to primary RMT set {2,5} and vice-versa, i.e. {0}{4,1,2} $\overset{[}{0}]2\leftrightarrows$ {2,5}. On the other hand, at least one RMT from both the RMT sets {1,4} and {3,6} should be active for transitions : {4,1,2} $\overset{[}{1}, 4]3, 6\leftrightarrows$ {5,3,6}; {0}{4,1,2} $\overset{[}{1}, 4]3, 6\leftrightarrows$ {4,1,3,6}; and {4,1,3,6} $\overset{[}{1}, 4]3, 6\leftrightarrows$ {7}{5,3,6}. To sumup, this discussion includes all possible primary RMT sets. Hence the proof.

Theorem 2 identifies 5 minimal representative ACAs (ACA 3, 11, 19, 25, 27) which are associated multi-length attractor of cycle length $2^n - 1$.

**Remark 2** ACA 3, 11, 19, 25 and 27, which are associated with cycle of length $2^n - 1$, are expected to be 'good' randomness enhancer.

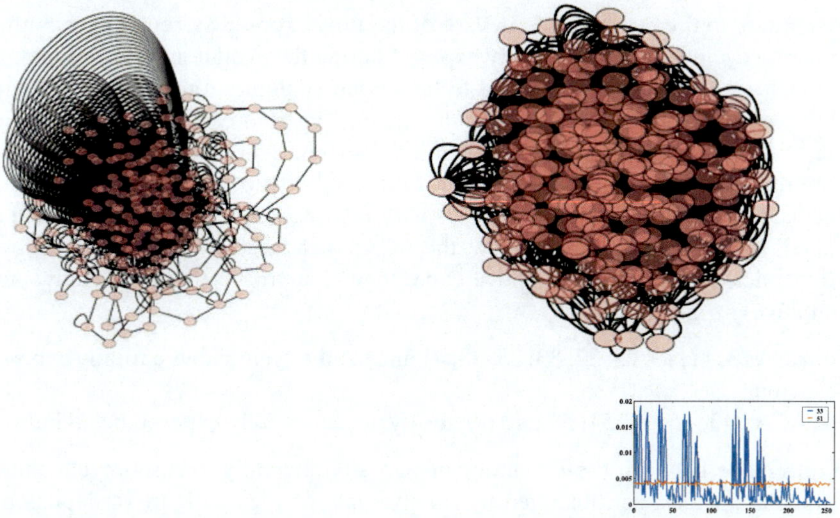


**Fig. 1** Dynamics of cycle **a** ACA 33 with partially exposed cycle; **b** ACA 51 with fully exposed cycle. The plot shows the profile of relative frequency as a function of the configurations. Here, we consider 8 size CA

## 4 ACAs with Large Cycles and Sense of Democracy

According to Remark 1 and 2, ACA 3, 11, 19, 25, 27, 33, 35, 38, 41, 43, 46, 51, 54, 57, 60 and 62 with large cycle of length $2^n$ or $2^n - 1$ are the candidate form the proposed randomness enhancer. However, large cycle length of a system sometimes induces some regularity. To avoid these circumstance, we study the dynamics of the cycle during evolution of the system.

In this experimental study, we express the ACA's evolution through graph where we need to follow the evolution of the system starting from a random initial configuration over a few time steps. During this evolution, we store the predecessor-successor relationship. Hereafter, we plot the predecessor-successor relationship where the configurations are the nodes of the graph. In this experiment, we evolve the CA for $n\times \mid \varrho \mid$ time steps where n is the size of the CA and $\mid \varrho \mid$ is the length of the corresponding $\varrho$ cycle. Note that, this number ($n\times \mid \varrho \mid$) is chosen to give a reasonable fair chance to every configurations of the cycle. Though, this approach is not a formal one to fulfil classification criterion, but it provides a good insight on the dynamics of the cycle.

To introduce a formal quantitative classification, we quantify the cycle dynamics by following the *relative frequency*, i.e. $f_r(\varrho_c) = \frac{f(\varrho_c)}{n\times|\varrho|}$, where $\varrho_c$ is a configuration associated with $\varrho$ cycle and f($\varrho_c$) is the (absolute) frequency for the $\varrho_c$ configuration.

**Definition 5** In this study, a cycle $\varrho$ is said to be fully exposed where $f_r(\varrho_i) \approx f_r(\varrho_j)$, $\forall i, j \in \{1, \cdots, \mid \varrho \mid\}$, otherwise the cycle $\varrho$ is said to be partially exposed.

According to the Definition 5, in case of the fully exposed system, every configuration of a cycle is (almost) equally exposed during the evolution of the system. In this direction, a configuration is said to be hidden configuration where it is "quite hard to reach" during the evolution of the system. Therefore, for partially exposed cycle, some configurations of the cycle are hidden configuration and some are not. In our previous work [18], we have studied this kind of behaviour for recurrent rule in details. Here, our hypothesis is, ACAs with fully exposed cycles are expected to be 'good' randomness enhancer. Now, the ACAs with large cycle show the following dynamics according to the above experimental approach (both qualitative and quantitative).

- For ACAs 3, 11, 19, 25, 27, 33, 35, 38, 41 and 62, the cycle shows partially exposed behaviour.
- For ACAs 43, 46, 51, 54, 57 and 60, the cycle shows fully exposed behaviour.

The difference between the dynamics of partially and fully exposed cycle shows remarkable changes, as illustrated by the example of Fig. 1. Figure 1a depicts the partially exposed cycle dynamics for ACA 33 where some configurations are associated with large number of self-loops. On the other hand, Fig. 1b shows the fully exposed cycle dynamics for ACA 51. Figure 1 also depicts the profile of relative frequency (blue line) for ECA 33 where the hidden configurations are associated with very low profile of relative frequency.

**Remark 3** Finally, ACA 43, 46, 51, 54, 57 and 60, which

- are associated with large cycle of length $2^n$ or $2^n - 1$; and
- shows fully exposed behaviour;

are expected to be 'good' randomness enhancer.

## 5 Empirical Verification of the Proposed ACAs-based Randomness Enhancer

It is a common practice of testing the quality of a PRNG by using empirical tests [23]. There are several empirical test beds available for this purpose [16, 24, 25]. Here, we use Diehard [16], a basic yet well-known test bed to see how our ACAs act as randomness enhancer. Diehard contains the following fifteen tests.

**Diehard Battery of Tests**

Birthday spacings, overlapping permutations, ranks of $31 \times 31$ and $32 \times 32$ matrices, ranks of $6 \times 8$ matrices, monkey tests on 20-bit Words, OPSO (Overlapping-Pairs-Sparse-Occupancy), OQSO (Overlapping-Quadruples-Sparse-Occupancy) and DNA tests, count the 1's in a stream of bytes, count the 1's in specific bytes, parking lot, minimum distance, random spheres, squeeze, overlapping sums, runs, and craps test.

To see how our ACAs act as randomness enhancer, we need to select a PRNG with poor randomness quality. One of the most common PRNGs, `rand` of GCC Library [26] satisfies our purpose. It has already been shown that, this PRNG fails to pass almost all tests of Diehard [27]. So, our target is to improve its randomness quality.

Recall that, for our fully asynchronous CA, one cell is updated in a single time step. To choose the particular cell, here we use `rand`, that is the update pattern is selected by `rand`. For each candidate ACAs of CA size $n$, $24 \leq n \leq 29$, 1000 initial configurations are considered. For each such run, a *.bin* file of size $10 - 12$ MB is generated using the whole configuration of the CA and is given as input to Diehard.

Table 3 depicts the results of Diehard battery tests for 16 candidate asynchronous CA rules with large cycle of length $2^n$ or $2^n - 1$. Recall that, we have claimed that ACA rules with fully exposed cycle dynamics are expected to be good randomness enhancer. In Table 3, these ACA rules are marked with colour. Observe that, ACA rules 43, 51, 54 and 57 with fully exposed cycle dynamics show extremely good result in Diehard battery tests. However, the proper theoretical understanding behind bad result of fully exposed ACA 46 and 60, is still open to us. Overall, to sum up, ACAs with large cycle and fully exposed cycle dynamics are 'good' randomness enhancer. Table 3 shows an evidence of that. Finally, to establish the result, Table 4 shows the comparison of existing PRNGs with our proposed ACA-based PRNG (best performing ACA 51) using Diehard battery tests for some fixed initial configuration. Table 4 also depicts the randomness improvement quality in comparison between `rand` and proposed ACA. Observe that, `rand` passes only one test out of 15 Diehard battery tests, on the other hand, the proposed ACAs-based randomness enhancer passes $8 - 10$ tests. Therefore, asynchronous cellular automata are able to improve randomness quality of a 'poor' random number generator.

## 6 Conclusion

In this work, we have explored the randomness enhancing capability of asynchronous cellular automata. Now, the natural extensions of this work on the sensitivity to asynchronous CAs and randomness include:

**Table 3** Results of Diehard battery tests for asynchronous cellular automata with large cycle of length $2^n$ or $2^n - 1$. The rules with fully exposed cycle are marked with colour. Note that, in this Table, max, min, avg respectively indicate maximum, minimum, average number of tests passed out of 15 tests

| Rule | CA Size | | | | | | | | | | | | | | | | | |
|---|---|---|---|---|---|---|---|---|---|---|---|---|---|---|---|---|---|---|
| | 24 | | | 25 | | | 26 | | | 27 | | | 28 | | | 29 | | |
| | Max | Min | Avg | Max | Min | Avg | Max | Min | Avg | Max | Min | Avg | Max | Min | Avg | Max | Min | Avg |
| 3 | 5 | 1 | 3.12 | 4 | 1 | 2.63 | 4 | 0 | 2.68 | 5 | 1 | 2.89 | 4 | 1 | 2.83 | 4 | 1 | 2.78 |
| 11 | 6 | 2 | 3.41 | 5 | 1 | 3.06 | 5 | 2 | 3.16 | 5 | 0 | 2.82 | 4 | 1 | 2.98 | 5 | 1 | 2.42 |
| 19 | 5 | 1 | 2.40 | 6 | 1 | 2.97 | 5 | 0 | 2.36 | 5 | 0 | 2.82 | 5 | 1 | 2.48 | 5 | 1 | 2.26 |
| 25 | 5 | 1 | 2.62 | 5 | 1 | 2.12 | 5 | 2 | 2.11 | 6 | 2 | 2.98 | 5 | 1 | 2.46 | 5 | 0 | 2.25 |
| 27 | 7 | 2 | 2.98 | 6 | 2 | 2.51 | 5 | 0 | 2.42 | 5 | 1 | 2.15 | 5 | 1 | 1.96 | 6 | 1 | 2.56 |
| 33 | 5 | 1 | 3.58 | 4 | 1 | 3.01 | 4 | 1 | 3.14 | 4 | 1 | 3.19 | 4 | 1 | 3.07 | 4 | 1 | 3.21 |
| 35 | 5 | 2 | 3.98 | 5 | 1 | 3.72 | 5 | 1 | 3.81 | 5 | 1 | 3.78 | 4 | 1 | 3.59 | 5 | 1 | 3.68 |
| 38 | 2 | 0 | 0.91 | 2 | 0 | 0.62 | 1 | 0 | 0.31 | 2 | 0 | 0.78 | 2 | 0 | 1.19 | 3 | 0 | 1.01 |
| 41 | 4 | 1 | 2.91 | 4 | 1 | 2.94 | 4 | 1 | 2.81 | 4 | 1 | 3.01 | 4 | 1 | 2.73 | 5 | 1 | 3.15 |
| 43 | 9 | 4 | 7.87 | 11 | 6 | 8.61 | 11 | 5 | 8.44 | 10 | 5 | 8.12 | 10 | 5 | 8.23 | 10 | 4 | 8.03 |
| 46 | 2 | 0 | 1.01 | 3 | 0 | 1.11 | 1 | 0 | 0.44 | 2 | 0 | 0.96 | 3 | 0 | 1.23 | 2 | 0 | 1.03 |
| 51 | 12 | 9 | 10.81 | 12 | 8 | 10.05 | 11 | 7 | 9.88 | 12 | 7 | 9.95 | 11 | 7 | 9.62 | 10 | 6 | 9.11 |
| 54 | 7 | 3 | 5.60 | 8 | 3 | 5.91 | 9 | 4 | 6.12 | 9 | 4 | 6.23 | 8 | 4 | 6.45 | 7 | 3 | 5.96 |
| 57 | 9 | 5 | 7.93 | 10 | 5 | 8.11 | 10 | 5 | 7.85 | 10 | 6 | 8.16 | 9 | 5 | 7.25 | 9 | 4 | 7.56 |
| 60 | 4 | 1 | 2.34 | 5 | 1 | 3.02 | 5 | 1 | 3.12 | 5 | 1 | 2.23 | 4 | 1 | 1.92 | 4 | 1 | 2.11 |
| 62 | 5 | 1 | 3.06 | 4 | 1 | 2.65 | 4 | 1 | 2.54 | 4 | 1 | 2.71 | 5 | 1 | 2.62 | 5 | 2 | 2.41 |

**Table 4** Comparison of existing PRNGs with our proposed ACA-based PRNG (best performing ACA 51) using Diehard battery of Tests for some fixed initial configuration (seed)

| Name of the PRNGs | | Seeds | | | | |
|---|---|---|---|---|---|---|
| | | 7 | 1234 | 12345 | 19650218 | 123456789123456789 |
| LCGs | MMIX | 6 | 5 | 6 | 4 | 5 |
| | **rand** | **1** | **1** | **1** | **1** | **1** |
| | lrand48() | 1 | 1 | 1 | 1 | 1 |
| | MRG31k3p | 1 | 1 | 1 | 1 | 0 |
| | PCG-32 | 9 | 9 | 11 | 10 | 9 |
| LFSRs | random() | 1 | 1 | 1 | 1 | 1 |
| | Tauss88 | 11 | 9 | 11 | 11 | 10 |
| | LFSR113 | 5 | 11 | 9 | 7 | 9 |
| | LFSR258 | 0 | 0 | 1 | 1 | 1 |
| | WELL512a | 9 | 10 | 10 | 8 | 7 |
| | WELL1024a | 9 | 10 | 9 | 9 | 9 |
| | MT19937-32 | 10 | 9 | 9 | 9 | 9 |
| | MT19937-64 | 10 | 10 | 8 | 11 | 10 |
| | SFMT19937-32 | 10 | 9 | 10 | 9 | 10 |
| | SFMT19937-64 | 11 | 10 | 10 | 9 | 10 |
| | dSFMT-32 | 7 | 8 | 11 | 11 | 10 |
| | dSFMT-52 | 5 | 5 | 7 | 6 | 7 |
| | xorshift32 | 4 | 4 | 4 | 0 | 4 |
| | xorshift64* | 10 | 10 | 8 | 7 | 8 |
| | xorshift1024* | 7 | 9 | 7 | 8 | 6 |
| | xorshift128+ | 9 | 9 | 10 | 10 | 8 |
| CAs | Hybrid CA with Rules 30 & 45 | 3 | 0 | 2 | 1 | 1 |
| | Maximal Length CA with $\gamma = 0$ | 2 | 0 | 1 | 1 | 2 |
| | Maximal Length CA with $\gamma = 1$ | 4 | 3 | 3 | 4 | 3 |
| | Non-linear 2-state CA | 6 | 8 | 5 | 5 | 7 |
| | 3-state CA | 3 | 3 | 3 | 2 | 3 |
| | **Proposed PRNG** | **10** | **8** | **9** | **8** | **9** |

- What can be said for bad results of fully exposed ACA 46 and 60?
- What can be said for ACAs with large cycle of length $l$, where $2^n - 1 > l > \frac{n}{2}$?
- What can be said for empirical verification of the proposed ACAs-based randomness enhancer with other popular testbeds like battery rabbit of TestU01 library, NIST testbeds?

**Acknowledgements** The authors are grateful to Prof. Sukanta Das and Dr. Kamalika Bhattacharjee for enlightening discussions on this topic. The authors acknowledge the anonymous reviewers for their comments and suggestions, which have helped to improve the quality and readability of the paper.

## References

1. Wolfram, S.: A New Kind of Science. Wolfram-Media Inc., Champaign, Ilinois, United States (2002)
2. von Neumann, J.: The Theory of Self-reproducing Automata, A. W. Burks ed. Univ. of Illinois Press, Urbana and London (1966)
3. Chaudhuri, P.P., Chowdhury, D.R., Nandi, S., Chatterjee, S.: Additive cellular automata—theory and applications. IEEE Comput. Soc. Press USA **1** ISBN 0-8186-7717-1 (1997)
4. Bhattacharjee, K., Naskar, N., Roy, S., Das, S.: A survey of cellular automata: types, dynamics, non-uniformity and applications. Natural Comput. **19**, 433–461 (2020)
5. Wolfram, S.: Cryptography with cellular automata. In: Williams, H.C. (Ed.) Advances in Cryptology—CRYPTO '85 Proceedings, pp. 429–432. Springer, Berlin, Heidelberg (1986)
6. Damgård, I.B.: A design principle for hash functions. In: Brassard, G. (Ed.) Advances in Cryptology—CRYPTO' 89 Proceedings, pp. 416–427. Springer, New York (1990)
7. Kurka, P.: Languages, equicontinuity and attractors in cellular automata. Ergodic Theory Dyn. Syst. **17**, 417–433 (1997)
8. Meier, W., Staffelbach, O.: Analysis of pseudo random sequence generated by cellular automata. In: Advances in Cryptology—EUROCRYPT '91, Workshop on the Theory and Application of of Cryptographic Techniques, Brighton, UK, April 8–11,: Proceedings, ser. Lecture Notes in Computer Science, vol. 547. Springer **1991**, 186–199 (1991)
9. Hortensius, P.D., McLeod, R.D., Card, H.C.: Parallel random number generation for vlsi systems using cellular automata. IEEE Trans. Comput. **38**(10), 1466–1473 (1989)
10. Adak, S., Bhattacharjee, K., Das, S.: Maximal length cellular automata in $gf(q)$ and pseudo-random number generation. Int. J. Modern Phys. C **31**(03), 2050037 (2020)
11. Cori, R., Metivier, Y., Zielonka, W.: Asynchronous mappings and asynchronous cellular automata. Inform. Comput. **106**(2), 159–202 (1993)
12. Droste, M., Gastin, P., Kuske, D.: Asynchronous cellular automata for pomsets. Theoret. Comput. Sci. **247**(1–2), 1–38 (2000)
13. Sethi, B., Roy, S., Das, S.: Asynchronous cellular automata and pattern classification. Complexity **21**(S1), 370–386 (2016)
14. Sethi, B., Das, S.: On the use of asynchronous cellular automata in symmetric-key cryptography. In: Mueller, P., Thampi, S.M., Alam Bhuiyan, M.Z., Ko, R., Doss, R., Alcaraz Calero, J.M. (Eds.) Security in Computing and Communications, pp. 30–41. Springer, Singapore (2016)
15. Manzoni, L., Mariot, L.: Cellular automata pseudo-random number generators and their resistance to asynchrony. In: Mauri, G., El Yacoubi, S., Dennunzio, A., Nishinari, K., Manzoni, L. (eds.) Cellular Automata, pp. 428–437. Springer International Publishing, Cham (2018)
16. Marsaglia, G.: DIEHARD: A battery of tests of randomness. http://stat.fsu.edu/geo/diehard.html (1996)

17. Li, W., Packard, N.: The structure of the elementary cellular automata rule space. Complex Syst. **4**, 281–297 (1990)
18. Roy, S., Das, S.: Asynchronous cellular automata that hide some of the configurations during evolution. Int. J. Modern Phys. C **32**(04), 2150054 (2021)
19. Fatès, N., Thierry, E., Morvan, M., Schabanel, N.: Fully asynchronous behavior of double-quiescent elementary cellular automata. Theoret. Comput. Sci. **362**(1), 1–16 (2006)
20. Sarkar, A., Mukherjee, A., Das, S.: Reversibility in asynchronous cellular automata. Complex Syst. **21**, 71–84 (2012)
21. Sethi, B., Fatès, N., Das, S.: Reversibility of elementary cellular automata under fully asynchronous update. In: Gopal, T.V., Agrawal, M., Li, A., Cooper, S.B. (eds.) Theory and Applications of Models of Computation, pp. 39–49. Springer International Publishing, Cham (2014)
22. Fatès, N., Sethi, B., Das, S.: On the Reversibility of ECAs with Fully Asynchronous Updating: The Recurrence Point of View, pp. 313–332. Springer International Publishing, Cham (2018)
23. Knuth, D.E.: The Art of Computer Programming, Volume 2: Seminumerical Algorithms, 3rd ed. Boston: Addison-Wesley (1997)
24. Rukhin, A., Soto, J., Nechvatal, J., Smid, M., Barker, E., Leigh, S., Levenson, M., Vangel, M., Banks, D., Heckert, A., Dray, J., Vo, S.: Statistical test suite for random and pseudorandom number generators for cryptographic applications, NIST special publication. revision 1a. National Institute of Standards and Technology, Technology Administration, U.S. Department of Commerce, vol. 800-22 (2010)
25. L'Ecuyer, P., Simard, R.: TestU01: a c library for empirical testing of random number generators. ACM Trans. Math. Softw. **33**(4), 22:1–22:40 (2007)
26. McGrath, R. et al.: GNU C Library. https://www.gnu.org/software/libc/manual/html_node/Pseudo_002dRandom-Numbers.html#index-pseudo_002drandom-numbers
27. Bhattacharjee, K., Maity, K., Das, S.: A search for good pseudo-random number generators : survey and empirical studies. CoRR vol. abs/1811.04035, 2018. [Online]. http://arxiv.org/abs/1811.04035

# Cellular-Automata Based Simulation of Dynamic Recrystallization and Statistical Analysis of Resulting Grain Growth

**R. S. Subramanian, Ramadev Sai Shree, and P. G. Kubendran Amos**

**Abstract** By employing cellular automata technique, in this work, a polycrystalline system corroborating experimentally observed microstructure is developed. Moreover, through appropriate transition functions, microstructural changes accompanying dynamic recrystallization, a manufacturing technique associated with the production of wide-range of components, is simulated. The grain growth that characterises this change is analysed to explicate the trend in the temporal evolution of mean grain size and its kinetics. As opposed to a progressive increase in mean grain size, which generally typifies a conventional grain growth, in dynamic recrystallization, it is observed that the mean grain size increases and decreases sequentially, thereby rendering a oscillating pattern. A perceptive investigation of the fluctuating trend unravels individual growth events, characterised by monotonic increase in mean grain size, whose kinetics follow the third-order power law.

**Keywords** Dynamic recrystallization · Grain growth · Kinetics · Cellular Automata

## 1 Introduction

When you consider an object made of metal, be it a routine kitchen utensil or meticulously designed component in an aerospace application, almost invariably its production involves a process wherein the desired shape is achieve by imposing appropriate mechanical forces [1]. For instances, sheets made of solid solution comprising of metal solvent and other solutes, which can be referred to as alloys, are generally fabricated by imposing a compressive load [2]. These techniques wherein considerable mechanical load is involved, the desired shape is render to the material (alloy)

R. S. Subramanian · R. S. Shree · P. G. K. Amos (✉)
Theoretical Metallurgy Group, Department of Metallurgical and Materials Engineering, National Institute of Technology, Tiruchirappalli, Tamil Nadu, India
e-mail: prince@nitt.edu



S. Das and G. J. Martinez (eds.), *Proceedings of First Asian Symposium on Cellular Automata Technology*, Advances in Intelligent Systems and Computing 1425,
https://doi.org/10.1007/978-981-19-0542-1_12

**Fig. 1** A *polycrystalline microstructure* comprising of numerous homogeneous regions called grains, which are separated by grain boundaries

by *deforming* it. Even though the act of deformation brings about the desired shape externally, it also introduces changes to the alloy which are not directly visible [3].

Metals and alloys that are employed in wide-range of application almost always share a common feature. When looked closely using a microscope, like a country encompassing different states, these materials comprise of numerous, yet definite, homogeneous regions separated by boundaries, together constituting a continuum. The homogeneous sections of the material, which are characterised by a specific 'directional' arrangement of the constituent atoms, are referred to as the grains, and the interface separating them as grain boundaries [4]. This structure of the metal (or alloy), which is only visible through a microscope, is termed as microstructure. And it is the microstructure that principally governs the behaviour of a material in a given *situation*, thereby dictating the properties. By assigning appropriate colours to the grains based on the directional arrangement of the atoms, the microstructure of a material is generally visualised as shown in Fig. 1.

The deformation that attempts to establish a desired shape externally, affects the internal microstructure of the alloy largely in two following ways [5]:

1. reducing the overall average size of the homogeneous regions i.e the grains,
2. disrupting the homogeneity within the grains by introducing defects referred to as dislocations.

The aforementioned changes made to the internal structure of the material are detrimental to further deformation. In other words, *the deformation imposed to bring about the desired shape alters the microstructure in a manner that hinders further deformation.* The hampered ability of the metal and alloy to deform ultimately leads to its failure. Therefore, in order to reverse the undesired changes, and to ensure that further deformation, either during the shaping or in an application, occurs with ease, the material is exposed to a befitting treatment called recrystallization [6].

The process of recrystallization (re-)establishes the initial microstructure by reducing the defects i.e dislocations, which disrupt the homogeneity of the grains, and increasing the reduced size of the grains (both ill-effects of deformation). In a commercial setting, a deformed material is generally recrystallized by adopting one of the two techniques. When the material assumes the desired shape, or is moving towards it with visible difficulties, the induced microstructural changes are reversed by heating it to a high temperature. The increase in temperature facilitates the nucleation and growth of new grains with reduced inhomogeneities. This technique is referred to as static recrystallization [7]. On the other hand, the deformation is performed at a pre-determined high temperature such that ill-effects are simultaneously reversed while the material is being shaped. This approach wherein the shape change is achieved while concurrently reversing the accompanying microstructural distortion is called as dynamic recrystallization [8]. Given the efficiency of the dynamic recrystallization over the static, on a industrial scale, the former is largely preferred over the latter. However, in order to devise the dynamic recrystallization, a comprehensive understanding of the microstructural changes induced by the technique is pivotal.

Gaining deeper insight on microstructural changes accompanying dynamic recrystallization through conventional experimental technique is an arduous task as it demands continual observation of the microscopic events that occur at high temperature [9]. Therefore, theoretical techniques, which are consistently formulated based on the experimental observations, are employed to model and analyse the evolution associated with dynamic recrystallization. Given its importance in manufacturing, the microstructural changes induced during the dynamic recrystallization have been modelled using various techniques [10–12]. In this work, an approach based on cellular automata is adopted to simulate the transformation of microstructure established during dynamic recrystallization [13, 14]. Furthermore, considering that kinetics of grain growth accompanying dynamic recrystallization is yet to be sufficiently analysed, the temporal change in the average grain-size is investigated.

The uniqueness of cellular automata essentially lies in it ability to spawn a temporally-varying dynamic system based on few straightforward *rule(s)* involving *neighbours* [15]. This technique begins with the discretization of the simulation domain into individual cells. Subsequently, the discretized domain is initialised by assigning specific, yet often not unique, states to these cells. A state of a given cell is allowed to change when a particular condition, imposed through an appropriate function, is satisfied. This function, referred to as the transition function or local rule, is applied parallely across all the cells in the simulation domain, thereby establishing a global evolution. In this work, a cellular automata technique with translation function suitably formulated based on the factors governing the dynamic recrystallization is employed to simulate the corresponding microstructural changes.

## 2 Model Description

The current investigation adopts the Cellular Automata model discussed in Refs [16, 17]. Even though an extensive description of the model, along with its relevance to physical system, can be found elsewhere [18], a concise and a contextual description is rendered here.

In order to model the characteristic microstructural changes associated with dynamic recrystallization, two different Cellular-Automata (CA) based approaches are employed. While the first establishes the initial *polycrystalline microstructure* comprising of numerous grains, it is the second approach which is used to simulate the evolution typifying dynamic recrystallization. The initial microstructure is generally established independent of the pre-existing state of the domain. Considering that this feature of the first CA approach is indicative of static recrystallization, wherein the evolution of grains are largely independent of the history of the microstructure, the corresponding model is referred to as the *static recrystallization model*. In other words, with static-recrystallization approach establishing the initial microstructure, its characteristic distortions are modelled by dynamic-recrystallization treatment, both operating in CA framework.

### *2.1 Static Recrystallization Model*

A polycrystalline microstructure characterised by numerous grains, under static recrystallization model, is developed by considering a homogeneous simulation domain with each discretized cell assuming an identical 'non-recrystallized' state. Several globular nuclei, with size typically greater than one cell and indicating 'recrystallized' state, are subsequently introduced in the domain. Besides recrystallized state, the cells constituting a globular nucleus take-up one of the 18 possible states ($\theta$) depicting the possible *orientation* or angular arrangement of atoms. Therefore, in addition to being recrystallized, a nucleus exhibit a specific orientation akin to physical grains. By assigning appropriate colour, nucleus of a grain characterised by definite orientation is distinguished from the other. These precursors to the polycrystalline system in the simulation domain are shown in the first two slides of Fig. 2. The randomly distributed nuclei are allowed to grow and consume the entire domain through a transition function that ensures that an initial non-recrystallized cell will

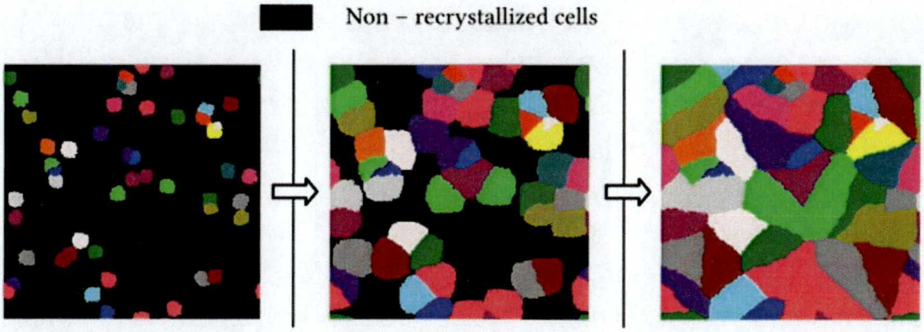


**Fig. 2** An initial polycrystalline microstructure established by CA-based static recrystallization treatment

recrystallize at a given time ($t$), when atleast one of its neighbours has already been recrystallized ($t - 1$). Owing to this rule, the growth of the nuclei stops when it impedes with another nuclei. Figure 2 shows the polycrystalline microstructure established by the static recrystallization approach.

In the microstructure established by the static recrystallization model, the edges of the cells of a given grain abutting the cells of another act as the grain boundary. By formulating a probabilistic movement for the grain boundary cells, microstructural features, particularly grain morphology, analogous to the experimental observations is achieved in the present approach [18].

## 2.2 *Dynamic Recrystallization Model*

For simulating the microstructural changes associated with dynamic recrystallization, a more rigorous treatment is rendered to the cells when compared to the static recrystallization model. The state of a cell in the dynamic recrystallization model is dictated by three distinct components. Correspondingly, the tuple indicating the state of the cell is expressed as

$$\mathbf{s} \in \{\theta, \rho, t_w\}, \tag{1}$$

where $\theta$ denotes the angular arrangement or orientation of the atoms in a grain. The degree of inhomogeneity in a grain, which is introduced during deformation through a defect called dislocation, is assigned to the corresponding cell as $\rho$, the dislocation density. The factor $t_w$ in Eq. 1 governs the probabilistic movement of the grain boundaries.

Dynamic recrystallization, as introduced in the introductory section, accompanies the deformation of the material. Therefore, it encompasses increase in the inhomogeneity of the grains, nucleation and growth of relatively homogeneous grains at

**Fig. 3** Formation of defect-free nuclei, and its subsequent growth during dynamic recrystallization

expense of the others. In other words, the material during dynamic recrystallization experiences *localised* increase dislocation density ($\rho$), formation of homogeneous nuclei and growth of grains with low dislocation density over inhomogeneous grains. For modelling such rather convoluted microstructural changes, two different transition functions are formulated and are imposed based on the nature of the cells.

Before imposing the local rule(s), a given cell is assessed to realise whether it is a part of the grain boundaries or an actual grain. Depending on this feature of the cell, specific transition function is imposed.

When a cell belongs to the grain boundary, its dislocation density ($\rho_i$) is compared to that of the neighbouring cell that belongs to the adjoining grain (($\rho_j$)). When the difference in the dislocation densities is beyond a critical value, $\rho_j - \rho_i > \rho_{crt}$, which is introduced separately in the formulation, the cell with higher degree of inhomogeneity ($\mathbf{s}_j$) transforms to the one with low dislocation density ($\mathbf{s}_i$) by assuming its orientation ($\theta_i$) along with the dislocation density ($\rho_i$). In other words, the grain with low dislocation density ($i$) grows into the grain with high inhomogeneity ($j$) by the probabilistic movement of the corresponding grain boundary. This evolution of grains establishes one characteristic feature of the dynamic recrystallization which is the reduction in inhomogeneity through grain growth.

As opposed to the grain boundaries, when a cell is located in with a grain, its dislocation density is allowed to progressively increase with time in accordance with a physically-relevant relation. This time-dependent increase in the dislocation density captures the progressive influx of inhomogeneities which is facilitated by the deformation associated with dynamic recrystallization. When the progressive raise in dislocation density reaches a critical value, a defect-free nucleus in introduced, and its allowed to the grow at the expense of the grain with increased dislocation density. Correspondingly, the ill-favoured effect of the deformation is reversed by the evolution of the nuclei. The nucleation and growth of defect-free grains modelled by dynamic recrystallization approach is shown in Fig. 3.

The combination of the above transition functions, which are imposed based on the nature of the cell, ultimately enable the modelling of microstructural changes characterised by dynamic recrystallization.

## 2.3 Summary of Rules

In the present work, considering that both static and dynamic recrystallizations are modelled using CA technique, characteristic rules are employed to simulate each of the microstructural changes.

In static recrystallization, the state of each discretized cell reflects its condition, which is recrystallized or non-recrystallized, and orientation. When the simulation domain is initialised with all cells assuming non-recrystallized condition and uniform orientation, the static recrystallization is achieved by introducing several nuclei and subsequently allowing them to grow. These nuclei, with the alternate recrystallized condition and specific orientation, grow by changing the state of the non-recrystallised matrix cells to its own when latter encounters the former as atleast one of its neighbours.

Once a polycrystalline microstructure is established through static recrystallization, different set of components are adopted to describe the state of cells and model dynamic recrystallization. In other words, after static recrystallization, each cell is associated to a grain, and its state is dictated by orientation ($\theta$), dislocation density ($\rho$) and probabilistic movement ($t_w$). Rule(s) that govern the transformation of a cell depends on its position. A cell along the grain boundary changes the state when the difference between its dislocation density and that of its neighbours is greater than a critical value. In such a case, the cell will adopt a state of its neighbours that is lowest dislocation density. On the other hand, for a cell within a grain, its dislocation density is allowed to progressively increase, till a critical value is reached. Upon reaching a critical value, a transformation is introduced which changes the state of cell one with zero dislocation density and different orienation.

## 2.4 Domain Setup

Since the present work is primarily geared towards realising the trend in grain growth associated with dynamic recrystallization, the corresponding simulations have been restricted to two-dimensional setup. Moreover, domains of identical size, discretized into 256 × 256 cells, is considered for all simulations. Periodic boundary condition that seemingly renders a toroidal shape to the domain, and offers no considerable boundary effects is invariably imposed. The transition functions are formulated by considering Moore's neighbourhood that includes eight nearest cells.

Through the static recrystallization model, a polycrystalline microstructure with 50 grains is established across the simulation domain. The critical dislocation density that dictates the growth of less homogeneous grains over others, in the present work, is set at $\rho_{crt} = 255$. Nucleation of a defect-free grain is induced when the dislocation density of the grains reach $\rho_{nucl} = 50$. The dislocation densities are appropriately normalised to render them dimensionless.

## 3 Result and Discussion

Grain growth is rather the most prevalent microstructural changes which is observed in wide-range of materials [19]. This phenomenon is primarily governed by the ability of the system to reduce its overall energy. A polycrystalline structure, generally, comprises of homogeneous grains and the separating grain boundaries. While grains are relatively low energy sections of the microstructure, the energy of the system is increased with raise in amount of grain boundaries. Therefore, when favoured, often at high temperatures, the system reduces the grain boundary area, thereby minimizing the overall energy, by allowing certain grains to grow at the expense of the others. However, the conventional grain growth is significantly different from the ones accompanying dynamic recrystallization. In conventional growth, the grains larger in size grow at the expensive of the smaller counterparts. In other words, a regular grain growth is fundamentally governed by curvature. Contrastingly, the evolution of grains during dynamic recrystallization, as modelled in the present work, is principally dictated by the difference in the degree of inhomogeneity or dislocation density. Accordingly, irrespective of the size, the grains with low dislocation density will grow by consuming the grains with higher degree of inhomogeneities. This primary difference in the governing mechanism indicates that the kinetics of conventional grain growth should be different from the one associated with dynamic recrystallization. To that end, in the present work, the grain growth is separately identified and its kinetics is analysed using statistical tools.

### *3.1 Realising Grain Growth in Dynamic Recrystallization*

Given that the principal focus of the present work is to understand grain growth accompanying dynamic recrystallization, the temporal change in the average grain size of the microstructure, which is the ratio of the sum of all the individual grain sizes and total number of grains, is plotted in Fig. 4. This representation indicates that the average grain size progressively decreases, without exhibiting an increase, thereby apparently suggesting that the grain growth is absent during this recrystallization process. One primary reason from this counterintuitive depiction is that the red curve in Fig. 4 begins with the average grain size of the polycrystalline microstructure rendered by the static recrystallization. However, the grain growth is introduced only through the evolution of the nuclei which are induced as a part of dynamic recrystallization. Furthermore, the size of these nuclei are considerably smaller than that of the original grains. Therefore, in order to realise the grain growth associated with dynamic recrystallization a section of the red curve is realised and separately illustrated in Fig. 4 as a subplot.

As opposed to a gradual increase in grain size, which is generally expected of grain growth, the subplot in Fig. 4 shows individual events of increase in grain size that are immediately followed by a corresponding reduction. In order to understand this

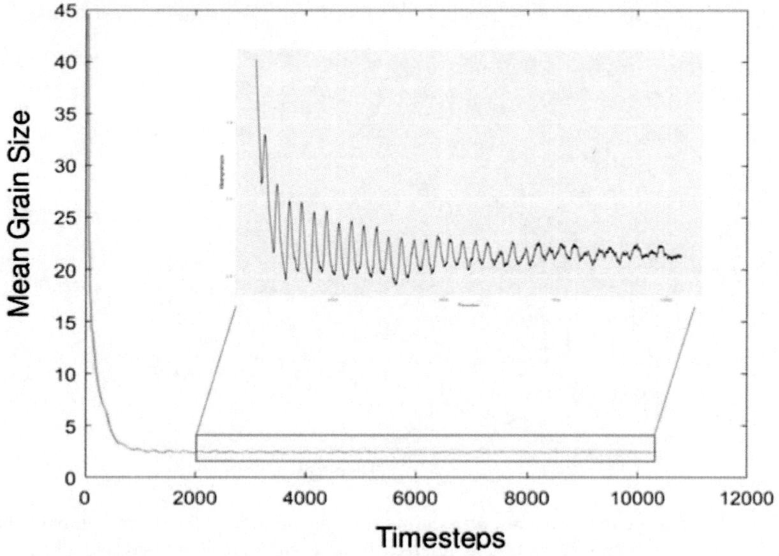


**Fig. 4** The change in the average grain size of the polycrystalline microstructure with time is plotted centrally, while a section of it is zoomed-in and presented as a subplot

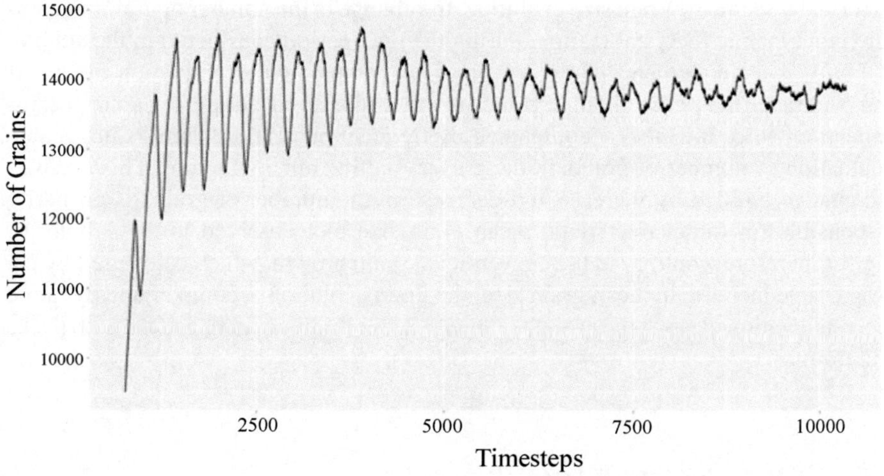


**Fig. 5** The number of grains in microstructure as the dynamic recrystallization progresses for the duration considered in the subplot of Fig. 4

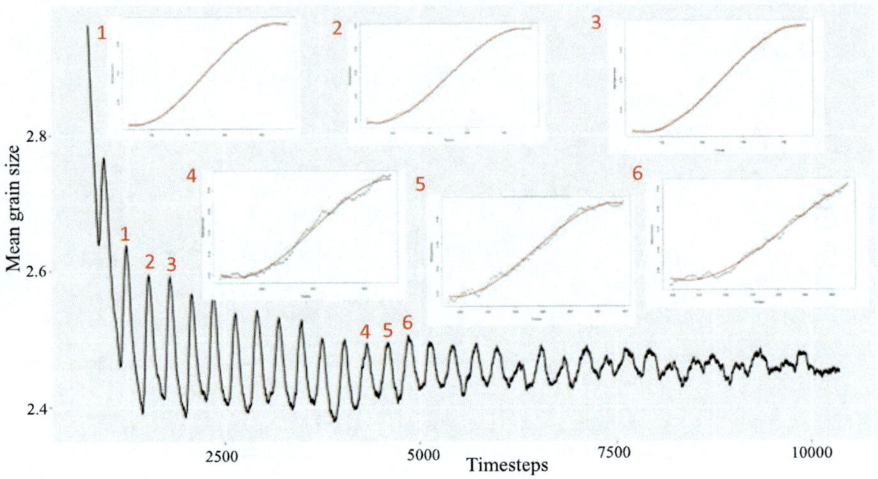


**Fig. 6** Temporal change in the average size during specific duration of the overall grain evolution, characterised by monotonic increase and referred to as *events*, are identified and illustrated as subplots with principal plot depicting the oscillated change in grain size

abnormal trend in the grain size evolution, the change in the number of grains accompanying dynamic recrystallization, during the time period considered in the subplot of Fig. 4, is estimated and plotted in Fig. 5. A comparative examination of Figs. 4 and 5 unravel that, even though grain growth is observed during dynamic recrystallization, since it is also accompanied by frequent nucleation, there is noticeable fluctuation of number of grains in the polycrystalline microstructure. This oscillating change typified by increase and decrease in the number of grains is primarily responsible for the characteristic mean grain size plot observed in the subplot of Fig. 4. Therefore, contrary to the conventional grain growth, which is featured by the progressive increase in mean grain size, the grain evolution accompanying dynamic recrystallization, despite the growth of nuclei, exhibit a fluctuating pattern with peaks and troughs.

## 3.2 Examining Grain Growth Kinetics

Owing to the unique trend in the grain size evolution associated with dynamic recrystallization, the conventional treatment of analysing the grain growth by considering the entire process cannot be adopted. Therefore, sections of the grain size evolution in Fig. 4, wherein the average size monotonically increases, is identified, and the size change corresponding to these sections are illustrated separately as different subplots in Fig. 6.

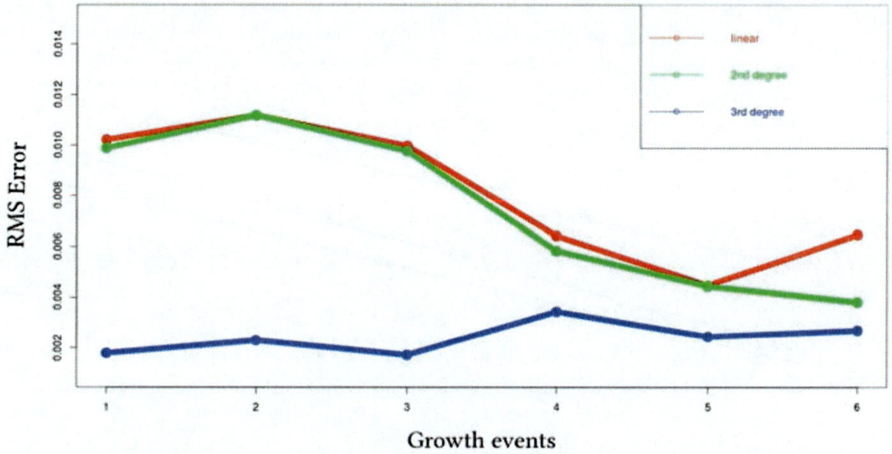


**Fig. 7** Root Mean Square Error (RMSE) for the linear and non-linear regression for datapoints associated with grain growth events exhibiting monotonic increase in average grain size

The subplots of individual *events* in Fig. 6 resemble, albeit qualitatively, the temporal change in average grain size observed during conventional grain growth. Therefore, although the overall evolution of average grain size during dynamic recrystallization adopts a unique pattern, a recognisable trend can be observed when considering the individual events, which are characterised by monotonic increase in average grain size.

The growth kinetics of the specific events typified by the monotonic increase in average size, shown in the subplots of Fig. 6, is quantitatively delineated by using statistical approach. Linear and non-linear regression techniques are adopted to realise the trend associated with the temporal change in the average grain size during the six growth events unravelled in Fig. 6. The Root Mean Square Error (RMSE) for different linear and non-linear fittings of the datasets pertaining to the six growth events are estimated in order to identify the suitable relation between the time and average grain size. The RMSE for linear, second and third order fittings of the events are graphically represented in Fig. 7.

The illustration in Fig. 7 indicates that, as opposed to linear and second order fitting, the monotonic increase in average grain size during specific events are accurately described by third order expression. In other words, the kinetics of the specific grain growth events accompanying dynamic recrystallization, adhere to power law, $\bar{R}^n \propto t$, with $n = 3$, where $\bar{R}$ and $t$ represent the average grain size and time, respectively.

The kinetics of conventional grain growth, wherein the evolution of the grains are primarily driven by curvature, largely follows second-order power law, $n = 2$ [20]. This growth rate indicates that the evolution of the grains, when driven by curvature, is primarily governed by migration of the grain boundaries facilitate to short-range diffusion of atoms [21]. Adherence to third-order power law is noticed when the

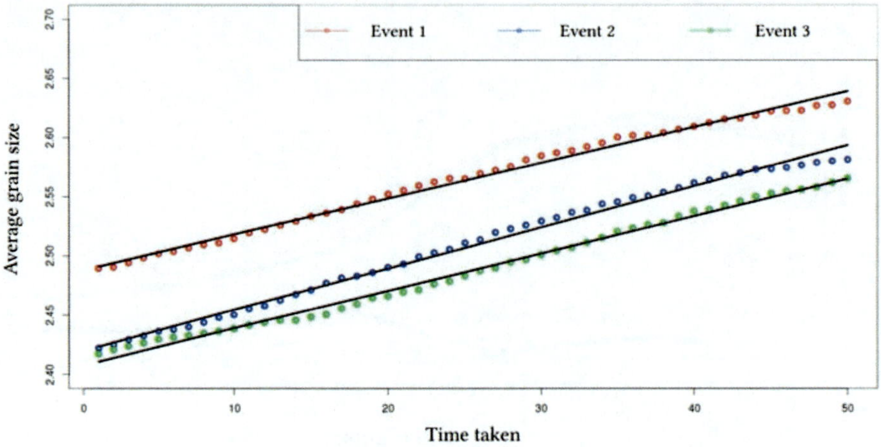


**Fig. 8** Monotonic increase in mean grain size during event 1, 2, and 3 which are identified in Fig. 6

curvature-driven grain growth is accompanied by coarsening [22]. In other words, when the grains constituting polycrystalline system, instead of exhibiting chemical uniformity, assume a distinct composition, their growth follow the power law characterised by exponent, $n = 3$ [23]. While the second-order power law indicates growth dictated by the migration of grain boundaries, the grain growth achieved by long-range diffusion adheres to third-order law. Therefore, Fig. 7 seemingly suggests that the kinetics of grain growth events associated with dynamic recrystallization resembles growth dictated by long-range diffusion rather than grain boundary migration.

Figure 6 unravels that the amount of fluctuation in the mean grain size decreases during the dynamic recrystallization. Correspondingly, the peak grain size attained by individual growth events gradually reduce with the progress of dynamic recrystallization. Besides this minimization of grain size oscillation, the impact of the degree of dynamic recrystallization on the kinetics of growth event is not adequately noticeable in Fig. 6. Therefore, the temporal change in the average grain size during the first three events are collectively presented in Fig. 8. It is evident from this illustration that, in addition to the degree of fluctuation, the kinetics of grain growth exhibited by the individual growth events steadily decreases as the dynamic recrystallization proceeds. Despite the reduction in the growth kinetics, all the growth event invariably adhere to the third-order growth law.

## 4 Conclusion

Grain growth during dynamic recrystallization, in contrary to the conventional phenomenon, is driven by the difference in the inhomogeneities across the various grains. In this study, the microstructural changes accompanying dynamic recrystallization has been modelled using cellular automata to analyse corresponding grain growth. Besides the general trend in mean grain size evolution, the kinetics of grain growth associated with dynamic recrystallization is realised to be different from that of curvature-driven microstructural changes. Furthermore, the present investigations unravel that the rate of growth event gradually decreases as the dynamic recrystallization proceeds. Considering that this work includes simulations of a specific condition, a statistically rigorous study will follow wherein material-specific critical dislocation and nucleation densities are involved.

## References

1. Wagoner, R.H., Chenot, J.-L.: Metal Forming Analysis. Cambridge University Press (2001)
2. Altan, T., Tekkaya, A.E.: Sheet metal forming: fundamentals. ASM Int. (2012)
3. Hu, J., Marciniak, Z., Duncan, J.: Mechanics of Sheet Metal Forming. Elsevier (2002)
4. Priester, L.: Grain boundaries: From Theory to Engineering. Springer Science & Business Media, p. 172 (2012)
5. Lin, J., Balint, D., Pietrzyk, M.: Microstructure Evolution in Metal Forming Processes. Elsevier (2012)
6. Humphreys, F.J., Hatherly, M.: Recrystallization and Related Annealing Phenomena. Elsevier (2012)
7. Doherty, R.D., Hughes, D.A., Humphreys, F.J., Jonas, J.J., Jensen, D.J., Kassner, M.E., King, W.E., McNelley, T.R., McQueen, H.J., Rollett, A.D.: Current issues in recrystallization: a review. Mater. Sci. Eng. A **238**, 219–274 (1997)
8. Huang, K., Logé, R.E.: A review of dynamic recrystallization phenomena in metallic materials. Mater. Design **111**(5), 548–574 (2016)
9. Liss, Kl.-D., Garbe, U., Li, H., Schambron, T., Almer, J.D., Yan, K.: In situ observation of dynamic recrystallization in the bulk of zirconium alloy. Adv. Eng. Mater. **11**,637–640 (2009)
10. Gourdet, S and Montheillet, F: A model of continuous dynamic recrystallization. Acta Materialia 51, 2685–2699 (2003)
11. Momeni, A., Dehghani, K., Ebrahimi, G.R.: Modeling the initiation of dynamic recrystallization using a dynamic recovery model. Journal of Alloys and compounds **509**, 9387–9393 (2011)
12. Takaki, T., Yoshimoto, C., Yamanaka, A., Tomita, Y.: Multiscale modeling of hot-working with dynamic recrystallization by coupling microstructure evolution and macroscopic mechanical behavior. International Journal of Plasticity **52**, 105–116 (2014)
13. Goetz, R.L., Seetharaman, V.: Modeling dynamic recrystallization using cellular automata. Scripta Materialia **38** (1998)
14. Sitko, M., Madej, Ł: Development of dynamic recrystallization model based on Cellular Automata approach. Key Engineering Materials **622**, 617–624 (2014)
15. Wolfram, S.: Cellular automata. Los Alamos. Science **9**, 42 (1983)
16. Kroc, J.: Application of cellular automata simulations to modelling of dynamic recrystallization. Int. Conf. Comput. Sci., 773–782 (2002)
17. Kroc, J., Paidar, V.: Modelling of recrystallization and grain boundary migration by cellular automata. Materials Science Forum **426**, 3873–3878 (2003)
18. Kroc, J.: Simulation of dynamic recrystallization by cellular automata. Univerzita Karlova Praha, Praha (2000)
19. Ralph, B.: Grain growth. Materials Science and Technology **6**, 1136–1144 (1990)
20. Hillert, M.: On the theory of normal and abnormal grain growth. Acta metallurgica **13**, 227–238 (1965)

21. Perumal, R., Amos, P.G.K., Selzer, M., Nestler, B.: Phase-field study on the formation of first-neighbour topological clusters during the isotropic grain growth. Comput. Mater. Sci. **140**, 209–223 (2017)
22. Amos, P.G.K., Perumal, R., Selzer, M., Nestler, B.: Multiphase-field modelling of concurrent grain growth and coarsening in complex multicomponent systems. J. Mater. Sci. Technol. **45**, 215–229 (2020)
23. Amos, P.G., Perumal, R., Koeppe, A., Nestler, B.: Data-driven analysis of grain-growth kinetics in duplex and triplex systems. arXiv preprint arXiv:2108.02236 (2021)

# A New Notion of Universality in Respect of Logic Gates Generation Capability of ECAs

**Raju Hazari**

**Abstract** This paper studies the gate generation capability of elementary cellular automaton (ECA). We first show that the basic logic gates, such as AND, OR, NOT, NAND, NOR, XOR, XNOR etc. can be realized by evolving ECAs. Since any logic circuit can be implemented by utilizing basic logic gates, the CA can implement all logic circuits. Therefore, we get a new notion of universality of CAs. We find out that there are some ECAs rules which have the universality property. Finally, we compare the gate generation capability of these CAs.

**Keywords** Cellular automata (CAs) · Elementary cellular automata (ECAs) · Universality · Logic gate

## 1 Introduction

Since early 1980s, elementary cellular automata (ECAs) have received wide attention from researchers due to their simplicity and capability of modeling physical systems [2, 14]. ECAs are the simplest cellular automata (CAs) where systems are binary and one-dimensional, and where the behavior of each cell depends only on itself and its two nearest neighbors. In this work, we explore the gate generation capability of all the ECAs.

The ECAs have been analysed for their capacity of producing interesting structures which can interact in order to compose more complex reactions. Stephen Wolfram in 1984, made a conjecture that ECA 110 is capable of universal computation [12]. Latter computational universality of ECA 110 has been proved [3, 9]. In [7], Martínez et al. have envisaged that the ECA 54 will make a good alternative to ECA 110 in a sense of logical universality as well as simulation of a Turing machine. Obviously, these CAs would be able to implement the logic gates. However, these ECAs are infinite CAs.

R. Hazari (✉)
Department of Computer Science and Engineering, National Institute of Technology Calicut, Kattangal, Kerala 673601, India

 
S. Das and G. J. Martinez (eds.), *Proceedings of First Asian Symposium on Cellular Automata Technology*, Advances in Intelligent Systems and Computing 1425,
https://doi.org/10.1007/978-981-19-0542-1_13

The designing of logic circuits using CAs have been initiated long before [10]. In 1996, Melanie Mitchell et al. have constructed AND, OR and NOT gates by the gliders guns [8]. According to that work, when two gliders collide, they annihilate. This is used to build a NOT gate. In the same way, AND and OR gates can also be constructed. Similarly, in [6], Martínez et al. have presented the designs of logical gates, which are implemented via collisions between gliders. In 2008, an evolutionary developmental technique has been introduced for the design of combinational circuits by exploring one-dimensional CAs [1]. The traditional concept of the cellular automata was modified in order to generate the circuits at the gate-level during the development of the CA.

In [4], we have already observe that ECA 184 can implement any logic circuits. In the work, we have shown that all the basic logic gates, such as AND, OR, NOT, NAND, NOR, XOR, XNOR etc. can be realized by evolving ECA 184 and called ECA 184 as a universal logic gate generator. This work is the extension of the previous work [4]. Here we explicitly study the gate generation capability of all the ECAs.

In this work, we consider an $n$-cell rule $\mathcal{R}$ CA as "$\mathcal{R}$ block"having $n$ inputs and $n$ outputs. The initial configuration of CA is the input to the block. To understand its gate generation capability, we evolve the CA. Obviously, each output of the block is a function of the inputs. After each evolution, we check if any known logic gate is realized by the output bits. To realize the logic gates, we need to set some input bits. In this approach, we can observe the gate generation capability of any CA.

To observe the universality property of ECAs, we first explore the gate generation capability of all the 88 equivalent ECAs. We find out that there are 28 ECAs, those have the universality property. That is, these 28 ECAs are capable to realize any logic gates. At the end of the study, we compare the gate generation capability of these 28 ECAs. We find out that, with respect to less number of bits required to generate all the logic gates, the ECA 57 is the best ECA rule. Similarly, ECAs 73, 106 and 134 are the best ECA rules in order to generate all the logic gates with respect to less time evolution.

The next section introduces the basics of ECAs, and consider an $n$-cell rule $\mathcal{R}$ CA as "$\mathcal{R}$ block". In Sect. 3.1, we explore the gate generation capability of all the ECAs. Finally, in Sect. 3.2 report a comparison of gate design capability of ECAs.

## 2 Elementary Cellular Automata

### *2.1 Basics*

An elementary cellular automaton is a one-dimensional array of finite-state automata. The automata take two states, 0 and 1, and update their states simultaneously in discrete time by a function, called next state function $\mathcal{R} : \{0, 1\}^3 \rightarrow \{0, 1\}$. Each automaton updates its state depending on its current state and the states of its two

**Table 1** Look-up table for ECA rules 57 and 184

| PS: | 111 | 110 | 101 | 100 | 011 | 010 | 001 | 000 | Rule |
|---|---|---|---|---|---|---|---|---|---|
| (RMT) | (7) | (6) | (5) | (4) | (3) | (2) | (1) | (0) | |
| (i) NS: | 0 | 0 | 1 | 1 | 1 | 0 | 0 | 1 | 57 |
| (ii) NS: | 1 | 0 | 1 | 1 | 1 | 0 | 0 | 0 | 184 |

closest neighbors (left and right). Therefore, in an ECA, next state of a cell $i$ is determined as:

$$S_i^{t+1} = \mathcal{R}(S_{i-1}^t, S_i^t, S_{i+1}^t) \tag{1}$$

where $\mathcal{R}$ is the next state function; $S_{i-1}^t$, $S_i^t$ and $S_{i+1}^t$ are the present states of the left neighbor, self and right neighbor of the $i^{th}$ cell at time $t$. The function $\mathcal{R} : \{0, 1\}^3 \mapsto \{0, 1\}$ can be expressed as a look-up table (see Table 1). A state of a cell with its three neighbors can form $2^3 = 8$ possible patterns. The decimal equivalent of the 8 possible outputs is call 'rule' of CAs [11]. There are $2^{2^3} = 256$ possible ECA rules. Two such rules 57 and 184 are shown in Table 1. First row of the table shows the possible combinations of present states (PSs) of left, self and right neighbors of a cell. Whereas, third and fourth rows show the next states (NSs) of corresponding PSs. The last column notes the rules.

Some researchers identified that CA rules have equivalences [5, 13]. Consequently, a similar behaviour can be projected and organised in set of rules, and it is usually convenient to consider only the minimal (lowest-numbered) representatives of each set. So, in ECAs, there are 88 minimal representative rules. In this work, we only consider these minimal rules.

## 2.2 *ECA as n-Input n-Output Block*

We have used CAs as a computational circuit block. We consider an $n$-cell rule $\mathcal{R}$ CA as "$\mathcal{R}$ block". The initial configurations of CAs are the input to the block. A CA block maps $n$ inputs $I_1, I_2, \cdots, I_n$ to $n$ outputs $O_1, O_2, \cdots, O_n$. A representative of $n$-input $n$-output CA block is shown in Fig. 1.

For example, let us consider a 3-input 3-output rule 57 block, shown in Fig. 2. Now we can express the output variables of 3-input 3-output rule 57 block as explicit

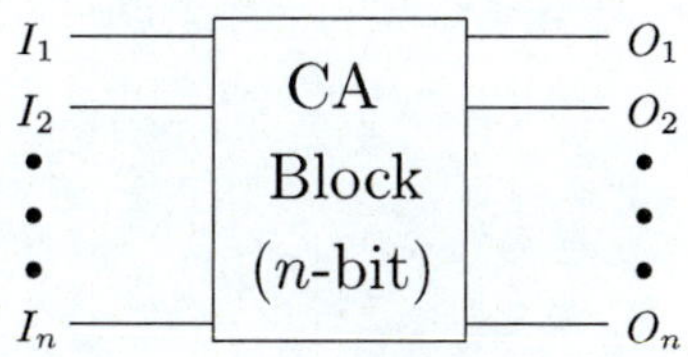


**Fig. 1** $n$-input $n$-output CA block

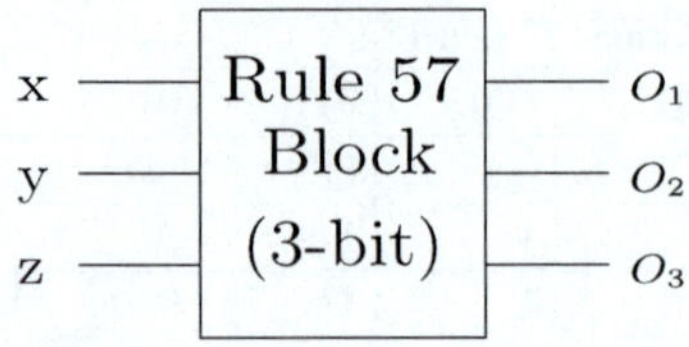


**Fig. 2** 3-input 3-output rule 57 block

| $I_1$ | $I_2$ | $I_3$ | $O_1$ | $O_2$ | $O_3$ |
|---|---|---|---|---|---|
| 0 | 0 | 0 | 1 | 1 | 1 |
| 0 | 0 | 1 | 1 | 0 | 0 |
| 0 | 1 | 0 | 0 | 0 | 1 |
| 0 | 1 | 1 | 1 | 1 | 0 |
| 1 | 0 | 0 | 0 | 1 | 0 |
| 1 | 0 | 1 | 0 | 1 | 1 |
| 1 | 1 | 0 | 1 | 0 | 1 |
| 1 | 1 | 1 | 0 | 0 | 0 |

Initial Input Output

(a) After $1^{st}$ evolution

| $I_1$ | $I_2$ | $I_3$ | $O_1$ | $O_2$ | $O_3$ |
|---|---|---|---|---|---|
| 0 | 0 | 0 | 0 | 0 | 0 |
| 0 | 0 | 1 | 0 | 1 | 0 |
| 0 | 1 | 0 | 1 | 0 | 0 |
| 0 | 1 | 1 | 1 | 0 | 1 |
| 1 | 0 | 0 | 0 | 0 | 1 |
| 1 | 0 | 1 | 1 | 1 | 0 |
| 1 | 1 | 0 | 0 | 1 | 1 |
| 1 | 1 | 1 | 1 | 1 | 1 |

Initial Input Output

(b) After $2^{nd}$ evolution

**Fig. 3** Truth table of rule 57 for 3-cell

functions of input variables. To do that, we first evolve the CA and build the truth table. After 1st evolution we get the truth table, which is shown in Fig. 3a and output can be written as following:

$$\begin{aligned} O_1 &= xyz' + x'z + x'y' \\ O_2 &= x'yz + xy' + y'z' \\ O_3 &= xy'z + yz' + x'z' \end{aligned} \tag{2}$$

Similarly, after $2^{nd}$ evolution again we get the truth table, which is shown in Fig. 3b and output can be written as following:

$$\begin{aligned} O_1 &= xz + x'y \\ O_2 &= xy + y'z \\ O_3 &= yz + xz' \end{aligned} \tag{3}$$

In this way we can find out the outputs of any $n$-input $n$-output CA block. These output functions are further used to design the logic gates.

## 3 Exploring Gate Design Capability of ECAs

In this section, we explore the logic gate generation capability through evolution of the ECAs rules. As we have discussed earlier, that the dynamic behavior of equivalent ECAs are same, so here we consider only 88 minimal rules. To explore the potential in logic gate design of all the minimal ECAs rules, we first evolve the CAs and build the truth tables for each evolution. After that we get the output function from those truth tables. Using the output functions, we realize the logic gates.

### *3.1 Realization of Basic Gates*

By proper selection of inputs, it might be possible to implement AND, OR, NOT, FAN-OUT, NAND, NOR, XOR and XNOR by an $n$-input $n$-output CA block. In this section, we target to realize the basic gates by all the 88 minimal ECA rules. During this realization, we take the desired output from output $O_1$. However, one can take the desired output from outputs other than $O_1$, but then she needs to set input variable in different way. In this section, the input settings are reported only to consider output from $O_1$.

During the realization of logic gates using rule 184 block (see [4]), we have seen that 3-input 3-output 184 block has been used to realize AND, OR, NOT and FAN-OUT. Again, to realize NOR, NAND, XOR and XNOR we have used 5-input 5-output 184 block. So, to realize different logic gates we may need to used different input size CA block. For 3-input CA block we marked the inputs as $x$, $y$ and $z$. Similarly, for 4-input CA block, the inputs are marked as $w$, $x$, $y$, $z$ and for 5-input CA block we marked the inputs as $v$, $w$, $x$, $y$, $z$ and so on.

However, here we present an example of implementation of logic gates considering ECA rule 57. To implement the logic gates using rule 57 block, we set some input in such a way that after some evolution get the desire output.

For realizing NOT, FAN-OUT, XOR and XNOR we consider 3-input 3-output 57 block and need to evolve the CA single time step. Let us set the input $x$ and $z$ with logic 0. Then after $1^{st}$ evolution, output $O_1$ results in $y'$. That is, under this setting $O_1$ is the output of NOT (NOT $y$). This realization is shown in Fig. 4a. Similarly, if we set input $x$ and $z$ with 1, 0 respectively then $O_1$ is the output of FAN-OUT. Fig. 4b shows the realization FAN-OUT. Again if we set input $y$ with logic 1 then output $O_1$ is the output of the XOR of two inputs $x$ and $z$ ($x \oplus z$). Similarly, if we set input $z$ with logic 0 then output $O_1$ is the output of the XNOR of the two inputs $x$ and $y$ ($x \odot y$). Fig. 4c shows the realization XOR and Fig. 4d shows the realization XNOR.

Apart from the above gates, we can also design AND and OR using 3-input 3-output 57 block. To realize AND and OR, we need to evolve the CA for two time steps. Let us set the input $y$ with logic 0 then after $2^{nd}$ evolution output $O_1$ becomes $xz$. That is, under this setting $O_1$ is the output of two input AND. Fig. 4e shows this

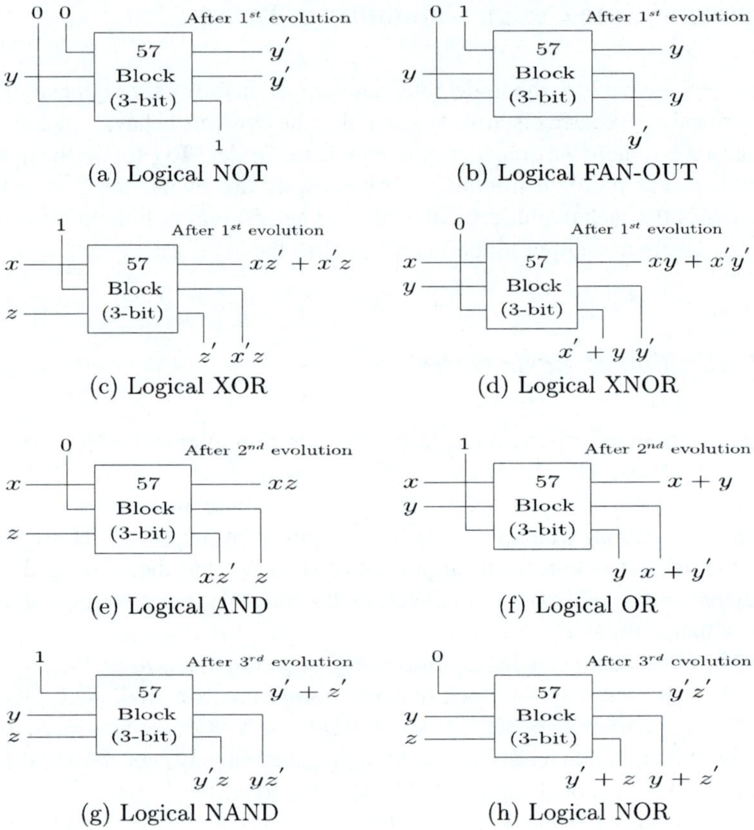


(a) Logical NOT (b) Logical FAN-OUT

(c) Logical XOR (d) Logical XNOR

(e) Logical AND (f) Logical OR

(g) Logical NAND (h) Logical NOR

**Fig. 4** Implementation of logic gates using 3-input 3-output 57 block

realization. Similarly, if we set input $z$ with logic 1 then output $O_1$ is the output of the OR of the two inputs $x$ and $y$. This realization is shown in Fig. 4f.

To realize the universal logic gates, the NAND and NOR, we also used 3-input 3-output 57 block but we need to evolve the CA for three time steps. Let us set input $x$ with logic 1 then after $3^{rd}$ evolution output $O_1$ results in $y^{'} + z^{'}$. That is, under this setting, $O_1$ is the output of two input NAND. This realization is shown in Fig. 4g. Similarly, if we set input $x$ with logic 0 then output $O_1$ is the output of the NOR of the two input $y$ and $z$. Fig. 4h shows this realization.

In this way, by proper selection of inputs and after some evolution we can realize all the logic gates. We observed that all the ECAs rules are not capable of design all the logic gates, such as NOT, FAN-OUT, AND, OR, NAND, NOR, XOR and XNOR. We find out that, out of 88 minimal ECA rules, there are 28 ECA rules, which are capable to realize any logic gates. These rules are listed in Table 2.

Next, in Table 3, we report the realization of basic gates function along with input setting for the ECA rules which are noted in Table 2.

**Table 2** List of rules those are capable to design any logic gate

| | | | | | | | | | |
|---|---|---|---|---|---|---|---|---|---|
| 14 | 25 | 26 | 28 | 30 | 35 | 37 | 41 | 43 | 45 |
| 54 | 57 | 58 | 62 | 73 | 74 | 94 | 104 | 106 | 110 |
| 122 | 126 | 134 | 142 | 146 | 154 | 156 | 184 | | |

### *3.2 Comparison of Gate Design Capability of ECAs*

In this section, we compare the logic gate generation capability of 28 minimal ECA rules which are reported in Table 2. Now, when we compare the logic gates generation capability of the rules, we consider two cases—*average case* and *worst case*. In average case, we consider two parameters—one is total number of bits required to realize all the logic gates and another one is the total number of time evolution needed to design all the logic gates. Similarly, in worst case, we also consider two parameter—one is the maximum number of bits required and another one is maximum number of evolution needed to design the gates. During the comparison of gate generation capability of ECAs, in average case, we find that with respect to less number of bits required to generate all the logic gates, the ECA 57 is the best ECA rule. Similarly, ECAs 73, 106 and 134 are the best ECA rules in order to generate all the logic gates with respect to less time evolution. On the other hand, in worst case, we find that with respect to maximum bits required to design logic gate, the ECA 57 is the best ECA rule. Similarly, ECAs 41, 62, 73, 94, 104, 106, 110, 122, 126, 134 and 156 are the best ECA rules in order to design logic gates with respect to less number of time evolution. The summary of average case and worst case for all the 28 ECA rules are reported in Table 4.

## 4 Conclusion

This work has shown the gate generation capability of all the elementary cellular automata rules. At the beginning of this study, we have explored the gate generation capability of 88 minimal ECAs rules. After that, we have compared the gate generation capability of these ECAs rules. We have shown that, out of 88 minimal rules, the 28 rules are capable to implement all the logic gates.

**Table 3** Realization of basic gates using ECA rules

| Rule | Block Size | # Evolution | Output function ($O_1$) | Realization of gates | |
|---|---|---|---|---|---|
| | | | | Gates | Input setting |
| **14** | 3 | 1 | $yz' + xz'$ | FAN-OUT | $y = 0, z = 0$ |
| | | | | NOT | $x = 0, y = 1$ |
| | | | | OR | $z = 0$ |
| | | 2 | $x'y + yz'$ | NAND | $y = 1$ |
| | 4 | 2 | $w'xz' + w'y + yz'$ | NOR | $x = 1, y = 0$ |
| | | 3 | $w'yz + x'z + x'y$ | AND | $w = 0, x = 1$ |
| | 5 | 5 | $v'wx'y + wx'z + vwy' + v'y'z + vx'y'$ | XOR | $w = 1, x = 0, z = 0$ |
| | | 6 | $v'wx'z + vx'y'z' + xy'z + wxz' + wy'z'$ | XNOR | $v = 1, w = 0, y = 0$ |

(continued)

**Table 3** (continued)

| Rule | Block Size | # Evolution | Output function ($O_1$) | Realization of gates | |
|---|---|---|---|---|---|
| | | | | Gates | Input setting |
| **25** | 3 | 1 | $xyz' + x'y'$ | NOT | $y = 0$ |
| | | | | FAN-OUT | $y = 1, z = 0$ |
| | | | | XNOR | $z = 0$ |
| | | | | NOR | $z = 1$ |
| | | 2 | $x'yz' + xz$ | AND | $y = 0$ |
| | 4 | 2 | $wxz + wx'y + w'y'z + w'xy'$ | OR | $w = 0, y = 0$ |
| | | 3 | $w'x'y'z' + w'yz + w'xy + wy'z + wxy'$ | XOR | $x = 1, z = 0$ |
| | 5 | 5 | $vw'xy + v'xz + v'y'z + v'xy' + v'w'y' + vyz'$ | NAND | $v = 1, x = 1, y = 1$ |
| **26** | 3 | 1 | $x'y'z + yz'$ | FAN-OUT | $z = 0$ |
| | | | | NOT | $y = 0, z = 1$ |
| | | | | XOR | $x = 0$ |
| | | | | NOR | $z = 1$ |
| | | 2 | $x'z + y'z$ | NAND | $z = 1$ |
| | 4 | 2 | $wxy'z + w'yz$ | AND | $w = 0$ |
| | 5 | 3 | $v'w'x'y'z + vw'xy + wx'yz + vwx'y' + v'xyz' + vwy'z' + wx'y'z' + w'yz'$ | XNOR | $x = 0, y = 0, z = 1$ |
| | | 4 | $v'wxy + x'yz + xy'z + w'y'z + v'z$ | OR | $v = 0, x = 1, y = 1$ |

(continued)

**Table 3** (continued)

| Rule | Block Size | # Evolution | Output function ($O_1$) | Realization of gates | |
|---|---|---|---|---|---|
| | | | | Gates | Input setting |
| **28** | 3 | 1 | $x'y'z + xz'$ | FAN-OUT | $z = 0$ |
| | | | | NOT | $y = 0, z = 1$ |
| | | | | NOR | $z = 1$ |
| | | | | XOR | $y = 0$ |
| | | 2 | $xy' + xz'$ | NAND | $x = 1$ |
| | 4 | 2 | $w'x'y + wxy' + wz'$ | AND | $y = 0, z = 1$ |
| | | 3 | $x'y + wz'$ | OR | $x = 0, z = 0$ |
| | 5 | 3 | $v'w'xz + vwx'y + v'w'x'y + w'xy'z' + vz'$ | XNOR | $x = 0, y = 1, z = 1$ |
| **30** | 3 | 1 | $x'y'z + yz' + xz'$ | OR | $z = 0$ |
| | | | | FAN-OUT | $x = 0, z = 0$ |
| | | | | NOT | $y = 0, z = 1$ |
| | | | | XOR | $y = 0$ |
| | | | | NOR | $z = 1$ |
| | | 2 | $x'yz + xy'z + xyz'$ | AND | $z = 0$ |
| | 4 | 2 | $wxy'z + w'xy'z' + w'yz + wyz'$ | XNOR | $w = 0, x = 1$ |
| | | 3 | $wx'y' + w'xz' + wx'z' + w'y$ | NAND | $w = 1, x = 0$ |

(continued)

**Table 3** (continued)

| Rule | Block Size | # Evolution | Output function ($O_1$) | Realization of gates | |
|---|---|---|---|---|---|
| | | | | Gates | Input Setting |
| **35** | 3 | 1 | $x'y + x'z'$ | FAN-OUT | $x = 0, z = 1$ |
| | | | | NOT | $y = 1, z = 1$ |
| | | | | NOR | $y = 0$ |
| | | 2 | $xz + y'z$ | AND | $y = 1$ |
| | 4 | 2 | $wx'y + wz + x'z$ | OR | $w = 1, x = 0$ |
| | | 3 | $w'xy' + wxz' + y'z'$ | NAND | $x = 1, y = 0$ |
| | 5 | 4 | $v'wy'z + vyz + v'xy+$ $w'xz + v'w'y$ | XNOR | $w = 1, x = 0, z = 1$ |
| | | | | XOR | $v = 0, x = 0, z = 1$ |
| **37** | 3 | 1 | $x'yz + y'z'$ | FAN-OUT | $x = 0, z = 1$ |
| | | | | NOT | $y = 1, z = 1$ |
| | | | | XNOR | $x = 0$ |
| | | | | NOR | $x = 1$ |
| | | 2 | $xy'z' + yz$ | AND | $x = 0$ |
| | 4 | 2 | $wxz + x'y'z + xy'z'+$ $w'y + wy'$ | XOR | $w = 0, y = 0$ |
| | | | | OR | $y = 0, z = 0$ |
| | | 3 | $w'x'y'z' + w'xz+$ $wx'y + xy'z + wyz'$ | NAND | $x = 1, z = 1$ |

(continued)

**Table 3** (continued)

| Rule | Block Size | # Evolution | Output function ($O_1$) | Realization of gates | |
|---|---|---|---|---|---|
| | | | | Gates | Input setting |
| **41** | 3 | 1 | $x'yz + xyz' + x'y'z'$ | FAN-OUT | $y = 1, z = 0$ |
| | | | | NOT | $y = 1, z = 1$ |
| | | | | XNOR | $z = 0$ |
| | | | | XOR | $y = 1$ |
| | | 2 | $x'y + xy' + z$ | OR | $y = 0$ |
| | 4 | 2 | $xyz + w'yz + wx'z'$ | AND | $w = 1, z = 1$ |
| | | | | NOR | $w = 1, y = 0$ |
| | 5 | 2 | $vw'xyz' + vw'x'y'z' + vwyz + vx'yz + v'wxy' + w'xy'z + v'x'yz' + v'xz$ | NAND | $w = 0, x = 1, z = 1$ |
| **43** | 3 | 1 | $x'y + yz' + x'z'$ | FAN-OUT | $x = 0, z = 1$ |
| | | | | NOT | $y = 1, z = 1$ |
| | | | | NOR | $y = 0$ |
| | | | | NAND | $y = 1$ |
| | 4 | 2 | $w'x'z + yz + w'y + x'y$ | AND | $w = 1, x = 1$ |
| | 5 | 2 | $vyz + v'xz + v'xy + vw'x + v'w'z + v'w'y$ | OR | $v = 1, w = 0, z = 1$ |
| | | 5 | $vx'yz' + v'xy + vw'x + v'w'x' + v'xz'$ | XOR | $w = 1, y = 1, z = 0$ |
| | | | | XNOR | $w = 0, y = 0, z = 1$ |

(continued)

**Table 3** (continued)

| Rule | Block Size | # Evolution | Output function ($O_1$) | Realization of gates | |
|---|---|---|---|---|---|
| | | | | Gates | Input setting |
| **45** | 3 | 1 | $x'yz + xz' + y'z'$ | FAN-OUT | $y = 1, z = 0$ |
| | | | | NOT | $y = 1, z = 1$ |
| | | | | XOR | $y = 1$ |
| | | | | XNOR | $x = 0$ |
| | | 2 | $yz + xy'$ | AND | $x = 0$ |
| | | 3 | $xy' + xz' + y'z'$ | NAND | $x = 1$ |
| | | | | NOR | $x = 0$ |
| | 4 | 2 | $xyz + w'x'z + w'x'y + xy'z' + wy'z'$ | OR | $y = 0, z = 0$ |
| **54** | 3 | 1 | $xy'z' + x'z + x'y$ | FAN-OUT | $y = 0, z = 0$ |
| | | | | NOT | $y = 0, z = 1$ |
| | | | | OR | $x = 0$ |
| | | | | NOR | $x = 1$ |
| | | | | XOR | $z = 0$ |
| | | 2 | $x'yz + xy'z + xyz'$ | AND | $z = 0$ |
| | 4 | 2 | $w'xyz + w'x'yz' + wx'z + wxz'$ | XNOR | $y = 1, z = 0$ |
| | | 4 | $wx'z + wxz' + wy'$ | NAND | $w = 1, z = 1$ |

(continued)

**Table 3** (continued)

| Rule | Block Size | # Evolution | Output function ($O_1$) | Realization of gates | |
|---|---|---|---|---|---|
| | | | | Gates | Input setting |
| **57** | 3 | 1 | $xyz' + x'z + x'y'$ | NOT | $x = 0, z = 0$ |
| | | | | FAN-OUT | $x = 1, z = 0$ |
| | | | | XOR | $y = 1$ |
| | | | | XNOR | $z = 0$ |
| | | 2 | $xz + x'y$ | AND | $y = 0$ |
| | | | | OR | $z = 1$ |
| | | 3 | $xy' + xz' + y'z'$ | NAND | $x = 1$ |
| | | | | NOR | $x = 0$ |
| **58** | 3 | 1 | $x'z + yz'$ | NOT | $z = 1$ |
| | | | | FAN-OUT | $z = 0$ |
| | | | | OR | $x = 0$ |
| | 4 | 2 | $w'yz + wy'z + x'y + wx'$ | AND | $w = 0, x = 1$ |
| | | | | NAND | $y = 1, z = 1$ |
| | | | | XOR | $x = 1, z = 1$ |
| | | 4 | $wx'z' + w'y + x'y + yz'$ | NOR | $w = 1, y = 0$ |
| | 5 | 3 | $vw'xy + v'x'z + vwz' + wx' + yz'$ | XNOR | $w = 0, y = 1, z = 1$ |

(continued)

**Table 3** (continued)

| Rule | Block Size | # Evolution | Output function ($O_1$) | Realization of gates | |
|---|---|---|---|---|---|
| | | | | Gates | Input setting |
| **62** | 3 | 1 | $x'z + yz' + xz'$ | FAN-OUT | $y = 0, z = 0$ |
| | | | | NOT | $z = 1$ |
| | | | | OR | $z = 0$ |
| | | | | XOR | $y = 0$ |
| | | 2 | $x'yz + xy'z$ | AND | $y = 0$ |
| | 4 | 2 | $w'xy'z' + w'yz+$ $wx'z + w'x'y+$ $wy'z$ | NOR | $x = 1, z = 0$ |
| | | | | NAND | $w = 1, z = 1$ |
| | | | | XNOR | $w = 0, x = 1$ |
| **73** | 3 | 1 | $xy'z + xyz' + x'y'z'$ | FAN-OUT | $y = 0, z = 1$ |
| | | | | NOT | $y = 0, z = 0$ |
| | | | | XNOR | $z = 0$ |
| | | | | XOR | $x = 1$ |
| | | | | NOR | $x = 0$ |
| | | 2 | $y'z + yz' + x$ | OR | $z = 0$ |
| | 4 | 2 | $wxyz + w'xy'z +$ $wx'y' + wx'z' +$ $wy'z'$ | AND | $y = 1, z = 1$ |
| | | | | NAND | $w = 1, z = 0$ |

(continued)

**Table 3** (continued)

| Rule | Block Size | # Evolution | Output function ($O_1$) | Realization of gates | |
|---|---|---|---|---|---|
| | | | | Gates | Input setting |
| **74** | 3 | 1 | $xy'z + yz'$ | NOT | $x = 1, z = 1$ |
| | | | | FAN-OUT | $y = 0, z = 1$ |
| | | | | AND | $y = 0$ |
| | | | | XOR | $x = 1$ |
| | | 3 | $xy' + xz'$ | NAND | $x = 1$ |
| | 4 | 2 | $wx'y'z + w'yz'$ | NOR | $w = 1, z = 1$ |
| | 5 | 2 | $vwx'y + vw'y'z + wxy'z' + v'xy' + v'xz'$ | XNOR | $v = 1, x = 0, z = 1$ |
| | | 3 | $v'w'xy + vxy'z + v'wxz' + wxy' + w'yz'$ | OR | $v = 1, x = 1, y = 0$ |
| **94** | 3 | 1 | $y'z + yz' + xz'$ | FAN-OUT | $y = 0, z = 0$ |
| | | | | NOT | $x = 1, y = 1$ |
| | | | | OR | $z = 0$ |
| | | | | XOR | $x = 0$ |
| | | 2 | $xy'z + xyz'$ | AND | $z = 0$ |
| | 4 | 2 | $w'x'y'z + w'xy'z' + wx'y + wyz'$ | NAND | $w = 1, y = 1$ |
| | | | | NOR | $y = 0, z = 1$ |
| | | | | XNOR | $x = 1, z = 0$ |

(continued)

**Table 3** (continued)

| Rule | Block Size | # Evolution | Output Function ($O_1$) | Realization of Gates | |
|---|---|---|---|---|---|
| | | | | Gates | Input Setting |
| **104** | 3 | 1 | $x'yz + xy'z + xyz'$ | FAN-OUT | $y = 0, z = 1$ |
| | | | | NOT | $y = 1, z = 1$ |
| | | | | AND | $z = 0$ |
| | | | | XOR | $z = 1$ |
| | 5 | 2 | $v'wx'yz + v'wxy'z+$ $vw'xy + vx'y'z+$ $vwx'y' + vxyz'$ | XNOR | $v = 1, w = 0, z = 1$ |
| | | | | OR | $v = 1, x = 0, y = 0$ |
| | | | | NAND | $v = 1, x = 1, y = 1$ |
| | | | | NOR | $w = 1, y = 1, z = 1$ |
| **106** | 3 | 1 | $xy'z + x'y + yz'$ | FAN-OUT | $y = 0, z = 1$ |
| | | | | NOT | $y = 1, z = 1$ |
| | | | | XOR | $z = 1$ |
| | | | | AND | $y = 0$ |
| | | | | NAND | $y = 1$ |
| | | 2 | $x'y'z$ | NOR | $z = 1$ |
| | 4 | 2 | $wy'z + x'y + yz'$ | OR | $x = 0, z = 1$ |
| | 5 | 2 | $wx'yz + vx'y'z +$ $w'xy + v'xy' + xz'$ | XNOR | $v = 1, w = 0, z = 1$ |

(continued)

**Table 3** (continued)

| Rule | Block Size | # Evolution | Output Function ($O_1$) | Realization of Gates | |
|---|---|---|---|---|---|
| | | | | Gates | Input Setting |
| **110** | 3 | 1 | $x'y + xy' + xz'$ | FAN-OUT | $y = 0, z = 0$ |
| | | | | NOT | $y = 1, z = 1$ |
| | | | | XOR | $z = 1$ |
| | | | | NAND | $x = 1$ |
| | | 2 | $x'y'z + x'yz' + xy'z'$ | NOR | $z = 1$ |
| | 4 | 2 | $wx'z + w'x'y + wy'z + wx'y' + xyz' + w'xz'$ | XNOR | $w = 1, z = 0$ |
| | 5 | 2 | $vw'yz + v'w'x + vx'z + vw'x' + wxz' + v'wz'$ | AND | $v = 1, w = 0, x = 1$ |
| | | | | OR | $w = 0, y = 1, z = 1$ |
| **122** | 3 | 1 | $x'y + y'z + yz'$ | FAN-OUT | $y = 0$ |
| | | | | NOT | $y = 1, z = 1$ |
| | | | | XOR | $x = 1$ |
| | | | | NAND | $y = 1$ |
| | | 2 | $x'y'z + x'yz' + xy'z'$ | NOR | $z = 1$ |
| | 4 | 2 | $wxy'z + wx'y + wyz' + x'yz' + wx'z'$ | AND | $y = 0, z = 1$ |
| | | | | XNOR | $w = 1, y = 0$ |
| | | | | OR | $x = 0, z = 0$ |

(continued)

**Table 3** (continued)

| Rule | Block Size | # Evolution | Output Function ($O_1$) | Realization of Gates | |
|---|---|---|---|---|---|
| | | | | Gates | Input Setting |
| **126** | 3 | 1 | $x'y + y'z + xz'$ | FAN-OUT | $y = 0, z = 0$ |
| | | | | NOT | $y = 1, z = 1$ |
| | | | | NAND | $z = 1$ |
| | | | | OR | $y = 0$ |
| | 4 | 2 | $wx'yz + wxy'z+$ $w'x'y'z + wxyz'+$ $w'x'yz' + w'xy'z'$ | XOR | $w = 1, z = 1$ |
| | | | | XNOR | $y = 0, z = 1$ |
| | | | | AND | $w = 1, x = 0$ |
| | | | | NOR | $w = 0, x = 1$ |
| **134** | 3 | 1 | $xyz + x'yz' + xy'z'$ | FAN-OUT | $y = 1, z = 1$ |
| | | | | NOT | $y = 1, z = 0$ |
| | | | | AND | $z = 1$ |
| | | | | XNOR | $x = 1$ |
| | | | | XOR | $z = 0$ |
| | 4 | 2 | $wxz + w'x'y + yz'$ | NAND | $w = 0, y = 1$ |
| | | | | OR | $w = 1, x = 1$ |
| | 5 | 2 | $v'wx'yz + vwx'y'z+$ $vwxy + v'w'xy+$ $vw'yz' + v'xz'$ | NOR | $x = 1, y = 0$ |

(continued)

**Table 3** (continued)

| Rule | Block Size | # Evolution | Output Function ($O_1$) | Realization of Gates | |
|---|---|---|---|---|---|
| | | | | Gates | Input setting |
| **142** | 3 | 1 | $xy + yz' + xz'$ | FAN-OUT | $y = 1, z = 1$ |
| | | | | NOT | $x = 1, y = 0$ |
| | | | | OR | $z = 0$ |
| | | | | AND | $z = 1$ |
| | 4 | 2 | $w'xz' + xy + w'y + yz'$ | NAND | $x = 0, y = 1$ |
| | | | | NOR | $x = 1, y = 0$ |
| | 5 | 5 | $v'wx'y + vyz + vwy' + v'y'z + vx'y'$ | XNOR | $w = 0, x = 1, z = 1$ |
| | | | | XOR | $w = 1, x = 0, z = 0$ |
| **146** | 3 | 1 | $xyz + x'y'z + x'yz'$ | FAN-OUT | $y = 1, z = 1$ |
| | | | | NOT | $y = 0, z = 1$ |
| | | | | XNOR | $z = 1$ |
| | | | | AND | $x = 1$ |
| | | | | XOR | $x = 0$ |
| | 4 | 2 | $wxy + wx'z + wxz'$ | OR | $w = 1, y = 1$ |
| | 5 | 2 | $vw'x'y'z + vwx'y'z' + vxyz + vwxy + w'x'yz' + w'xy'z'$ | NOR | $v = 1, y = 0, z = 1$ |
| | | 3 | $vwyz + vwxz + v'w'x'z + v'w'y'z + v'wx'z' + v'wy'z'$ | NAND | $v = 0, w = 1, z = 0$ |

(continued)

**Table 3** (continued)

| Rule | Block Size | # Evolution | Output Function ($O_1$) | Realization of Gates | |
|---|---|---|---|---|---|
| | | | | Gates | Input Setting |
| **154** | 3 | 1 | $x'y'z + xy + yz'$ | FAN-OUT | $y = 1, z = 1$ |
| | | | | NOT | $y = 0, z = 1$ |
| | | | | XOR | $x = 0$ |
| | | | | XNOR | $z = 1$ |
| | 4 | 2 | $yz + xy$ | OR | $y = 1$ |
| | | | | AND | $z = 0$ |
| | 5 | 3 | $v'w'x'y'z + v'wyz + vwx'y' + wx'y'z' + vw'y + w'yz' + xy$ | NOR | $v = 0, y = 0, z = 1$ |
| | 6 | 3 | $u'vxyz + u'v'w'x'z + u'vw'y'z' + uvw'x' + v'xy'z + v'xyz' + vw'x'z' + uv'x + wx$ | NAND | $u = 0, v = 1, w = 0$<br>$z = 0$ |
| **156** | 3 | 1 | $x'y'z + xy + xz'$ | FAN-OUT | $y = 1, z = 1$ |
| | | | | NOT | $y = 0, z = 1$ |
| | | | | XOR | $y = 0$ |
| | | | | XNOR | $z = 1$ |
| | 4 | 2 | $wy + x'y + wz'$ | AND | $x = 1, z = 1$ |
| | | | | OR | $x = 0, z = 0$ |
| | 5 | 2 | $v'w'y + vx + vz'$ | NOR | $x = 0, y = 1, z = 1$ |
| | | | | NAND | $w = 0, x = 0, y = 1$ |

(continued)

**Table 3** (continued)

| Rule | Block Size | # Evolution | Output Function ($O_1$) | Realization of Gates | |
|---|---|---|---|---|---|
| | | | | Gates | Input Setting |
| **184** | 3 | 1 | $xy + x^{'}z$ | FAN-OUT | $y = 1, z = 0$ |
| | | | | NOT | $y = 0, z = 1$ |
| | | | | AND | $z = 0$ |
| | | | | OR | $y = 1$ |
| | 5 | 2 | $wxz + vwx + vw^{'}z + v^{'}yz^{'} + w^{'}yz^{'}$ | NAND | $x = 0, y = 1, z = 0$ |
| | | | | NOR | $w = 1, x = 0, y = 1$ |
| | | 4 | $v^{'}w^{'}xy + vwy^{'}z^{'} + vyz + vxz + v^{'}wz + v^{'}wx^{'}$ | XOR | $v = 0, y = 1, z = 0$ |
| | | | | XNOR | $v = 1, w = 1, x = 0$ |

**Table 4** Comparison of ECAs in respect of average case and worst case

| | Average case | | Worst case | |
|---|---|---|---|---|
| ECAs | Total bits | Total Evolution | Maximum bits | Maximum Evolution |
| **14** | 30 | 21 | 5 | 6 |
| **25** | 28 | 16 | 5 | 5 |
| **26** | 29 | 15 | 5 | 4 |
| **28** | 28 | 14 | 5 | 3 |
| **30** | 26 | 12 | 4 | 3 |
| **35** | 30 | 18 | 5 | 4 |
| **37** | 27 | 13 | 4 | 3 |
| **41** | 28 | 12 | 5 | 2 |
| **43** | 31 | 18 | 5 | 5 |
| **45** | 25 | 14 | 4 | 3 |
| **54** | 26 | 13 | 4 | 4 |
| **57** | 24 | 14 | 3 | 3 |
| **58** | 30 | 16 | 5 | 4 |
| **62** | 27 | 12 | 4 | 2 |
| **73** | 26 | 11 | 4 | 2 |
| **74** | 29 | 14 | 5 | 3 |
| **94** | 27 | 12 | 4 | 2 |
| **104** | 32 | 12 | 5 | 2 |
| **106** | 27 | 11 | 5 | 2 |
| **110** | 29 | 12 | 5 | 2 |
| **122** | 27 | 12 | 4 | 2 |
| **126** | 28 | 12 | 4 | 2 |
| **134** | 28 | 11 | 5 | 2 |
| **142** | 30 | 18 | 5 | 5 |
| **146** | 29 | 12 | 5 | 3 |
| **154** | 31 | 14 | 6 | 3 |
| **156** | 30 | 12 | 5 | 2 |
| **184** | 32 | 16 | 5 | 4 |

# References


1. Bidlo, M., Vašícek, Z.: Gate-level evolutionary development using cellular automata. In: NASA/ESA Conference on Adaptive Hardware and Systems, AHS'08. pp. 11–18. IEEE (2008)
2. Chopard, B., Droz, M.: Cellular Automata Modelling of Physical Systems. Cambridge University Press, Cambridge (1998)
3. Cook, M.: Universality in elementary cellular automata. Complex Syst. **15**(1), 1–40 (2004)
4. Hazari, R., Kundu, S., Bhardwaj, M., Das, S.: ECA 184 can implement any logic circuits. J. Cell. Autom. **13**(4), 359–371 (2018)

5. Li, W., Packard, N.: The structure of the elementary cellular automata rule space. Complex Syst. **4**(3), 281–297 (1990)
6. Martínez, G.J., Adamatzky, A., Morita, K.: Logical gates via gliders collisions. In: Reversibility and Universality, pp. 199–220. Springer, Berlin (2018)
7. Martínez, G.J., Adamatzky, A., McIntosh, H.V.: Phenomenology of glider collisions in cellular automaton rule 54 and associated logical gates. Chaos, Solitons Fractals **28**(1), 100–111 (2006)
8. Mitchell, M.: Computation in cellular automata: A selected review. In: Non-Standard Computation. Wiley-VCH Verlag GmbH & Co. KGaA, Weinheim (1996)
9. Neary, T., Woods, D.: P-completeness of cellular automaton rule 110. In: Proceedings of the 33rd International Colloquium on Automata, Languages and Programming - Volume Part I, ICALP 2006, pp. 132–143. Springer, berlin (2006)
10. Pries, W., Thanailakis, A., Card, H.C.: Group properties of cellular automata and VLSI applications. IEEE Trans. Comput. **C-35**(12), 1013–1024 (1986)
11. Wolfram, S.: Statistical mechanics of cellular automata. Rev. Mod. Phys. **55**(3), 601–644 (1983)
12. Wolfram, S.: Universality and complexity in cellular automata. Phys. D **10**(1–2), 1–35 (1984)
13. Wolfram, S.: Cellular Automata and Complexity: Collected Papers. Westview Press (1994)
14. Wolfram, S.: A New Kind of Science. Wolfram-Media Inc., Champaign (2002)

# Impact of Genetic Algorithm on Low Power QCA Logic Circuit with Regular Clocking

**Amit Kumar Pramanik, Jayanta Pal, and Bibhash Sen**

**Abstract** Quantum-dot Cellular Automata (QCA) is the alternative approach to synthesize the logic circuit with high density and low power dissipation to overcome the limitation of current VLSI technology. The underlying regular clocking scheme plays a significant role in the systematic cell layout, scalability, and reliability of the QCA circuit. This work analyzes the impact of genetic algorithm (GA) on regular clock based QCA circuits for less power dissipation. The elitism-based methodology is utilized to effectively realize multi-output boolean functions, embedding regular clocking schemes. A detailed analysis of power dissipation with different regular clocking schemes is reported. QCADesigner is used for logic synthesis, whereas QCAPro and QCADesignerE have been utilized for energy dissipation analysis.

**Keywords** QCA · Genetic algorithm · Regular clocking · Power analysis · QCAPro · QCADesignerE

## 1 Introduction

In the last few decades, CMOS technology has improved extremely; as a result, it has reached its ultimate limit in terms of its size, and at the same time, it faces significant problems. Quantum-dot Cellular Automata (QCA) technology is one of the alternative technology of CMOS [1], as it has low power consumption and high computation speed. The primary element of QCA is the cell. Each cell contains four quantum dots. Due to the coulombic interaction between the cells, information

A. K. Pramanik (✉)
Department of CSE, Dumka Engineering College, Dumka, Jharkhand, India

J. Pal
Department of Information Technology, Tripura University, Suryamaninagar, India
e-mail: jayantapal@tripurauniv.ac.in

B. Sen
Department of CSE, National Institute of Technology Durgapur, Durgapur, India
e-mail: bibhash.sen@cse.nitdgp.ac.in

 
S. Das and G. J. Martinez (eds.), *Proceedings of First Asian Symposium on Cellular Automata Technology*, Advances in Intelligent Systems and Computing 1425,
https://doi.org/10.1007/978-981-19-0542-1_14

propagates from one part to another. The basic QCA logic comprises of QCA wire, inverter, and majority voter. Conventional AND, OR gate can be formed from majority voter.

One of the emerging minimization technique is the Genetic Algorithm (GA). The Natural Evolution Theory influences GA. In previous researches, GA was used to minimize QCA circuits [2, 3], but none of them has investigated to the fact that clocking scheme is appropriate for these minimized circuits. Moreover, no previous work has considered elitism-based optimization methods for QCA circuit synthesis. Elitism preserves the best possible solutions, which plays a vital role in optimal logic realization [4, 5].

A proper clocking layout has a significant role in realizing logic design in QCA technology. Few of them are reported in [6–9], where the first-mentioned three do not follow the universal shape and bound. Support of feedback path for the sequential circuit is one of the major pitfalls. A clocking with feedback path is reported in [6] compromising the complexity. This issue is resolved in [7, 8], and it also has the advantage of three directional information flow.

Power dissipation is one of the significant challenges in the development of digital circuits. Bennet showed that the power dissipation could be reduced (tends of zero) by realizing reversible circuits [10]. Bennett clocking is energy efficient but cannot be implemented in the QCA simulator due to its reversibility nature. A power loss may occur in depolarize state, switching state, and when the clock changes from high to low, to hold a new state [11]. It becomes of utmost importance to investigate the power dissipation of a proposed logic.

Inspired by the above factors, a GA is utilized for QCA circuit synthesis, embedding a regular clocking scheme. The primary focus of this work is to analyse the power dissipation of the circuits before and after the application of GA. The paper contributes to the research as follows:

- Power dissipation of the functions (before and after application of GA) are investigated.
- QCAPro and QCADesignerE tool are utilized for the power analysis of the circuits.

The rest of the article is structured as follows: Sect. 2 presents a brief discussion about the basics of QCA, Genetic Algorithm, and Regular Clocking. Sect. 3 contains a detailed discussion about the proposed method. Section 4 presents power analysis of the five multi-output functions before and after utilization of GA. Section 5 concludes the article.

## 2 Background and Related Work

This section deal with the fundamental concepts of different aspects used in this research work.

## 2.1 QCA Basic

The fundamental units are depicted in Fig. 1. A cell with four quantum dots contains two electrons placed diagonally opposite each other (Fig. 1(a)). The Coulombic repulsion forces the electrons to form two possible placements (polarization (P) equal to +1 and -1) (Fig. 1(b)). The electrons are free to move within the dots in a cell via tunneling. The majority gate is the most important unit in QCA, shown in Fig. 1(c)). It can implement AND, OR basic logics by simply fixing one of the input cells only.

The information propagates from input to output through the QCA wire. QCA cells are placed adjacent to each other to form a wire. The information flowing vertically should not interact with the information flowing horizontally, is realized with different techniques of crossover (coplanar and multi-layer). Sometimes, it is required to flip the input at the other end, which can be realized in QCA by an inverter. The implementation of the inverter can be done in two different ways. Any of them can be used alternatively based on the uses and application.

There are four phases of the QCA clocks which control the flow of information throughout the circuit. The phases are: Switch (the potential barrier between the dots are at least energy, allowing the electrons to tunnel between the dots and propagating data), Hold (tunneling of electrons prevented as the clock signal remains high), Release (cell starts to lose its stored value) and Relax (cell became un-polarized and lost its stored value) as shown in Fig. 1(d). There is simultaneous modulation of inter dot potential barrier for the cells in a row, which can be partitioned by dividing it into sub row and achieving the multi-phase clocking (sub-array of cells), which are known as the clock zones [7].

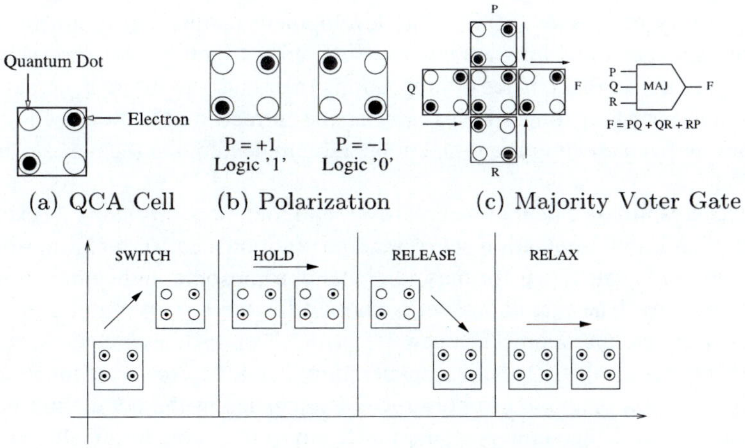


(a) QCA Cell (b) Polarization (c) Majority Voter Gate

(d) Four Phases of QCA Clock

**Fig. 1** Basic structural components of QCA

## 2.2 Regular Clocking

The concept of QCA clocking was first proposed by [12] with a uni-dimensional placement of the clocking zones, also known as One-Dimensional clocking. Later, in [13] came up with a suggestion that it should be regular, bounded shape, and uniform. A formal two-dimensional clocking was presented with bi-directional input and output in [6]. However, the challenge of realizing the feedback path remains the same as it was in [13].

An effective realization of feedback path was reported in [7], termed as USE clocking scheme. The interacting clock zones are kept together (e.g., zone 0 propagates data to zone 1, zone 2 propagates to zone 3, and so on). Four grids with the facility of loop maintain the clock zones in a regular manner. It also maintains the opposite directional information flow in the adjacent rows and columns. In another research, a Robust, Efficient, and Scalable clocking scheme is proposed [8] with coplanar wire crossing. Additionally, three directional information flow can also be achieved. Both USE and RES proposed an electric field generation circuitry for the clock zones. In achieving the advantages mentioned above, this scheme lacks a uniform underlying layout for the particular clock zone 3. It is responsible for unnecessary wire crossing under the QCA layer, and at the same time, the complexity of fabrication also increases.

## 2.3 Genetic Algorithm

The concept of Genetic Algorithm was first proposed in the mid 70's. The evolutionary theory of species inspires the development of this algorithm. In nature, weak and unfit species within their environment face extinction by natural selection. The stronger has a higher chance of propagating their genes to future generations via reproduction. For a long time, species carrying the correct combination become dominant, starting from creating the initial population to developing the next-generation population.

GA executes fitness calculation, crossover, and mutation. GA starts its execution by creating an initial population set of random solutions, called a chromosome. A fitness function is calculated for the individual chromosome, indicating feasibility and sustainability. The population set is formed for the next generation by selection, crossover, and mutation. Then new offspring chromosomes are generated from the parent chromosomes for the next generation using the crossover function. The mutation operation is applied to preserve discrepancies in the population pool. In the process of mutation, some portion of a chromosome is randomly altered with a specified probability to introduce new characteristics.

All the operations are repeated until the intended goal is realized or reaches the maximum steps. Whereas elitism-based GA considers less number of chromosomes with the most suitable value and transfers to the next generation [4, 5]. A Multi-objective genetic algorithm considers more than one objective to be optimized. Several multi-objective-based genetic algorithms have been reported, suitable for different applications [14–16].

## 3 Proposed Method

An elitism-based genetic algorithm is utilized to synthesize functions. In this article, GA is used to minimize the multi-output functions in QCA. As under laying clocking scheme plays a vital role in the efficient designing of QCA circuits, we have used USE [7], and RES [8] clocking scheme for realization. Our main objective is to analyze the power dissipation for the realized circuits. However, a brief discussion on the genetic algorithm to synthesize the QCA circuit is done.

The first step for the GA-based approach is to create the initial population set. This set contains randomly selected chromosomes. The next step is to calculate fitness values to find the best chromosome for the next generation. Here two different fitness functions are used to compute the fitness value. The first fitness function is applied for the initial output function and it is calculated as the ratio of the number of total min-terms and the number of min-term matches with the output function. The second fitness function is applied for all the remaining outputs. It differs from the first, where the previous output is also considered to optimize the current output function.

The elite chromosome is considered to find the better fitness value to be migrated to the next generation without changing the logical output of the function. The traditional mutation operation can not be applied in QCA technology due to the function variation of a primitive logic gate regarding the number of inputs. Therefore, the alternative approach is used here to maintain the variation in the population pool. Here the chromosomes with equal number of mutation children are selected randomly, which is redirected to the succeeding generation population pool to be considered as mutation children. Finally, chromosomes with the least fitness values are discarded totally, and new random chromosomes are added at their place. All these newly generated chromosomes start from elitism, then crossover, mutation up to the arbitrary add-on; all these are put together as a pool of the next-generation population. The steps involved, from the calculation of fitness value to creating the next-generation population pool, are repeated several times until the specified termination condition is not reached.

### 3.1 Case Study

The following functions are considered for performance evaluation (Eqs. 1–5). One from each group of available multi-output functions is chosen to verify its impact on power dissipation.

$$P_1 = \Sigma(m_0, m_2, m_4, m_7) \tag{1}$$

$$P_2 = \Sigma(m_1, m_3, m_4, m_5, m_7, m_{12}, m_{13}, m_{15}) \tag{2}$$

$$P_3 = \Sigma(m_2, m_4, m_6) \tag{3}$$

$$P_4 = \Sigma(m_4, m_6, m_7) \tag{4}$$

$$P_5 = \Sigma(m_3, m_6, m_7, m_{11}, m_{13}, m_{14}, m_{15}) \tag{5}$$

Generally, the Karnaugh map is used to minimize any logic circuit. However, in QCA, logic can only be represented using its primitive gates. The functions are presented using QCA gates, in Eqs. 6–10, where A, B, and C are the inputs to the circuit.

$$F_1 = M(M(M(A, B, 0)', C', 0), M(M(A, B, 0), C, 0), 1) \tag{6}$$

$$F_2 = M((M(M(A', D, 0), M(B, D, 0), 1), M(B, C', 0), 1)) \tag{7}$$

$$F_3 = (M(M(A, C', 0), M(B, C, 0), 1) \tag{8}$$

$$F_4 = M(M(B, C', 1), A, 0) \tag{9}$$

$$F_5 = (M(B, D, 0), A, 0), M(M(B, D, 1)C, 0), 1) \tag{10}$$

This paper aims to investigate the impact of GA in low power circuit realization. In this regard, all the functions are scrutinized to prove the viability of the proposed method. The functions (Eqs. 6–10) are reproduced using GA and represented in Eqs. 11–15.

$$G_1 = M(M(B, 0, C), M(A, C', 1), M(A, C, B)') \tag{11}$$

$$G_2 = M(M(C', 0, B), D, M(1, A', B)) \tag{12}$$

$$G_3 = M(M(1, B, A), 0, C') \tag{13}$$

$$G_4 = M(M(0, C', A), A, M(B, C, 0)) \tag{14}$$

$$G_5 = M(C, M(B, C, D), M(A, D, 0)) \tag{15}$$

The renowned clocking schemes USE and RES have been utilized for both the pre (Eqs. 6–10) and post GA (Eqs. 11–15) circuit realization. All the functions have been verified with USE and RES clocking schemes, utilizing the proposed method. In this regard, to show the buildability of the proposed method, the circuit synthesis of the function correspond to Eq. 3 is shown in Fig. 2. The renowned clocking scheme USE

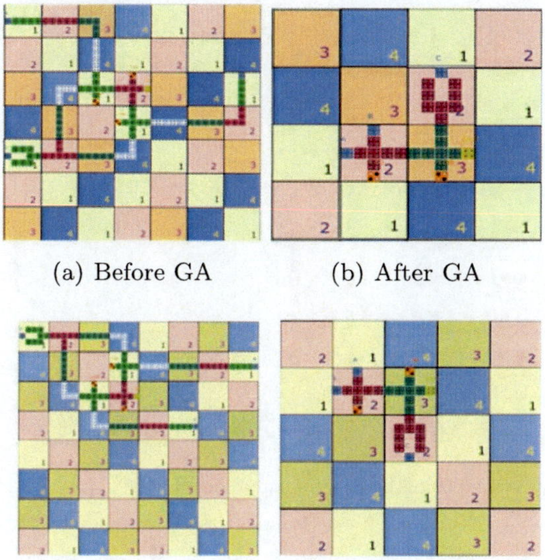

(a) Before GA (b) After GA

(c) Before GA (d) After GA

**Fig. 2** (a-b) RES clock and (c-d) USE based realization of $P = \sum(m_2, m_4, m_6)$

and RES have been utilized for the same. The figure depicts the function synthesis in both pre and post-application of GA. Similarly, other functions have also been verified with this method.

## 4 Power Analysis

The estimation of power consumption is also an important factor, attributed in accordance with the integration levels increment [17]. In QCA technology, the upper limit power estimation model, as proposed in [11, 18], is widely accepted, which can be analyzed by applying the QCAPro simulation tool [19]. Additionally, the QCAesignerE tool can also be used for energy analysis [20] as it also calculates the energy dissipation of QCA circuits. This section measures the power dissipation for the proposed circuits using both QCAPro and QCADesignerE.

**Using QCAPro**: The power dissipation approximation tool QCAPro is used for calculating the approximate value of average leakage and switching energy dissipation of the proposed design. The energy dissipation map is created in three different tunneling energy levels (0.5Ek, 1Ek, and 1.5Ek). Figure 3 represents the power dissipation map for five multi-output functions by using the USE before applying GA. In contrast, Fig. 4 illustrates the power dissipation map after applying GA. In the same way, Fig. 5 represents the power dissipation map for five multi-output functions by

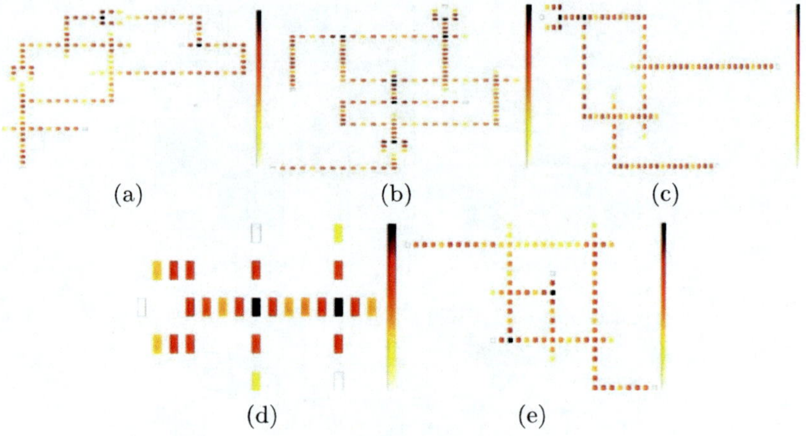

**Fig. 3** Power dissipation map using USE before applying GA

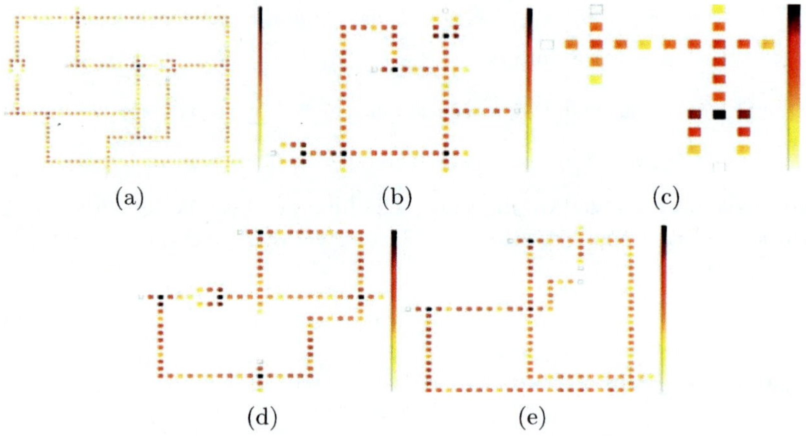

**Fig. 4** Power dissipation map using USE after applying GA

using the RES before applying GA, while Fig. 6 represents the power dissipation map after applying GA. The calculated values at three tunneling energy levels are tabulated in Table 1 at temperature 2K.

**Using QCADesignerE:** The energy dissipation of QCA circuits can also be calculated using QCADesignerE. QCADesignerE is the advanced version of the QCADesigner tool. It has an extra advantage of calculating energy dissipation compared to the QCADesigner tool. Energy dissipation is calculated for the functions in the coherence vector (w/Energy) simulation engine of QDE with default parameters at 2K temperature. The calculated values are tabulated in Table 2.

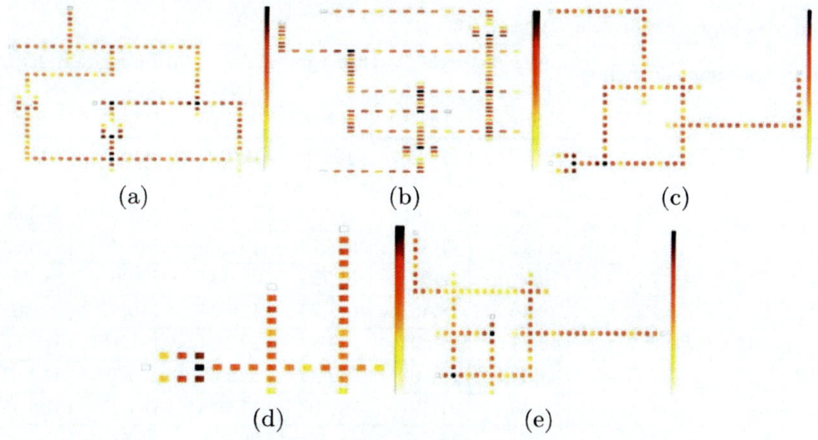

Fig. 5 Power dissipation map using RES before applying GA

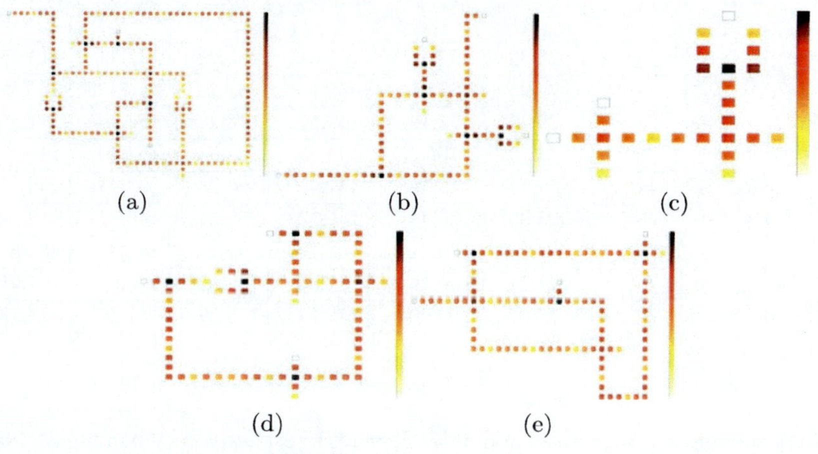

Fig. 6 Power dissipation map using RES after applying GA

## 4.1 Discussion

It is clear from Table 1 that if the energy level increases from 0.5 to 1.5 EK, then average leakage energy also increases, but average switching energy decreases. Average leakage energy and switching energy for the USE clocking scheme are 47% and 52%–53% less, respectively, when the $P_1$ function is realized without GA. Similarly, the average leakage energy and switching energy for the RES clocking scheme are 28% and 36%–37% less, respectively, when the $P_1$ function is realized without GA. In the case of $P_2$ function, the average leakage energy and switching energy for the USE clocking scheme are 44%–45% and 41%–42% less, respectively, when realized with GA. Similarly, the average leakage energy and switching energy for the RES

**Table 1** Power dissipation using QCAPro (in meV)

| Status | Scheme | Function | Avg. leakage power dissipation | | | Avg. switching power dissipation | | |
|---|---|---|---|---|---|---|---|---|
| | | | $0.5E_k$ | $1E_k$ | $1.5E_k$ | $0.5E_k$ | $1E_k$ | $1.5E_k$ |
| Before GA | USE [7] | $P_1$ | 36.07 | 114.98 | 209.29 | 196.39 | 169.56 | 143.39 |
| | | $P_2$ | 44.46 | 140.48 | 255.33 | 259.85 | 225.00 | 190.84 |
| | | $P_3$ | 28.17 | 91.37 | 167.13 | 173.86 | 149.94 | 126.63 |
| | | $P_4$ | 7.47 | 22.57 | 40.33 | 42.54 | 37.06 | 31.75 |
| | | $P_5$ | 24.46 | 78.46 | 142.74 | 128.04 | 110.47 | 93.40 |
| | RES [8] | $P_1$ | 42.51 | 134.22 | 243.62 | 222.34 | 192.15 | 162.66 |
| | | $P_2$ | 41.99 | 132.08 | 239.59 | 248.67 | 215.30 | 182.63 |
| | | $P_3$ | 28.14 | 91.29 | 167.02 | 173.18 | 149.27 | 125.95 |
| | | $P_4$ | 11.71 | 36.25 | 65.01 | 59.92 | 51.58 | 43.59 |
| | | $P_5$ | 25.82 | 82.95 | 151.02 | 135.57 | 116.95 | 98.84 |
| After GA | USE [7] | $P_1$ | 68.39 | 217.80 | 396.94 | 415.76 | 360.01 | 305.37 |
| | | $P_2$ | 24.29 | 77.41 | 141.01 | 152.99 | 132.98 | 113.37 |
| | | $P_3$ | 7.02 | 22.21 | 40.21 | 38.12 | 33.14 | 28.26 |
| | | $P_4$ | 29.42 | 93.15 | 169.61 | 178.28 | 154.76 | 131.61 |
| | | $P_5$ | 33.52 | 108.26 | 198.36 | 194.04 | 168.40 | 143.10 |
| | RES [8] | $P_1$ | 59.15 | 187.68 | 341.47 | 350.28 | 302.95 | 256.65 |
| | | $P_2$ | 25.39 | 80.41 | 146.32 | 162.27 | 140.77 | 119.74 |
| | | $P_3$ | 7.02 | 22.21 | 40.21 | 38.12 | 33.14 | 28.26 |
| | | $P_4$ | 22.21 | 70.26 | 127.86 | 137.62 | 119.64 | 101.91 |
| | | $P_5$ | 29.07 | 93.20 | 170.14 | 175.27 | 151.50 | 128.28 |

clocking scheme are 39%–40% and 34%–35% less, respectively, when the $P_2$ function is realized with GA. The average leakage energy and switching energy for the USE clocking scheme are 75%-76% and 77%-78% less, respectively, when function $P_3$ is realized with GA. The average leakage energy and switching energy for USE clocking scheme are 74%-76% and 75%-76% less, respectively, when $P_4$ function is realized without GA. In the same way, the average leakage energy and switching energy for RES clocking scheme are 47%–49% and 56%–57% less, respectively, when $P_4$ function is realized without GA. The average leakage energy and switching energy for USE clocking scheme are 27%-28% and 34%-35% less, respectively, when function $P_5$ is realized without GA.

It is observed from Table 2 that the total energy dissipated by $P_1$ using USE and RES clocking schemes are 38% and 21% less, respectively, when the function is realized without GA. In the case of $P_2$ the total energy dissipation using USE and RES clocking schemes are 34% and 34% less, respectively, when the function is realized with GA. The total energy dissipated by $P_3$ using USE and RES clocking schemes are 56% and 58% less, respectively, when the function is realized with GA.

**Table 2** Energy dissipation using QCADesignerE simulation tool

| Clocking scheme | Before GA | | After GA | |
|---|---|---|---|---|
| | Total dissipation ($Sum\_E_{bath}$) | Dissipation Per Cycle ($Avg\_E_{bath}$) | Total dissipation ($Sum\_E_{bath}$) | Dissipation per cycle ($Avg\_E_{bath}$) |
| $P_1 = \sum(m_0, m_2, m_4, m_7)$ | | | | |
| USE [7] | 3.78e-002 eV | 3.43e-003 eV | 6.08e-002 eV | 5.53e-003 eV |
| RES [8] | 4.10e-002 eV | 3.73e-003 eV | 5.19e-002 eV | 4.72e-003 eV |
| $P_2 = \sum(m_1, m_3, m_4, m_5, m_7, m_{12}, m_{13}, m_{15})$ | | | | |
| USE [7] | 4.09e-002 eV | 3.72e-003 eV | 2.69e-002 eV | 2.44e-003 eV |
| RES [8] | 4.03e-002 eV | 3.67e-003 eV | 2.67e-002 eV | 2.43e-003 eV |
| $P_3 = \sum(m_2, m_4, m_6)$ | | | | |
| USE [7] | 2.79e-002 eV | 2.54e-003 eV | 1.24e-002 eV | 1.13e-003 eV |
| RES [8] | 2.95e-002 eV | 2.68e-003 eV | 1.24e-002 eV | 1.13e-003 eV |
| $P_4 = \sum(m_4, m_6, m_7)$ | | | | |
| USE [7] | 1.41e-002 eV | 1.29e-003 eV | 2.88e-002 eV | 2.62e-003 eV |
| RES [8] | 1.33e-002 eV | 1.21e-003 eV | 2.55e-002 eV | 2.32e-003 eV |
| $P_5 = \sum(m_3, m_6, m_7, m_{11}, m_{13}, m_{14}, m_{15})$ | | | | |
| USE [7] | 3.18e-002 eV | 3.01e-003 eV | 3.35e-002 eV | 3.14e-003 eV |
| RES [8] | 3.14e-002 eV | 3.11e-003 eV | 3.37e-002 eV | 3.26e-003 eV |

In the case of $P_4$ the total energy dissipation using USE and RES clocking schemes are 51% and 48% less, respectively, when the function is realized without GA. The total energy dissipated by $P_5$ using USE and RES clocking schemes are 5% and 7% less, respectively, when the function is realized without GA.

From Tables 1, 2, it is observed that Power dissipation and average leakage power are high when the function $P_1$ is realized using GA. Because regular clocking schemes are used to fabricate the functions, it increases the circuit area though the majority voter count is the same before and after applying GA. In the case of $P_2$ and $P_3$, as the majority voter count is reduced after applying GA, power dissipation and average leakage power are also lowered. Due to the expansion in majority voter count, total power dissipation and average leakage energy are high when $P_4$ is realized using GA. The majority voter count is reduced when $P_5$ is realized using GA. Still, the power dissipation is slightly high because the regular clocking schemes are used here to fabricate the functions which increase the circuit area.

## 5 Conclusion

This article applies the genetic algorithm (GA) based on elitism to realize the QCA circuit. The significance of the underlying regular clocking schemes is discussed. The multi-output logical functions are examined and tested for the proposed job. The effect of GA on regular clocking to synthesis the low power circuit is investigated. The experimental result signifies that after applying GA, the energy dissipation of the QCA circuit is reduced to 30–55% with few exceptions. Moreover, it is observed that the circuits realized with different regular clocking schemes dissipate different amounts of energy. An automatic cell layout of QCA with GA for low power dissipation is targeted for future endeavors.

**Acknowledgements** This work is sponsored by the Young Faculty Research Fellowship (YFRF) of Visvesvaraya Ph.D. scheme through the grant number MLA/MUM/GA/10(37)B.

## References

1. Karim, F., Walus, K.: Efficient simulation of correlated dynamics in quantum-dot cellular automata (qca). IEEE Trans. Nanotechnol. **13**(2), 294–307 (2014)
2. Houshmand, M., Saleh, R.R., Houshmand, M.: Logic Minimization of QCA Circuits Using Genetic Algorithms, pp. 393–403. Springer, Berlin (2011)
3. Tehrani, M.A., Navi, K., Kia-kojoori, A.: Multi-output majority gate-based design optimization by using evolutionary algorithm. Swarm Evol. Comput. **10**, 25–30 (2013)
4. Vasconcelos, J.A., Ramirez, J.A., Takahashi, R.H.C., Saldanha, R.R.: Improvements in genetic algorithms. IEEE Trans. Magn. **37**(5), 3414–3417 (2001)
5. Ahn, C.W., Ramakrishna, R.S.: Elitism-based compact genetic algorithms. IEEE Trans. Evol. Comput. **7**(4), 367–385 (2003)
6. Vankamamidi, V., Ottavi, M., Lombardi, F.: Two-dimensional schemes for clocking/timing of qca circuits. IEEE Trans. Comput. Aided Des. Integr. Circuits Syst. **27**(1), 34–44 (2007)
7. Campos, C.A.T., Marciano, A.L., Neto, O.P.V., Torres, F.S.: Use: a universal, scalable, and efficient clocking scheme for qca. IEEE Trans. Comput. Aided Des. Integr. Circuits Syst. **35**(3), 513–517 (2015)
8. Goswami, M., Mondal, A., Mahalat, M.H., Sen, B., Sikdar, B.K.: An efficient clocking scheme for quantum-dot cellular automata. Int. J. Electron. Lett. 1–14 (2019)
9. Wang, L., Xie, G., Zhu, R., Yu, C.: An optimized clocking scheme for nanoscale quantum-dot cellular automata circuit. In: 2019 IEEE 14th International Conference on Nano/Micro Engineered and Molecular Systems (NEMS), pp. 336–341. IEEE (2019)
10. Bennett, C.: Logical reversibility of computation. IBM J. Res. Devel. 525–532 (1973)
11. Srivastava, S., Sarkar, S., Bhanja, S.: Power dissipation bounds and models for quantum-dot cellular automata circuits. In: 2006 Sixth IEEE Conference on Nanotechnology, vol. 1, pp. 375–378. IEEE (2006)
12. Lent, C.S., Tougaw, P.D.: A device architecture for computing with quantum dots. Proc. IEEE **85**(4), 541–557 (1997)
13. Hennessy, K., Lent, C.S.: Clocking of molecular quantum-dot cellular automata. J. Vacuum Sci. Technol. B: Microelectron. Nanometer Struct. Process., Measur., Phenom. **19**(5), 1752–1755 (2001)
14. Konak, A., Coit, D.W., Smith, A.E.: Multi-objective optimization using genetic algorithms: a tutorial. Reliab. Eng. Syst. Safety **91**(9), 992–1007 (2006)

15. Deb, K.: Multi-Objective Evolutionary Algorithms, pp. 995–1015. Springer, Berlin (2015)
16. Coello, C.A.C., Lamont, G.B., Van Veldhuizen, et al. D.A.: Evolutionary Algorithms for Solving Multi-objective Problems, vol. 5. Springer (2007)
17. Liu, W., Lu, L., O'Neill, M., Swartzlander, E.E.: A first step toward cost functions for quantum-dot cellular automata designs. IEEE Trans. Nanotechnol. **13**(3), 476–487 (2014)
18. Timler, J., Lent, C.S.: Power gain and dissipation in quantum-dot cellular automata. J. Appl. Phys. **91**(2), 823–831 (2002)
19. Srivastava, S., Asthana, A., Bhanja, S., Sarkar, S.: Qcapro-an error-power estimation tool for qca circuit design. In: IEEE International Symposium of Circuits and Systems (ISCAS), pp. 2377–2380. IEEE (2011)
20. Torres, F.S., Wille, R., Niemann, P., Drechsler, R.: An energy-aware model for the logic synthesis of quantum-dot cellular automata. IEEE Trans. Comput. Aided Des. Integrated Circ. Syst. **37**, 3031–3041 (2018)

# Cellular Automata Based Model for Finding Minimum Spanning Tree in Wireless Sensor Networks

**Parijat Sen and Debopriya Barman**

**Abstract** This work presents a cellular automata (CA) based model for finding minimum spanning tree in wireless sensor networks (WSN) with distinct edge weights. A regular grid topology network is created by the sensor nodes where every sensor node can have maximum four neighbor nodes in the tree. To perform the computation every sensor node communicates with its immediate neighbour nodes which are within its communication range . Nodes use simple CA based localized algorithms which can optimize the performance. By limiting the maximum number of neighbour nodes it also minimizes the redundancy in the network which can increase the lifetime of the network.

**Keywords** Cellular automata (CA) · Wireless sensor networks · Minimal spanning tree

## 1 Introduction

Graphs are quite commonly used for modelling computer networks. Spanning tree is a similar kind of structure. Spanning tree of a graph can be defined as a sub graph containing all the vertices and is a tree. The sum of all edge weights is called weight of the spanning tree. The smallest weighted spanning tree, among all the possible spanning trees of a graph, is called minimum spanning tree (MST). MST problem plays an important role in network optimization.

Applications of wireless sensor networks have increased exponentially due to the advancements of sensor technology. WSNs are used for environmental monitoring [2], battlefield surveillance, healthcare sector [1] etc.

P. Sen (✉) · D. Barman
Department of Information Technology, Indian Institute of Engineering Science and Technology, Shibpur Howrah, West Bengal 711103, India



S. Das and G. J. Martinez (eds.), *Proceedings of First Asian Symposium on Cellular Automata Technology*, Advances in Intelligent Systems and Computing 1425,
https://doi.org/10.1007/978-981-19-0542-1_15

Cellular Automata (CA) is really useful for solving many computational problems like Leader election problem [3], Mutual exclusion problem [4], Shortest path problem [5] etc. All this CAs use different local rules and neighborhood structure for computation which is quite similar with the locally decision making nature of sensor nodes. In this work we are trying to explore this similarities of WSN and CA for finding Minimum Spanning Tree.

One of the major benefits of using a cellular automaton based model is that it needs very limited local information to compute the solution. Best feature of cellular automata model is that usually cellular automata models can solve much complex solutions in a comparatively simpler way so it becomes very much practical to implement those models in WSNs.

## 2 Preliminaries

### 2.1 *Wireless Sensor Networks*

Wireless Sensor networks are built with set of nodes which consist of a radio transceiver, microcontroller and sensors. Sensors are the main components of any WSN. Sensors with different applications [6] are used so that it can perform well in any area. A sensor is a very low cost device with limited processing power, short communication range, limited battery power and limited memory. Generally WSN consists of several number of sensors. Sensors are deployed in the monitored area but there are not any centralized controls on these sensors. The sensors use their immediate neighbors for communication. Two metrics are considered for measuring the performance of the network. First is coverage and the second one is network lifetime. Coverage is defined as the area monitored by the network. Network lifetime can be defined as the time elapsed until all the sensors are dead.

### 2.2 *Cellular Automata*

The Cellular Automaton is a biologically inspired model that has been widely used to model different physical systems. von Neumann constructed a two-dimensional cellular automaton which can simulate arbitrary Turing machines. So von Neumann's cellular automaton is computation universal. Many simpler and more special cellular automaton can also be computation universal.

**Definition 1** Cellular automaton is a quadruple $(\mathscr{L}, S, \mathscr{N}, \mathscr{R})$

1. $\mathscr{L} \subseteq \mathbb{Z}^d$ is a d-dimensional space consists of cells, where $\mathbb{Z}$ is the set of integers
2. S is the finite set of states

3. $\mathscr{N} = (\mathbf{x_1}, \mathbf{x_2}, ..., \mathbf{x_k})$ N is the neighborhood vector of k distinct elements of the Lattice $\mathscr{L}$. Basically it represents all those surrounding cells on which the next state value of a cell depends.
4. $\mathscr{R}$ is the local rule. If $S_1, S_2, ..., S_k$ is the present state of neighbors then $R(S_1, S_2, ..., S_k)$ is the value of the next state.

We can model a two dimensional cellular automaton as a two dimensional grid. Where the cells assume their next states in each specific time by following its local rule $\mathscr{R}$. Although classically all cells are updated simultaneously and follows same local rule for changing the cell states, but from 1986 many researchers had worked on the idea that the cells may follow different rules with different neighborhood dependencies in order to relate with some real-life scenarios.This class of non-classical CA is called *non-uniform CA*. There exist various different types of CAs [13] depending upon dimensions, neighborhood, lattice size, set of states, boundary conditions, local rules etc.

A *configuration* can be explained as a snapshot of states that are assigned to the cells of a CA at any time of evolution. Let $\mathcal{C} = S^{\mathscr{L}}$ be the set of all configurations. Then a CA can be described as a function $G : \mathcal{C} \to \mathcal{C}$, this is called as Global Transition Function. If we have two configurations $x$ and $y$ such that, $x, y \in \mathcal{C}$ and after $t > 0$ time $y$ is reached from $x$ then $y = G^t(x)$. During evolution a CA can also reach a special configuration, called *fixed point*. When a CA reaches to a fixed point, then the evolution terminates and no further change in configuration happens.

## 2.3 Survey of Existing Work

Among all the existing approaches one of the most famous one is known as GHS algorithm. GHS [7] algorithm finds the minimum spanning tree in a graph with bidirectional edges.In GHS, each node starts as a different fragment. The core idea is to join fragments progressively by joining minimum weighted outgoing edge until there is only one fragment remaining. For merging different fragments this algorithm uses level and fragment id. The user identity of last added node is used as fragment id in order to identify different fragments. Algorithm proposed by Lavalle et al. [8] also finds the minimum spanning tree like GHS Algorithm but the difference is in this algorithm nodes don't need to wait for achieving maturity at each level unlike the previous one.

Both the above mentioned algorithms extensively depend on the unique node ids and use those unique ids of every root node of a subtree (fragment) for distinguishing different subtrees (fragments) while merging the fragments. But in wireless sensor network it can be difficult for every sensor node to maintain the details of all sensors present in that network.

In [10] authors proposed a Cluster based method for finding MST for Heterogeneous Sensor Networks with Mobile Sink. Every cluster head nodes have their unique identification number and the sink node is aware of location information of every cluster head node. Here the sink node computes the MST path centrally by applying greedy algorithm and not by using any distributed approach.

In [11] authors proposed a Dynamic minimum spanning tree construction Algorithm called DGHS based on GHS [7] Algorithm for finding MST and maintenance for WSN. As they used GHS algorithm itself for finding the MST so the problem of level matching before merging the fragments persist.

In this [9] recent work authors presented an idea where they merged Artificial Bee Colony algorithm with quantum computing to find Spanning Tree in Industrial Wireless Sensor networks.But this algorithm is not suitable for finding Minimum Spanning Tree, it is suitable for finding Spanning Tree.

### *2.4 Contribution of This Work*

Here we are implementing a CA based algorithm where subtrees (fragments) don't need to manage the level of the tree for merging fragments. The MST finding approach taken in this work is a little different from most commonly used Kruskal's or Prim's algorithm. In this approach initially multiple nodes get connected simultaneously and then each and every small fragment starts increasing the size of the subtree by merging with other fragments. Here we are also not using any unique fragment id for distinguishing different fragments rather we are using a discard state so that if two nodes of the same fragment intend to connect, then it is prevented by identifying the same minimum edge value twice in the searching phase. In Wireless sensor Network it can be very difficult for every sensor node to maintain all such information about the sensor nodes present in the network.

## 3 Proposed Work

### *3.1 Computation Using Cellular Automaton*

Here is the 2D non-uniform Cellular Automaton that solves the minimum spanning tree. We consider a 2D cellular automaton where the cells are used to indicate both **edges** and **nodes**. Since the graph is weighted, the cells which are used for modelling edges, store corresponding weights. As here we take 2D grid graphs, our lattice $\mathscr{L} = \mathbb{Z}^2$. Every node can have at most four neighbours. If our given grid graph has p rows and q columns where p and q are the maximum number of nodes present in a row and a column. To represent the grid graph on the cells of lattice, the cells are used for the nodes and the edges in an alternative manner. We need $(2p - 1) \times (2q - 1)$ cells of $\mathscr{L}$ to represent the grid graph. There are three possible types of cell in the $(2p - 1) \times (2q - 1)$ rectangle,

(1) Cell (i, j) is a **node** cell if i and j are both odd.

(2) Cell (i, j) is an **edge** cell if any one of i and j is odd and the other one is even.
(3) Cell (i, j) is redundant if i and j are both even.

## 3.2 Set of States

At any point every node cell of the 2D CA can be in any one of the 10 possible states as discussed below.
1. Initial State: In the initial configuration all the node cells are which are present in the network are in this state. In this state node cells are represented using $[N, \infty]$
2. Disjoint State: Just after the first generation every node cells evolves to this state. In this state every nodes are represented using $[N, MinEdgeVal]$ where MinEdgeVal is the minimum outgoing edge value of that node.
3. Search State: This state is used whenever a node cell is searching the minimum outgoing edge value among all its adjacent nodes.
4. Search Return State: This state is used whenever a node cell has completed the searching and returned the corresponding minimum outgoing edge value to its parent node.
5. Claim State: This state is used after the completion of minimum outgoing edge value searching phase of any subtree. Node cells evolve to this state whenever they are performing the search operation for claiming the minimum edge value.
6. Claim Return State: This state is also used after the completion of searching phase. A node cell evolves to this state to indicate the information that none of its adjacent node has the minimum edge of that subtree.
7. Final State: Whenever a node has the minimum outgoing edge of that whole subtree (fragment) then it evolves to this state.
8. Neutral state: Any node cell is evolved to this state when that node does not need to perform in the operation of that subtree (fragment). Any node can evolve to this state from search return state or discard state.
9. Discard State: This state is used whenever there is a chance of forming a cycle in the tree. If any node during the search phase encounters a

## 3.3 State Transition

As the CA evolves, the Cells update their states under a local rule and as a result of this continuous evolution the cellular automaton achieves the minimum spanning tree of the given grid graph. As we are assuming that the graph has only unique weighted edges so the MST will also be unique.

- If a node of the rectangular grid is not present then the corresponding cell of the CA is represented using Redundant state. A cell in redundant state always remains in the same state.

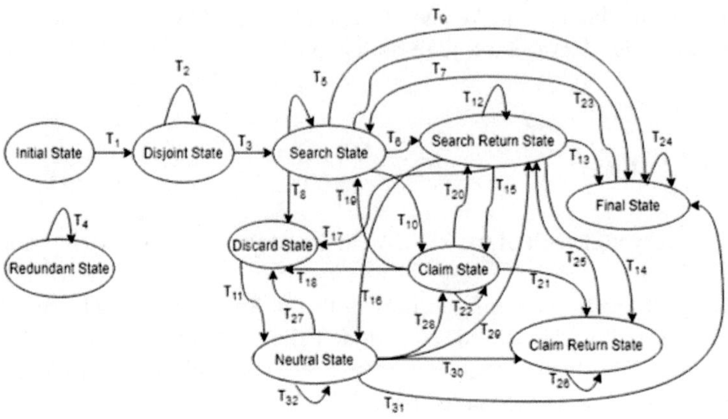


**Fig. 1** State transition diagram

- In initial state all the nodes have any minimum weighted outgoing edge value. In the next generation each and every node cell that are in initial state goes to disjoint State using transition $T_1$. Every cell updates their minimum edge weight value after obtaining the minimum weighted outgoing edge from that particular node, it updates the value and changes their state to disjoint state.
- Now when all the nodes are in disjoint state, if the corresponding edge value of a node matches with any of its neighbor node's edge value then those two nodes get connected and form a subtree. As we can see in Fig. 1 there are three possible transitions from Disjoint state.
- If any node cell in disjoint state fails to connect with any other neighbor node then it remains in the same state but in case the pair of disjoint nodes are merged then one node goes to search state but the other node of the pair remains in the same state.
- By changing its state a node which is in search state points to its different neighbors sequentially to find the minimum outgoing edge value of all the nodes present in that subgraph. Here in this CA preference is given to the neighbors in these order left node, Upper Node, Right Node, Down node. The node remains in search state until it gets the minimum value.
- When a node which is in disjoint state, and the neighbour node sends search message then it changes it goes to search return state to return the minimum outgoing edge value.
- During the searching phase when a node encounters that the previously found minimum weight edge values matches with its own minimum outgoing edge then it understands that this edge can create cycle, it changes it state to discard state and send the message to the node corresponding to that edge, then the corresponding node starts the search again without considering that edge as both the nodes are already part of the same sub graph.

- Once the search is completed then it evolves to Search Return State and returns the minimum outgoing edge value to its parent node. When a sub graph node is done with the Search Return phase and it does not have action to perform then it changes to Neutral State so that its state value does not affect any other neighbour.
- Once all the nodes complete their search operation then the root node evolves to Claim State to ask the corresponding sub graph node who has the minimum outgoing edge to claim the value and evolve to the Final State
- a node evolves to Final state when it has the minimum outgoing in the whole subgraph. The node which is in Final state requests the node on the other side of the edge to get connected. When both the nodes are in Final state with same minimum outgoing edge value then they get connected and after merging of two sub graphs again the searching starts.
- when the neighbor node evolves to Claim state and basically asks this node to claim the token if it has the minimum outgoing edge then in case it does not have that minimum weighted edge then it changes its state to claim state to pass on the claim message further to its immediate neighbors.
- If a node which is in Claim state does not have any neighbor who can claim the minimum value the it changes its state to Claim Return to lets its root node understand that it does not have the minimum edge.
- The evolution terminates when after completing one round of searching the root node receives no valid minimum outgoing edge value.
- After the completion of computation the root node of the spanning tree sends the tree information to the sink node of that Wireless sensor network.

### *3.4 Example*

Here in Fig. 2 on the left side we have the 2D 4 $\times$ 4 input grid graph and its corresponding initial configuration in the CA, as well as on the right side we can see the required MST of the 4 $\times$ 4 grid graph and its corresponding final configuration in the CA.

1. As we can see in the initial configuration that all the node cells i.e $V_{1,5}$, $V_{3,1}$, $V_{3,3}$, $V_{3,5}$, $V_{3,7}$, $V_{5,3}$, $V_{5,5}$, $V_{5,7}$, $V_{7,3}$ are represented using $[N, \infty]$, where initially no one aware of their minimum outgoing edge value, edges are represented using E$x$ where $x$ is the edge weight. and redundant cells are represented using B.
2. In the next generation of the evolution all the nodes will simultaneously update their minimum outgoing edge and replace the infinity.
3. Once the minimum outgoing edge values are updated then all the nodes with same minimum edge value will merge and create subtrees with two vertices. In this example we will get three different subtrees ($V_{1,5}$ , $V_{3,5}$), ($V_{3,3}$ , $V_{5,3}$) and ($V_{3,7}$ , $V_{5,7}$).

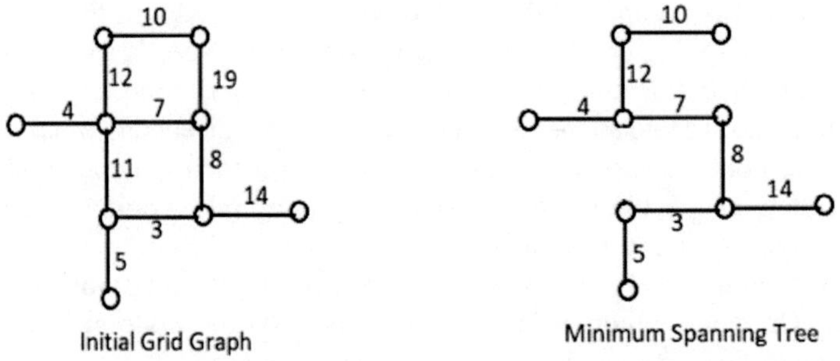


| | 1 | 2 | 3 | 4 | 5 | 6 | 7 |
|---|---|---|---|---|---|---|---|
| 7 | B | B | [N,∞] | [E,10] | [N,∞] | B | B |
| 6 | B | B | [E,12] | B | [E,19] | B | B |
| 5 | [N,∞] | [E,4] | [N,∞] | [E,7] | [N,∞] | B | B |
| 4 | B | B | [E,11] | B | [E,8] | B | B |
| 3 | B | B | [N,∞] | [E,3] | [N,∞] | [E,14] | [N,∞] |
| 2 | B | B | [E,5] | B | B | B | B |
| 1 | B | B | [N,∞] | B | B | B | B |

Initial Configuration of the CA

| | 1 | 2 | 3 | 4 | 5 | 6 | 7 |
|---|---|---|---|---|---|---|---|
| 7 | B | B | [N,10] | [E,10] | [N,10] | B | B |
| 6 | B | B | [E,12] | B | [E,19] | B | B |
| 5 | [N, 4] | [E,4] | [N, 4] | [E,7] | [N, 7] | B | B |
| 4 | B | B | [E,11] | B | [E,8] | B | B |
| 3 | B | B | [N,3] | [E,3] | [N,3] | [E,14] | [N,14] |
| 2 | B | B | [E,5] | B | B | B | B |
| 1 | B | B | [N,5] | B | B | B | B |

Configuration in generation 1

| | 1 | 2 | 3 | 4 | 5 | 6 | 7 |
|---|---|---|---|---|---|---|---|
| 7 | B | B | [F,12] | [P,10] | [CL,12] | B | B |
| 6 | B | B | [E,12] | B | [E,19] | B | B |
| 5 | [R,7] | [P,4] | [F,7] | [E,7] | [N, 7] | B | B |
| 4 | B | B | [E,11] | B | [E,8] | B | B |
| 3 | B | B | [R,5] | [P,3] | [CL,5] | [E,14] | [N,14] |
| 2 | B | B | [E,5] | B | B | B | B |
| 1 | B | B | [N, 5] | B | B | B | B |

Configuration in generation 4

| | 1 | 2 | 3 | 4 | 5 | 6 | 7 |
|---|---|---|---|---|---|---|---|
| 7 | B | B | T | P10 | T | B | B |
| 6 | B | B | [P,12] | B | B | B | B |
| 5 | T | [P,4] | T | [P,7] | T | B | B |
| 4 | B | B | B | B | [P,8] | B | B |
| 3 | B | B | T | [P,3] | T | [P,14] | T |
| 2 | B | B | [P,5] | B | B | B | B |
| 1 | B | B | T | B | B | B | B |

Final Configuration of the CA

**Fig. 2** Grid Graph and their corresponding CA configurations

4. As already mentioned in the state transions now this subtrees will start searching the minimum outgoing among all the nodes of these subtrees and join those edges like here in this example minimum outgoing edge $V_{3,1}$ is 5 so it returns 5 to node $V_{3,3}$. but $V_{3,3}$ has minimum edge value as 8 so it will evolve to claim state with value 5.
5. When $V_{3,3}$ evolves to claim state with token value 5 then $V_{3,1}$ claims that value as the value matches with its own min edge value 5 and evolves to final state with min edge value 5, similarly other subtrees also follow the same rule to reduce the number of subtrees to 1.
6. Computation terminates when the root nodes does not get any valid minimum outgoing edge value to connect.
7. In Fig. 2 we can see the final configuration of the CA where the evolution terminates and we have the minimum spanning tree

## *3.5 Implementation of CA Based Model in WSN*

Here in the proposed work we are trying to use a 2D cellular automaton based algorithm for finding minimum weight spanning tree in a 2D regular grid wireless network where edges have distinct weights. Wireless network will be of 2D regular grid topology where every node can have maximum four neighbour nodes for constructing the minimum spanning tree. Depending upon the location of surrounding nodes Four possible neighbours are Left, Upper, Right, Down. Every node can communicate with all the nodes that are available within its communication range, but considers only those edges which are within a range of distance $D_{max}$ and $D_{min}$. By limiting the total number of neighbor node in every co-ordinate and by maintaining the threshold distance it creates a virtual grid topology network.

There are two stages in this work

a. Maintaining virtual grid network

b. Finding MST in the Grid Network

**Maintaining virtual grid network** Initially all the sensors are not in awakened state. Random sensors spontaneously becomes active. Active nodes sends pairing request to other active nodes to form a virtual grid. Sensors use Receive Signal Strength Indicator to find the distance between two nodes. $D_{max}$ is considered as maximum Threshold Distance. If the distance between two nodes is more than $D_max$ then those nodes will not consider each other as their neighbour, similarly if the distance is less than $D_{min}$ then also that node is not considered to avoid redundancy. Sensor will use angle of arrival method for identifying the co-ordinate of a surrounding node. Sequentially every sleeping node wakes up spontaneously to change its state from undiscovered to paired state. NeighborCount is used to denote the total number of neighbors that are paired with that sensor.

Here in this Algorithm 1 S($N_i$) means the state of a requesting sensor. If a sensor already have a left neighbor then Left_Node value will be 1 else 0. Similarly values are assigned to Upper_Node, Right_Node, Down_Node. If the neighbor node is $N_i$ the direction($N_i$) is used to denote its location. $D_{max}$ and $D_{min}$ are used for denoting the maximum and minimum threshold distance respectively.

**Algorithm 1** Algorithm for Maintaining Grid

```
procedure for sending paring request
begin
   if status = active then
      if (available node >0 and NeighborCount < 4) then
         send paring request
      else
         if available node < 1 or S(Self)!=Paired then
            status ← sleep
end
procedure on receiving paring request from N_i
begin
   if status = active and S(N_i) != Paired then
      if (direction(N_i)=Left and Left_Node=0 and D≤ D_max and D≥ D_min) then
         S(N_i) ← Paired
         Left_Node ← 1
      if (direction(N_i)=Upper and Upper_Node=0 and D≤ D_max and D≥ D_min) then
         S(N_i) ← Paired
         Upper_Node ← 1
      if (direction(N_i)=Right and Right_Node=0 and D≤ D_max and D≥ D_min) then
         S(N_i) ← Paired
         Right_Node ← 1
      if (direction(N_i)=Down and Down_Node=0 and D≤ D_max and D≥ D_min) then
         S(N_i) ← Paired
         Down_Node ← 1
end
```

**Finding MST in the Grid Network** Here in this next stage of our work we are applying the CA based distributed Algorithm in a virtual regular grid Wireless sensor network for finding the Minimum Spanning tree.Every sensor node can have atmost four neighbors, sensors use the intermediate distance from its immediate neighbor as the edge weight. Sequentially whenever any node becomes inactive then other nearby node takes its place to maintain the virtual grid network and provide better lifetime. We have already mentioned in the state transition Function how every sensor node sends different messages to its immediate neighbors for during the computation to find the MST. Here we have given the pseudo code (Algorithms 2 and 3) of the Distributed Algorithm.

**Algorithm 2** Algorithm to find MST in WSN Part 1

```
procedure for sending join nodes
begin
   if status = active  then
      if (NeighborCount >0 and Weight[E]=MinEdgeValue then
         send Join(MinEdgeValue) over edge E
end
procedure Response to Join(Val) request on edge E
begin
   if status = active and val= MinEdgeValue then
      Send Return(MinEdgeValue except Weight[E])
end
procedure Response to Return(Val) request on edge E
begin
   if NodeState = Disjoint then
      if val< MinEdgeValue then
         EdgeState[E]← path
         NodeState ← ClaimReturn
         Send Claim(Val) on edge m [If the node doesn't have the min outgoing edge]
      else
         EdgeState[E] ← path
         NodeState ← Final
         Send Final(MinEdgeValue) on edge E
         Send Add[MinEdgeValue] on MinEdge [If the node has min outgoing edge]
   else
      if EdgeState[E]=path and NodeState=Search then
         ReturnCounter ← ReturnCounter +1
      if MinEdgeValue < Val then
         Val ← MinEdgeValue
      if ReturnCounter ≠ NeighborCounter then
         Send Search(Val) to other Neighbors
      else
         if Parent ≠ null then
            Return(Val) to Parent
         else
            if Val ≠ MinEdgeValue then
               for All edge in NeighborCounter do
               Send Claim(Val) on edge
               NeighborCounter gets NeighborCounter-1
            else
               if Val = MinEdgeValue = Null then
                  End Execution
               else
                  NodeState ← Final
                  Add[MinEdgeValue] on MinEdge
end
procedure Response to ClaimReturn(Val) on E
begin
   EdgeState = ClaimReturn
   if NeighborCounter > 0 then
      for All edge in NeighborCounter do
      NodeState
      NeighborCounter gets NeighborCounter-1
      Send Claim[Val] on edge
   else
      NodeState gets ClaimReturn
      Send ClaimReturn[Val] on E
end
```

**Algorithm 3** Algorithm to find MST in WSN Part 2

```
procedure Response to Add(Val) on E
begin
  if NodeState =Disjoint and Val=MinEdgeValue then
    Send Return(MinEdgeValue except Weight[E])
  else
    if NodeState =Final and Val=MinEdgeValue then
      EdgeState[E] ← path
      NodeState ← Search
      Parent ← Null State Send Search(MinEdgeValue) on E
end
procedure Response to Claim(Val) on E
begin
  if NodeState = Disjoint then
    EdgeState[E] ← path
    NodeState = Final
    Send Add[MinEdgeValue] on MinEdge
  else
    if EdgeState[E] = path then
      Parent ← Sender
      if Val = MinEdgeValue then
        for All edge in NeighborCounter do
        if Weight[edge]=MinEdgeValue then
          Send Add[MinEdgeValue] on edge
          NodeState = Final
      else
        if NeighborCounter > 0 then
          for All edge in NeighborCounter do
          Send Claim[Val] on edge
          NodeState = Claim
          NeighborCounter gets NeighborCounter-1
        else
          Send ClaimReturn[Val] on E
          NodeState = ClaimReturn
end
procedure Response to Search(Val) on E
begin
  if EdgeState[E] = path then
    Parent ← Sender
    ReturnCounter gets 0
    if Val = MinEdgeValue then
      for All edge in NeighborCounter do
      if Weight[edge]=Val then
        EdgeState[edge] gets Discard
        NodeState =Neutral
        Send Discard[Val] on edge [To prevent any Cycle]
      if NeighborCounter -1 =0 then
        NodeState gets Return
        Send Return(Minimum(Val,MinEdgeValue)) on E
      else
        NodeState gets Search
        for All edge in NeighborCounter do
        NeighborCounter gets NeighborCounter-1
        Send Search(Minimum(Val,MinEdgeValue)) on edge
end
procedure Response to Discard(Val) on E
begin
  EdgeState = Search
  Send Search(MinEdgeValue except Val)
end
```

## 4 Simulation Results

We have already mentioned how the Algorithm works. This section evaluates the effectiveness of the proposed model in real life scenario. In this simulation process we have considered $50m \times 50m$ square area which was divided in $5 \times 5$ regions. We deployed 40, 50, 60 and 70 sensor nodes within the same monitoring zone to understand the utilisation of sensor nodes and check the feasibility of this CA based model. As it is not feasible to explicitly choose the locations for deployment of the sensor nodes, here we have taken a randomized approach for deploying all these sensor nodes. All throughout our experiment we had identical initial energy levels for every sensor nodes. In Fig. 3. we can see how the sensors are deployed randomly. In Fig. 4 we can see various snapshots of the grid network to understand how the Minimum Spanning Tree formation takes place from the randomly deployed sensors as depicted in Fig. 3. Here in Table 1. we can see how the initial active node count changes when different number of sensors are deployed.

In Fig. 5. We can see the final configuration after completion of the Minimum Spanning tree finding computation. We can also see how a virtual grid network is formed by the sensors. It is also noticeable that all the sensors are not needed to remain active simultaneously when the coverage area is sensor in that to provide the coverage.

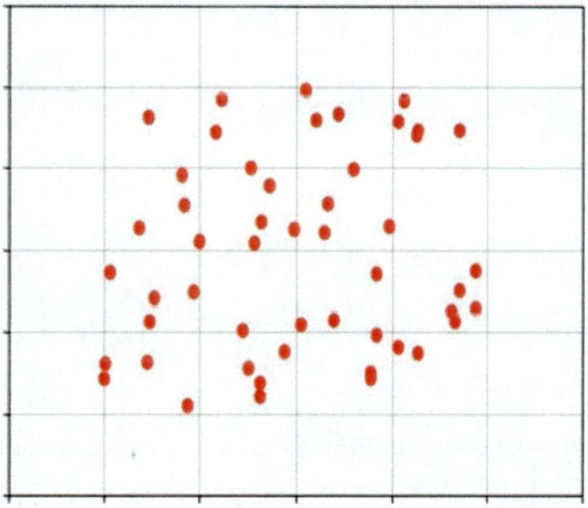

**Fig. 3** Randomly deployed sensor nodes

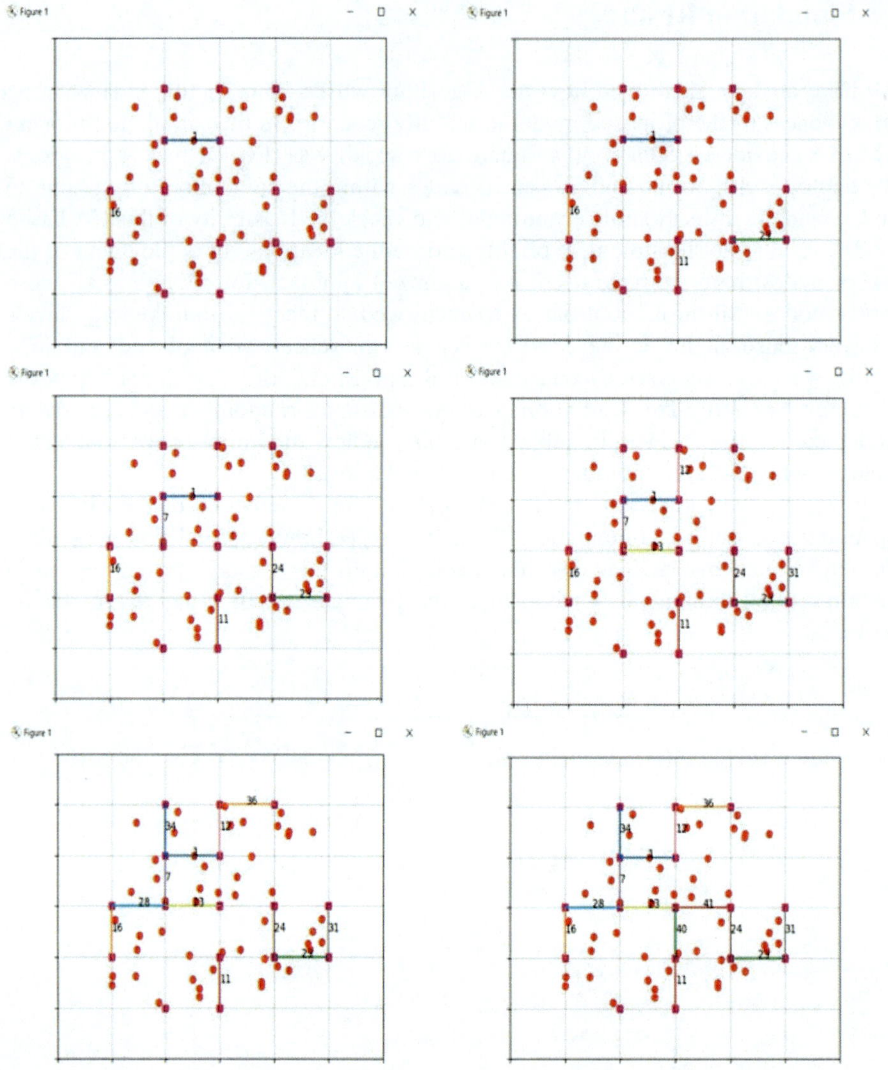

**Fig. 4** Snapshots of the Grid network at different time intervals in a sequential order

**Table 1** Mean of active node count after deploying different number of sensors

| Deployed node count | Mean of initial active node count |
|---|---|
| 40 | 14.2 |
| 50 | 15.3 |
| 60 | 17.7 |
| 70 | 19 |

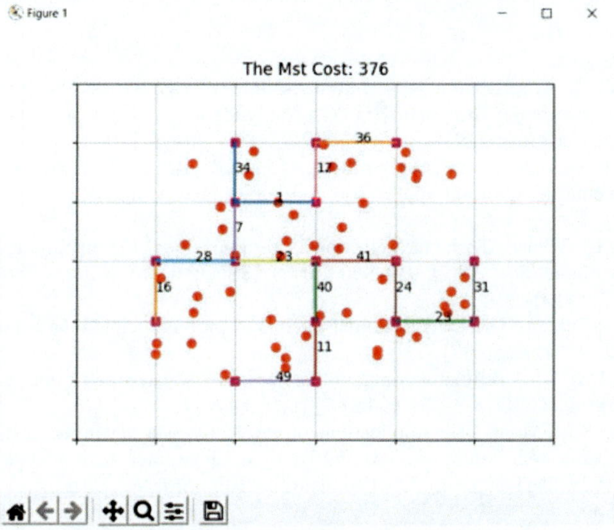


**Fig. 5** Final Minimum Spanning Tree

# 5 Conclusion

In this work we tried exploring how computational problems like finding minimum Spanning Tree can be solved in Wireless sensor networks using cellular automaton based models. The work also revealed how we can more efficiently use the sensors by reducing the number of active sensors at any single instance of time. We tried reviewing the existing literatures in this domain and how we can achieve feasible practical solutions. Wireless sensor networks can be well modeled using cellular automata. A cell can represent a sensor. There are enough results suggest that cellular automata can be used to model such problems of the sensor networks. As WSN and Cellular automata both are very interesting and unconventional computing models, there are various areas which are still unexplored. Here in this work we are using square grid cellular automata model for finding the MST, It will be interesting to solve different kinds of other computational problems also, moreover that hexagonal cellular automatons can also be interesting in future.

**Acknowledgements** This work is supported by the SERB, Govt. of India sponsored project titled "Computational Problems and Cellular Automata" (File No: EMR/2017/001571)

## References


1. Elhayatmy, G., Dey, N., Ashour, A.S.: Internet of things based wireless body area network in healthcare. In: Internet of Things and Big Data Analytics Toward Next-generation Intelligence, pp. 3–20. Springer, Cham (2018)
2. Valverde, J., Rosello, V., Mujica, G., Portilla, J., Uriarte, A., Riesgo, T.: Wireless sensor network for environmental monitoring: application in a coffee factory. https://doi.org/10.1155/2012/638067
3. Stratmann, M., Worsch, T.: Leader election in d-dimensional CA in time diam log(diam). Fut. Gener. Comput. Syst. **18**(7), 939–950 (2002). ISSN 0167-739X. https://doi.org/10.1016/S0167-739X(02)00073-0
4. Roy, S.: Das, Sukanta Distributed mutual exclusion problem in cellular automata. J. Cell. Autom. **12**(6), 493–512 (2017)
5. Barman, D., Das, S.: A cellular automaton that computes shortest paths in grid graph (2021). https://doi.org/10.1007/978-3-030-69480-7_1
6. Jino Ramson, S.R., Moni, D.J.: Applications of wireless sensor networks - a survey. In: 2017 International Conference on Innovations in Electrical, Electronics, Instrumentation and Media Technology (ICEEIMT), pp. 325-329. Coimbatore (2017). https://doi.org/10.1109/ICIEEIMT.2017.8116858
7. Gallager, R.G., Humblet, P.A., Spira, P.M.: A distributed algorithm for minimum-weight spanning trees. ACM Trans. Program. Lang. Syst. **5**(1), 66–77 (1983)
8. Lavallee, I., Roucairol, G.: A fully distributed (minimal) spanning tree algorithm. Information processing letter. **23**, 55–62 (1986)
9. Li, Y., Zhao, Y., Zhang, Y.: A spanning tree construction algorithm for industrial wireless sensor networks based on quantum artificial bee colony. J. Wirel. Comput. Netw. **2019**, 176 (2019). https://doi.org/10.1186/s13638-019-1496-z
10. Sudarmani, R., Vanithamani, R.: Minimum spanning tree for clustered heterogeneous sensor networks with mobile sink. In: 2015 IEEE International Conference on Computational Intelligence and Computing Research (ICCIC), pp. 1–6. Madurai (2015). https://doi.org/10.1109/ICCIC.2015.7435631
11. Diaz, S., Mendez, D., Scholzel, M.: Dynamic Gallager-Humblet-Spira algorithm for wireless sensor networks. In: 2018 IEEE Colombian Conference on Communications and Computing (COLCOM), pp. 1–6. Medellin (2018). https://doi.org/10.1109/ColComCon.2018.8466704
12. Choudhury, S.: Cellular automata and wireless sensor networks. In: Adamatzky, A., (eds.), Emergent Computation. Emergence, Complexity and Computation, vol. 24. Springer, Cham (2017). https://doi.org/10.1007/978-3-319-46376-6_14
13. Bhattacharjee, K., Naskar, N., Roy, S., Das, S.: A survey of cellular automata: types, dynamics, nonuniformity and applications. Nat. Comput. **19**(2), 433–461 (2020). Springer

# A Novel Hash Function Based on Hybrid Cellular Automata and Sponge Functions

**Anita John, Alan Reji, Ajay P Manoj, Atul Premachandran, Basil Zachariah, and Jimmy Jose**

**Abstract** Hash functions serve as the fingerprint of a message. They also serve as an authentication mechanism in many applications. Nowadays, hash functions are widely used in blockchain technology and bitcoins. Today, most of the work concentrates on the design of lightweight hash functions which needs minimal hardware and software resources. This paper proposes a lightweight hash function which makes use of Cellular Automata (CA) and sponge functions. This hash function accepts arbitrary length message and produces fixed size hash digest. An additional property of this function is that the size of the hash digest may be adjusted based on the application because of the inherent property of varying length output of sponge function. The proposed hash function can be efficiently used in resource constraint environments in a secure and efficient manner. In addition, the function is resistant to all known generic attacks against hash functions and is also preimage resistant, second preimage resistant and collision resistant.

**Keywords** Cryptographic Hash functions · Cellular automata · Sponge functions · Omega-flip permutation

A. John (✉) · J. Jose
Department of Computer Science and Engineering, National Institute of Technology Calicut, Kozhikode, Kerala, India
e-mail: anita_p170007cs@nitc.ac.in

J. Jose
e-mail: jimmy@nitc.ac.in

A. Reji · A. P. Manoj · A. Premachandran · B. Zachariah
Department of Computer Science and Engineering, Rajagiri School of Engineering and Technology, Kochi, Kerala, India

 
S. Das and G. J. Martinez (eds.), *Proceedings of First Asian Symposium on Cellular Automata Technology*, Advances in Intelligent Systems and Computing 1425,
https://doi.org/10.1007/978-981-19-0542-1_16

# 1 Introduction

Any digital document needs a fingerprint of itself to avoid misuse of the document known as the message digest or hash digest. Cryptographic hash functions are one-way functions that take an arbitrary length input and produce a fixed length hash digest as output. Different hash functions were developed so far which made use of Merkle-Damgård [11, 22] and Davies-Meyer [21] schemes. Today, most of the applications need lightweight hash functions which can be used in resource constraint environments. But, the lightweight property of the hash functions must not affect the security of the function. So this is a trade off between efficiency and security. In order to accomplish this, new cryptographic primitives like CA and sponge functions can be used in their design. This paper proposes a hash function that makes use of both CA and sponge function.

The rest of the paper is organized as follows. Section 2 gives a brief description about CA and sponge functions. Section 3 gives literature survey on the hash functions based on CA and sponge. Section 4 explains the detailed design of the proposed hash function followed by Sect. 5 that discusses the design rationale. Security analysis is done in Sect. 6 followed by conclusion.

# 2 Preliminaries

In this section, we discuss some basic definitions and terminologies related to CA, sponge functions and cryptographic hash functions.

## *2.1 Basics of CA*

CA are a discrete lattice of cells. Each cell can store a value and a next state computation function or rule which is a Boolean function associated with it. The value of the cells get updated based on the CA rule at every clock cycle and this happens in parallel. The transformation of the value in a particular cell is influenced by the state of the neighbouring cells. The rules of CA are developed based on the number of neighbours involved to update the values of a cell. There are two types of CA rules—linear rules and non-linear rules. Linear rules are those rules that use only XOR in their combinational logic and non-linear rules are rules that do not follow the conditions of linear rules. In general, the number of cells $n$ that participate in a CA cell update is given by $n = 2r + 1$, where $r$ is the radius of the neighbourhood. All the cells in a CA can update their states using the same CA rules or different CA rules. If they use the same rules, then they are called *uniform CA* and if they use more than one rule, then they are called *hybrid CA*. The neighbors of the cells at the extreme left and right are decided based on whether the CA follows periodic or null boundary conditions. In addition to these conditions, three more boundary conditions are discussed in [8] namely *one null*, *null one* and *one* boundary conditions. Here, we make use of *one null boundary* where the left most cell takes the value 1 as its left neighbour and rightmost cell takes the value 0 as it right neighbour. This makes CA rules cryptographically robust [8]. CA have recently evolved as a good cryptographic primitive due to its chaotic nature and parallel execution. It is also efficient in both hardware and software applications due to its simple logic operations. Three-neighbourhood CA developed by Stephen Wolfram [32] had been very powerful in developing cryptographically secure pseudorandom number generators. Later, 3-neighbourhood CA have been widely used in many cryptographic applications like stream ciphers [30], block ciphers [24] and hash functions [23].

## 2.2 *Sponge Functions*

A sponge function [3] is a simple iterative function that takes a variable length input, processes it and generates an infinite length output. Sponge functions have an internal function F that can either permute or transform the state bits. Sponge function is the basic technique used in Keccak[5] which was selected as the new SHA-3 standard by the National Institute of Standards and Technology (NIST) in October 2012. The function operates on a state of bits of width $b$, where $b = r + c$. Here $r$ is called the *bit rate* and $c$ is called the *capacity* of the sponge. The values of $r$ and $c$ can be varied. The security level that can be attained using sponge is dependent on $c$ [4].

Working of sponge function-

**Absorbing phase**-In this phase, each message block after XORing with $r$ bits of the $b$ state bits is subject to an internal transformation function. The output of $n_r$ rounds of transformation will be the state bits used for the next message block. This continues until all message blocks are exhausted.
**Squeezing phase**-In this phase, the $r$ bits from the output of absorption phase are extracted and it forms the first part of the final hash digest. This state again undergoes transformation and the next $r$ bits from its output are again squeezed out. This continues until the needed length hash digest has been squeezed out. If the length of the squeezed out bits is more than the required hash digest length, it must be truncated to match the needed length. The strength of the sponge function lies in the internal transformation function.

## 2.3 *Cryptographic Hash Functions*

A cryptographic hash function accepts an arbitrary length input and outputs a fixed length output. Earlier, all known hash functions made use of a compression function which processes every message block in the same manner. They were called iterated hash functions. The properties of a good hash function are (i) Collision Resistance (ii) Preimage resistance (iii) Second preimage resistance. The most popular standard hash functions were MD5 [26] and SHA [29] (Secure Hashing Algorithm) family of hash functions. SHA-256 which produces a 256-bit hash digest is considered to be secure and is now used in Blockchain. Even though MD5 and SHA were considered to be the standard hash functions, by 2004 and 2005, Wang et al. [31] had shown that finding collisions for MD5 and SHA were relatively easy. This made NIST to make an announcement for an open competition in 2007 for a stronger hash function design to be selected as SHA-3. In 2012, Keccak [5] was declared as the winner. Both Keccak and SHA-3 used a new design strategy called ***sponge functions***. The main advantage of using sponge functions in hash design is that we could get a hash digest of any desired length.

### 2.4 Omega-Flip Permutations

Several bit level permutation functions for fast software cryptography were discussed in [19]. Bit level permutations play an important role in introducing diffusion in a block of data. Omega-flip (OMFLIP) permutation permutes $n$ bits in at most $log(n)$ permutation instructions. The instruction used for performing this operation is *omflip*. For an Omega-flip network, there are 2 basic operations- omega and flip, the order of which are decided by a 2-bit opcode. In order to permute $n$ bits, we make use of an $n$-bit control string. More detailed description about the working of this instruction is available in [20].

## 3 Related Work

### 3.1 Hash Functions Based on Cellular Automata

Ivan Damgard was the first to propose a collision-free hash function based on CA [11]. Damgard proposed a hash function that computes a 128-bit hash digest from a 256-bit string. This made use of a binary CA rule $Rule120$ with periodic boundary conditions. The CA evolved through 384 iterations and the first bits of the strings from $257^{th}$ to $384^{th}$ iterations were taken as the final hash digest. But his proposal was cryptanalyzed by Daemen [10], who in turn proposed another hash function called Cell Hash. Cell Hash made use of non-linear and linear CA rules with periodic boundary conditions which was also cryptanalyzed later [9].

Another hash function based on CA was proposed by Mihaljevic et al. [23]. This paper proposes a dedicated one way hash function based on programmable CA. They have used an iterative hash function where the compression function and output function made use of CA. Another lightweight hash function based on CA developed by Hanin et al. is LCAHASH [15]. An extension of LCAHASH is proposed in LCAHASH-1.1 [27] which made use of non-uniform CA with rule $< 30, 90 >$ for 128 or 256 iterations and the required output is taken as the hash digest. Jamil et al. [17] had proposed a hash function based on hybrid CA that made use of linear and non-linear rules together in a CA along with a customized omega-flip permutation. They made use of Rules 30 and 134 and they were the first to employ such elements into the compression function which showed good avalanche effect. Another recent work based on CA was done by Sadak et al. in [28] where they made use of a carefully chosen cryptographically secure hybrid rule set which contains both linear and non-linear CA rules. The rules are selected based on Cryptographically Suited Hybrid Rules (CSHR) algorithm proposed in [8].

### *3.2 Hash Functions Based on Sponge Functions*

A lightweight hash function based on sponge construction with permutation is Quark [2]. The hash function made use of shift registers and Boolean functions. The permutation used in QUARK was done by using stream ciphers based on shift registers like Grain [16] and block cipher like KATAN [7]. SPONGENT [6] and PHOTON [13] were the other lightweight hash functions that made use of sponge functions in their domain extension algorithm. Keccak, the winner of NIST competition for a new hash function was also based on sponge construction. A detailed description of all lightweight cryptographic hash functions is found in B.Hammad et al. in [14].

### *3.3 Hash Functions Based on CA and Sponge Functions*

As part of the research to find lightweight and efficient cryptographic hash functions, new designs that used compression functions using CA and sponge functions were developed. The use of CA and sponge as individual primitives in these functions produced hash digests that satisfied the properties of collision resistance, preimage and second preimage resistance. The strength of a hash function that uses sponge function is the strength of the internal transformation function used in it. This inspired researchers in the field of cryptographic hash functions to use CA in the transformation function of sponge.

Keccak [5] had used a CA transformation that affects the two neighbourhood cells of each cell in its $\chi$ transformation. CASH [18] is a family of hash functions based on CA and sponge function. This hash function made use of tweakable parameters which allows the user to choose suitable parameters depending upon the level of security and efficiency required for a particular application. They made use of both permutation and transformation in the internal transformation function of the sponge function. The hash function provided diffusion and non-linearity with the help of both cryptographic primitives. Very few works have been done that incorporated hybrid CA in the internal function of sponge.

# 4 Proposed Hash Function

This paper proposes a new hash function which makes use of hybrid CA in the internal function of sponge function. The internal transformation function has two layers- the permutation layer and the transformation layer. The transformation layer makes use of CA rules for their transformation. Here we have combined linear and non-linear 3-neighbourhood CA rules in a single CA which is supposed to give good characteristics to the hash function. The linear CA rules offer good randomness, but lack security due to linear cryptanalysis. The non-linear CA rules offer high algebraic degree and non-linearity, but lack correlation immunity. So, the use of uniform linear and non-linear CA rules does not ensure good cryptographic properties. Kaushik et al. [8] proposed hybrid CA rule sets which combined the linear and non-linear CA rules. The combination of both types of rules made the rule set ideal for cryptographic applications by exhibiting good cryptographic properties. In this work, we have selected the rule vector for the hybrid CA as $< 30, 90, 150, 30, 210, 30, 90, 150 >$. The permutation layer makes use of Omega-Flip permutation [19]. The strength of the hash function depends on the strength of the internal function used and hence we have used hybrid CA together with permutation function.

## *4.1 Description of the Hash Function*

The hash function accepts an arbitrary length input message and produces 256-bit hash digest. The size of the hash digest can be varied which is an inherent property of the sponge function. The sponge function is implemented on a state bit of width $b$, where $b = r + c$. Here, we have taken $r$ as 92 and $c$ as 132. Another combination is r=92 and c=258 which gives good results and is resistant to attacks at the cost of more execution time. The block diagram for the proposed hash function is shown in Fig. 1.

The input message must be padded to make the length of the message a multiple of $r$. This is done by padding the message with '10' followed by the 64- bit representation of the message length. Padding step is mandatory even if the original message is a multiple of $r$ even before padding. This ensures the collision resistance property of the hash digest. The padded message is then divided into $r$ bit blocks $M_1$, $M_2$,..., $M_i$. Initially, $b$ is set to all zeros. Each message block is XORed with the $r$ bits of the state bits of $b$. The XORed $r$ bits together with $c$ bits will be transformed by the internal transformation function. The block diagram of the internal function $f$ of the sponge function is shown in Fig. 2.

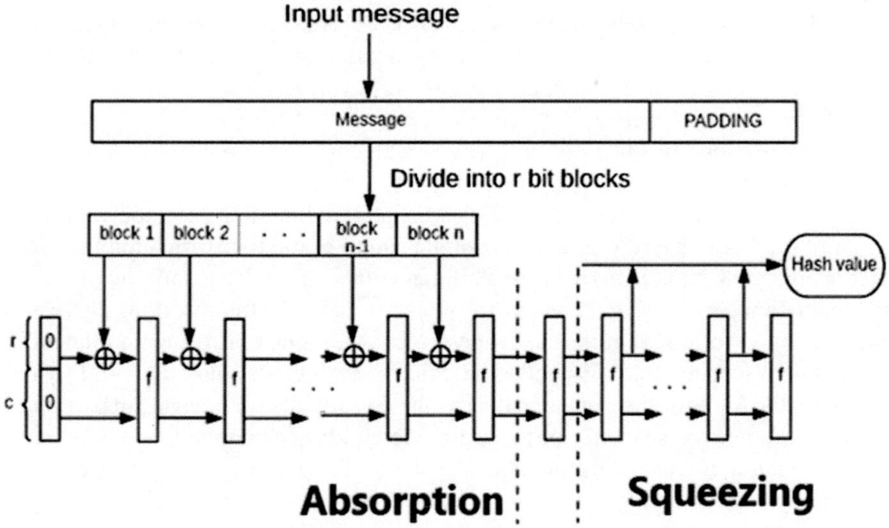


**Fig. 1** Block diagram for Hash Function

**Fig. 2** Internal function

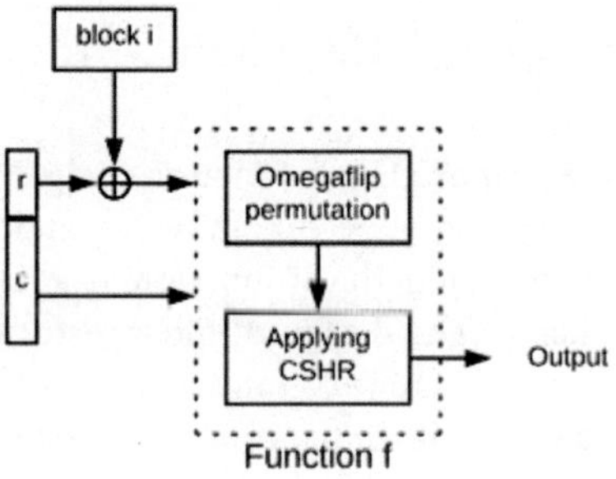


## 4.2 *Single Round of the Internal Function*

The state bits in $b$ undergo an internal transformation function F which proceeds in 2 steps- permutation and transformation. To permute the state bits, we have used Omega-Flip permutation. We have used the bits of the message block itself as the control bits for permuting them. In order to permute 224 bits, we need eight omflip instructions. The 2-bit opcode needed in each of these 8 instructions can be obtained from the first 16 bits of the $b$ state bits. i.e., the first 16 bits of the message block itself. So, the order of performing omega or flip operation is decided by the first 16 bits of the message block $M_i$. The entire omega-flip permutation is dependent on the message. After performing the permutation, the state will be subjected to a transformation phase. In the transformation phase, we make use of cryptographically secure hybrid linear and non-linear 3-neighbourhood CA rules. The order of rules to be used in each of the cells is decided by an algorithm called Cryptographically

Secure Hybrid Rules (CSHR) discussed in [8]. These rules are expected to give good results with respect to algebraic degree, non-linearity and period. The rule set that we use here is $< 30, 90, 150, 30, 210, 30, 90, 150 >$ where each component in the rule set represents an elementary CA rule. This rule set was proved to have good cryptographic properties like non-linearity and algebraic degree [8]. The CA is run for 13 clock cycles.

There will be two rounds (i.e., $n_r = 2$) for the internal function. After the last transformation function, a null round, i.e., a single round of the transformation function without any XOR input is performed. This marks the end of the absorption phase of the sponge function. During the squeezing phase, the $r$ bits of the absorbing phase are squeezed out to form part of the hash digest. Now, the same transformation function used in the absorbing phase is applied and the next $r$ bits are squeezed out. During this phase, the $b$ bits of the state are taken as the control bits for omega-flip permutation. This interleaved sequence of transformation and squeezing will continue until we get the required hash digest length. The size of the hash digest can be varied as per user needs by adjusting the number of times $r$ bits are squeezed out. If the length exceeds the required hash digest length, the remaining bits can be truncated.

## 5 Design Rationale

The proposed hash function design makes use of CA and sponge function. The emergence of sponge functions as random oracle encourages their use in the design of cryptographic hash function as a replacement to the conventional Merkle-Damgård scheme. The design of the transformation function has a crucial role in the strength of the sponge function which increases the resistance of the hash function to known attacks on it. In the proposed design, the internal transformation consists of permutation and transformation. Since we combine both operations into a single function, we have combined the best possible among them to retain the lightweight nature of the hash function. The design rationale for each of the steps in the hash function are discussed below.

**1. Padding the input**: Here we pad the message to make the length of the message a multiple of $r$, the bit rate. This padding scheme ensures that the last block never contains all zeros which is one of the criteria for using sponge functions. This padding rule also enhances the security of the hash function design.

**2. Sponge functions**: Merkle-Damgård scheme suffered from length extension attacks and multicollision attacks. The sponge function is known to be resistant to these attacks. This motivated us to use sponge function in our design. Our next aim was to strengthen the sponge function by designing a strong internal function. We have used a hybrid rule set that has a combination of linear and non-linear 3-neighbourhood rules. The use of hybrid CA enhances the security of the function in 2 ways- linear hybrid rules ensure diffusion of bits and non-linear hybrid rules introduce non-linearity to the function. Both non-linearity and diffusion are important factors that determine the strength of the function. The position of the CA rules in the selected rule set is based on the selection procedure followed in the *CSHR* algorithm proposed in [8]. We have added an additional step of permutation with the implementation of a faster method of Omega-flip [19] so as to distribute the bits throughout the state. The addition of null round in between the absorbing and squeezing phases helps to avoid sliding attacks on the sponge functions. So, we have tailored a CA with hybrid rules along with omega- flip permutation into the sponge function.

The application of hybrid CA rules and permutation function in a single internal round of sponge function ensures the dependency of all the bits in a single block of the message through simple logical operations. At the end of the absorption phase, all the blocks of the message will be thoroughly mixed. The squeezing phase again mixes the bits and squeezes out the required bits of the hash digest. This helps to strengthen the avalanche property of the proposed hash function.

# 6 Security and Performance Analysis

## *6.1 Security Analysis*

**1. Avalanche effect**: Avalanche effect is a desirable property for hash functions. This property states that, a small change in the input should bring a great change in the output bits of the hash digest. To evaluate the Avalanche effect, we have taken a sample of 100 varying length messages. For each message, we changed the original message by one bit and calculated the Hamming distance between the hash values of the original message and the corresponding modified message. The values for Hamming distance range from 117 to 145 where the size of the hash digest is 256. This shows that even a single bit change in the input has varied around half of the output bits which proves that the proposed hash function exhibits good avalanche effect. Figure 3 illustrates the Avalanche effect.

**2. Pre-image resistance and Second pre-image resistance**: The security of the proposed system is based on the internal transformation function of the sponge construction. This function is composed of hybrid CA and permutation. Given a

**Fig. 3** Avalanche Effect

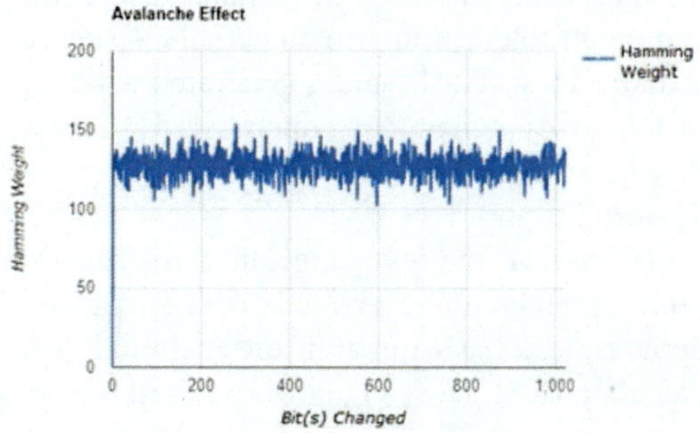

hash digest output of size $n$, we need to perform $2^n$ operations to find a preimage or a second pre image and $2^{n/2}$ operations to find a collision. Here, the internal transformation has transformed input message block in an efficient manner. Each message block first goes through a permutation followed by CA. In CA, each bit is dependent on 27 neighbouring bits at the end of 13 CA cycles, since we are using 3-neighbourhood CA and each bit is dependent on 2q+1 neighbouring bits where $q$ is the number of cycles of CA. The hash digest is produced after many rounds of internal transformation and hence is dependent on all bits of the input message. So the function is resistant to pre-image, 2nd pre-image and collision attacks.

**3. Sliding attacks**: One of the attacks on hash functions that made use of sponge function is the sliding attack [12]. Sliding attacks occur when we are able to find a *"slid pair"* of messages. In sponge functions, the first $r$ bits of the hash digest are obtained from the output of the last internal transformation of the absorbing phase, So, there is a chance that an attacker gets some hint regarding the last block of the input message or about the internal transformation. In order to avoid this, we run an additional null round which hides the last message block.

**4. Randomness Test using NIST Statistical Test Suite** The randomness of hash digest is a property that makes it suitable for cryptographic applications. So to test the randomness of the hash digest output, we make use of NIST Statistical Test Suite [25]. This contains a battery of 15 tests. Each test computes the chi-square statistic of a particular parameter. This is done by comparing this parameter for the generated bit stream with the ideal value of this parameter. The ideal value will be the one that is obtained from the theoretical results of such an identical sequence of bits. This chi-square value is converted to a random probability value called P-value. Each test is based on a calculated test statistic value which is a function of the input data. This value is compared with a critical value that is taken from the reference distribution. These are compared and a P-value is computed that summarizes the strength of evidence against the null hypothesis. Each P-value is the probability that a perfect random number generator would have produced a sequence less random than the sequence that was tested. A P-value=0 indicated non randomness and a P-value =1 indicates perfect randomness [1].

The tests in the test suite, in general, checks for the proportion of zeros and ones in the entire bit stream as well as in subblocks of the stream, checks whether there are any repeating patterns, non-periodic patterns or too many occurrences of ones in the bitstream. Based on these analyses, an output file will be generated by the test suite with relevant intermediate values such as test statistics and P-values for each statistical test. The P-values generated will help to analyze the randomness properties of the newly generated sequence [25]. The randomness test of the generated hash digests using NIST test suite affirms the cryptographic strength of the hash function against attacks.

To execute the test suite, the hash function has been used as a pseudorandom number generator to create a data stream of 10 Mb. The stream is generated by applying the hash function on an initial seed, $S$ which is incremented by 1. We calculate H(S), H(S+1) and so on until we are able to concatenate them to get a 10

**Table 1** NIST test results

| SI.no | Test name | P-value | Status |
|---|---|---|---|
| 1 | Frequency test | 0.419021 | Pass |
| 2 | Block frequency test | 0.030806 | Pass |
| 3 | Cumulative sums test | 0.935716 | Pass |
| 4 | Runs test | 0.236810 | Pass |
| 5 | Longest runs test | 0.017912 | Pass |
| 6 | Rank test | 0.616305 | Pass |
| 7 | FFT test | 0.897763 | Pass |
| 8 | Non overlapping template test | 0.616305 | Pass |
| 9 | Overlapping template test | 0.514124 | Pass |
| 10 | Random excursions | 0.911413 | Pass |
| 11 | Random excursions variant | 0.637119 | Pass |
| 12 | Linear complexity | 0.719747 | Pass |

Mb sequence. This acts as input to the NIST statistical test suite. The binary stream generated by the hash function showed good P-values and pass rates for the relevant tests for hash functions as shown in Table 1.

## 6.2 Performance Analysis

The proposed hash function has been implemented in C language, while the performance experiments were done on an Intel Core i5 (2.3 GHz) microprocessor. The speed of the hash function was calculated with a test file of size 10097 bytes and was found to be 1.1 MB/s.

The randomness property of the hash digest generated by the hash function is important to make it cryptographically secure. In order to get good P-values for the tests in NIST, we have tested the bit stream by increasing the number of cycles sequentially from 2 and found that 13 cycles gave good results. We have also implemented the same hash function without using omega-flip permutation, i.e., the internal function of the sponge function has only the transformation step which made use of hybrid 3-neighbourhood CA rules. The speed achieved during that design was 6.84 MB/s which is comparable or even more than other popular hash functions like SHA. The design also showed good avalanche effect, but it did not give good P-values for some of the tests in NIST test suite. While designing the new hash function, we have considered the randomness property and avalanche effect with equal priority and this accounts for the decreased speed of the hash. The size of the internal state bits of the sponge function, the values of $r$ and $c$ are also factors affecting the speed of the hash function. So, here we have a trade-off between security and speed. But, the reward for slow speed is the adjustable size of the hash digest due to the use of sponge functions and additional security due to CA and omflip permutation.

## 7 Conclusion and Future Work

In this paper, we propose a new hash function based on Cellular Automata and sponge functions which takes an arbitrary length input and produces a 256-bit hash digest. The proposed function has tailored the best cryptographic primitives into the design and hence shows good Avalanche effect. We have used the CSHR algorithm to develop a 3-neighbourhood CA rule set that uses both linear and non-linear rules. In addition, we have also used Omega-Flip permutation which helps in permuting the bits in a fast and efficient manner. The proposed hash function produces a hash digest that resembles a random number, which makes it resistant to all known attacks on hash functions. We are still investigating the possibility of using other rule sets under CSHR algorithm. We are also analysing the impact of using CA rules of higher radii like 5-neighbourhood CA in place of 3-neighbourhood.

## References

1. A statistical test suite for random and pseudorandom number generators for cryptographic applications. Tech. Rep. NIST Special Publication (SP) 800-22, Rev. 1a, Includes updates as of April 2010, National Institute of Standards and Technology, Gaithersburg (2010)
2. Aumasson, J.P., Henzen, L., Meier, W., Naya-Plasencia, M.: QUARK: A lightweight hash. In: Mangard, S., Standaert, F.X. (eds.) Cryptographic Hardware and Embedded Systems, CHES 2010, pp. 1–15. Springer, Berlin (2010)
3. Bertoni, G., Daemen, J., Peeters, M., Van Assche, G.: Sponge functions. In: ECRYPT hash workshop, vol. 2007. Citeseer (2007)
4. Bertoni, G., Daemen, J., Peeters, M., Van Assche, G.: On the Indifferentiability of the sponge construction. In: Smart, N. (ed.) Advances in Cryptology - EUROCRYPT 2008, pp. 181–197. Springer, Berlin (2008)
5. Bertoni, G., Daemen, J., Peeters, M., Van Assche, G.: Keccak. In: Johansson, T., Nguyen, P.Q. (eds.) Advances in Cryptology - EUROCRYPT 2013, pp. 313–314. Springer, Berlin (2013)
6. Bogdanov, A., Knežević, M., Leander, G., Toz, D., Varıcı, K., Verbauwhede, I.: SPONGENT: A lightweight hash function. In: Preneel, B., Takagi, T. (eds.) Cryptographic Hardware and Embedded Systems - CHES 2011, pp. 312–325. Springer, Berlin (2011)
7. Cannière, C., Dunkelman, O., Knežević, M.: Katan and ktantan – A family of small and efficient hardware-oriented block ciphers. In: Proceedings of the 11th International Workshop on Cryptographic Hardware and Embedded Systems, CHES '09, pp. 272-288. Springer, Berlin (2009)
8. Chakraborty, K., Chowdhury, D.R.: CSHR: Selection of cryptographically suitable hybrid cellular automata rule. In: Sirakoulis, G.C., Bandini, S. (eds.) Cellular Automata, pp. 591–600. Springer, Berlin (2012)
9. Chang, D.: Preimage attacks on Cellhash, Subhash and Strengthened versions of Cellhash and Subhash. IACR Cryptol. ePrint Arch. **2006**, 412 (2006)

10. Daemen, J., Govaerts, R., Vandewalle, J.: A framework for the design of one-way hash functions including cryptanalysis of Damgård's one-way function based on a cellular automaton. In: International Conference on the Theory and Application of Cryptology, pp. 82–96. Springer (1991)
11. Damgård, I.B.: A design principle for hash functions. In: Conference on the Theory and Application of Cryptology, pp. 416–427. Springer (1989)
12. Gorski, M., Lucks, S., Peyrin, T.: Slide attacks on a class of hash functions. In: Pieprzyk, J. (ed.) Advances in Cryptology - ASIACRYPT 2008, pp. 143–160. Springer, Berlin (2008)
13. Guo, J., Peyrin, T., Poschmann, A.: The PHOTON family of lightweight hash functions. In: Rogaway, P. (ed.) Advances in Cryptology - CRYPTO 2011, pp. 222–239. Springer, Berlin (2011)
14. Hammad, B.T., Jamil, N., Rusli, M.E., Reza, M.: A survey of lightweight cryptographic hash function. In: International Journal of Scientific and Engineering Research, vol. 8 (2017)
15. Hanin, C., Echandouri, B., Omary, F., Bernoussi, S.E.: L-CAHASH: A novel lightweight hash function based on cellular automata for RFID. In: UNet (2017)
16. Hell, M., Johansson, T., Maximov, A., Meier, W.: A stream cipher proposal: Grain-128. In: 2006 IEEE International Symposium on Information Theory, pp. 1614–1618 (2006)
17. Jamil, N., Mahmood, R.: A new cryptographic hash function based on cellular automata rules 30 134 and omega-flip network (2012)
18. Kuila, S., Saha, D., Pal, M., Chowdhury, D.R.: CASH: Cellular automata based parameterized hash. In: International Conference on Security, Privacy, and Applied Cryptography Engineering, pp. 59–75. Springer (2014)
19. Lee, R.B., Shi, Z., Yang, X.: Efficient permutation instructions for fast software cryptography. IEEE Micro **21**(6), 56–69 (2001)
20. Lee, R.B., Shi, Z.J., Yin, Y.L., Rivest, R.L., Robshaw, M.J.B.: On permutation operations in cipher design. In: Proceedings of the International Conference on Information Technology: Coding and Computing (ITCC'04), vol. 2, p. 569. IEEE Computer Society, USA (2004)
21. Matyas, S.M.: Generating strong one-way functions with cryptographic algorithm. IBM Tech. Disclos. Bull. **27**, 5658–5659 (1985)
22. Merkle, R.C.: One way hash functions and DES. In: Conference on the Theory and Application of Cryptology, pp. 428–446. Springer (1989)
23. Mihaljevic, M., Zheng, Y., Imai, H.: A fast cryptographic hash function based on linear cellular automata over GF(q) (1998)
24. Mukhopadhyay, D., RoyChowdhury, D.: Cellular automata: an ideal candidate for a block cipher. In: Ghosh, R.K., Mohanty, H. (eds.) Distributed Computing and Internet Technology, pp. 452–457. Springer, Berlin (2005)
25. NIST Statistical Test Suite. https://csrc.nist.gov/projects/random-bit-generation/documentation-and-software. Accessed 01 Nov 2021
26. Rivest, R.: The MD5 message-digest algorithm (1992)
27. Sadak, A., Echandouri, B., Ziani, F.E., Hanin, C., Omary, F.: LCAHASH-1.1: A new design of the lcahash system for iot. Int. J. Adv. Comput. Sci. Appl. **10**(11) (2019)
28. Sadak, A., Ziani, F.E., Echandouri, B., Hanin, C., Omary, F.: HCAHF: A new family of ca-based hash functions. Int. J. Adv. Comput. Sci. Appl. **10**(12) (2019)
29. Secure Hash Standard, FIPS publication 180-1. National Institute of Standards and Technology (1995)
30. Tomassini, M., Perrenoud, M.: Stream ciphers with one-and two-dimensional cellular automata. In: International Conference on Parallel Problem Solving from Nature, pp. 722–731. Springer (2000)
31. Wang, X., Yu, H.: How to break MD5 and other hash functions. In: Cramer, R. (ed.) Advances in Cryptology - EUROCRYPT 2005, pp. 19–35. Springer, Berlin (2005)
32. Wolfram, S.: Cryptography with cellular automata. In: Williams, H.C. (ed.) Advances in Cryptology – CRYPTO '85 Proceedings, pp. 429–432. Springer, Berlin (1986)

# Synthesis of Single Length Cycle Two Attractor CA Using NSRT Diagram

**Suvadip Hazra and Mamata Dalui**

**Abstract** Cellular Automata (CA) is a discrete model that is increasingly being employed in scientific study and simulations. It has several applications in the domains such as VLSI design, error-correcting codes, test pattern generation, and cryptography, etc. The majority of these applications make use of three-neighbourhood 1-dimensional CA. In this work, a comprehensive study on 3-neighborhood CA with 2-states per cell has been reported. The Next State RMT Transition Diagram (NSRTD), a graph-based tool, is presented to analyse CA's state transition behaviour with fixed points. A methodology has been proposed for synthesising Single Length Cycle Two Attractor CA (TACA).

**Keywords** Cellular automata · TACA · NSRTD · Fixed points

## 1 Introduction

The introduction of cellular automata (CA) is an important milestone in the history of computing [11]. It provides an abstract model of concrete computers. It is found to be more expressive than Turing machines and provides means for modelling parallel computation [11]. The 3-neighborhood 2-state CA has simple, elegant structure [11]. In recent years, it is proved to be a powerful modeling tool [4–6]. However, the characterization of such a machine is still an area of research.

While characterizing the state space of 3-neighborhood CA, the researchers identified a set of states called attractor towards which neighboring states asymptotically approach in the course of dynamic evolution. A fixed point is an attractor with only one state [2]. The CA that has only fixed points is of prime interest in devising schemes for different applications, especially in authentication and cryptography [2,

S. Hazra (✉) · M. Dalui (✉)
NIT, Durgapur, West Bengal 713209, India
e-mail: sh.16cs1102@phd.nitdgp.ac.in

M. Dalui
e-mail: mamata.dalui@cse.nitdgp.ac.in



S. Das and G. J. Martinez (eds.), *Proceedings of First Asian Symposium on Cellular Automata Technology*, Advances in Intelligent Systems and Computing 1425,
https://doi.org/10.1007/978-981-19-0542-1_17

10]. Other potential applications include—verification of cache coherence in Chip Multiprocessors [4], pattern classification [5], diagnosis of faulty nodes in Wireless Sensor Network [6], language recognition [1] etc. This work reports the characterization of such special class of CA in null-boundary, based on the analysis of individual rules and their potential to form the fixed points. The tool referred to as Next State RMT Transition Diagram (NSRTD) [3] is used to provide the theoretical framework for the identification of such rules.

The identification of attractors, for linear/additive CA, is explored in [2]. The graph-based scheme for the identification of attractors is proposed in [9]. These works employ the de bruijn diagram and its subdiagrams to study the CA and propose a quadratic time algorithm to determine whether a linear CA is reversible. An alternative graph-based scheme, referred to as RVG (rule vector graph), is also proposed in [8] for null-boundary CA. It targets a linear time solution to check the invertibility of CA. A similar solution for periodic boundary CA is reported in [7]. However, the characterization of CA with fixed points beyond linear domain is yet to be explored. This motivates us to develop CA theory with the target to characterize rules that can form a specified number of fixed points in null-boundary.

In the above background, this paper is organized as follows. A comprehensive study on 3-neighborhood CA with 2-states per cell is presented in Sect. 2. The description of NSRTD in Sect. 3 followed by the presentation of the proposed methodology for the synthesis of CA with fixed points in Sect. 4 is provided. Section 5, concludes the paper.

## 2 CA Preliminaries

A cellular automaton consists of a number of cells organized in the form of lattice. It evolves in discrete space and time, and can be viewed as an autonomous finite state machine. Each cell of a CA stores a discrete variable at time $t$ that refers to the present state (PS) of the cell. The next state (NS) of the cell at $t+1$ is affected by its PS and the PSs of its neighbors. In this work, we concentrate on 3-neighborhood CAs (self, left and right neighbors) having two states—0 or 1 in null-boundary condition, where left (right) neighbor of the leftmost (rightmost) CA cell is fixed to 0 state (Fig. 1). The NS of $i$th-cell of such a CA is

$$S_i^{t+1} = f_i(S_{i-1}^t, S_i^t, S_{i+1}^t)$$

where $S_i^t$, $S_{i-1}^t$, and $S_{i+1}^t$ are the PSs of itself and the left and right neighbors of the $i$th cell and $f_i$ is the next state function. The set of states $S^t = (S_1^t, S_2^t, \ldots, S_n^t)$ of the cells at $t$ is the PS of the CA. Therefore, the NS of an $n$-cell CA is

$$S^{t+1} = (f_1(S_0^t, S_1^t, S_2^t), f_2(S_1^t, S_2^t, S_3^t), \ldots, f_n(S_{n-1}^t, S_n^t, S_{n+1}^t))$$

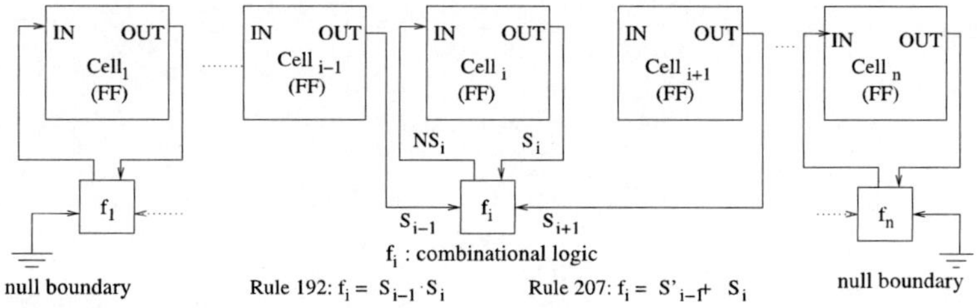


**Fig. 1** Null boundary CA

**Table 1** *RMT*s of the rules

| PS | 111 | 110 | 101 | 100 | 011 | 010 | 001 | 000 | Rule |
|---|---|---|---|---|---|---|---|---|---|
| RMT | T(7) | T(6) | T(5) | T(4) | T(3) | T(2) | T(1) | T(0) | |
| NS | 0 | 0 | 0 | 0 | 1 | 1 | 1 | 1 | 15 |
| NS | 1 | 1 | 1 | 1 | 1 | 1 | 1 | 0 | 254 |
| NS | 1 | 1 | 1 | 1 | 1 | 1 | 1 | 1 | 255 |

PS: present state; NS: next state, RMT: rule min term

The $f_i$ is expressed in the form of a truth table (Table 1). The decimal equivalent of the 8 outputs of $f_i$ is called 'Rule' $R_i$. In a 2-state 3-neighborhood CA, there can be $2^8$ (256) rules. Three such rules 15, 254, and 255 are illustrated in Table 1. The first row lists the possible $2^3$ (8) combinations of PSs of $(i-1)$th, $i$th and $(i+1)$th cells. The next three rows indicate the NSs of the $i$th cell, forming the rules 15 ($f_i = S'_{i-1}$), 254 ($f_i = S_{i-1} + S_i + S_{i+1}$) and 255 ($f_i = 1$). The following terminologies are relevant for ease of understanding of the CA theory developed.

**Definition 1** *(Rule Vector)*
The set RV = $\langle R_1, R_2, \ldots, R_i, \ldots, R_n \rangle$ of rules which configures the cells of a CA is called the rule vector of the CA.

**Definition 2** *(Uniform and Non-uniform CA)*
If $R_1 = R_2 = \cdots = R_n$ in RV = $\langle R_1, R_2, \ldots, R_i, \ldots, R_n \rangle$, then the CA is a uniform CA; otherwise it is a non-uniform CA.

**Definition 3** *(RMT)*
A combination of present states $(S^t_{i-1}, S^t_i, S^t_{i+1})$, shown in the 1st row of Table 1, is referred to as the Rule Min Term (RMT).

The column 100 of Table 1 is the 4th RMT. The next states corresponding to this RMT are 0 for the rules 15, and 1 for 254. An RMT is represented as T($m$), $m$ = 0, 1, 2, 3, 4, 5, 6, 7; and T($m$) ∈ {T}, where T = T(0), T(1), ..., T(7). The RMT for $i$th cell at time $t$ is denoted as $T^t_i(m)$. However, in the diagrams, an RMT is represented only by the corresponding decimal number ($m$). A rule divides the RMTs into two subsets, 0-RMT and 1-RMT, and are denoted as T/0 and T/1. An RMT String (RS) is defined as a sequence of consecutive RMTs which appear in a state of CA. It is represented as $(T_1, T_2, \ldots, T_n)$ for an $n$-cell CA, where $T_i \in$ T.

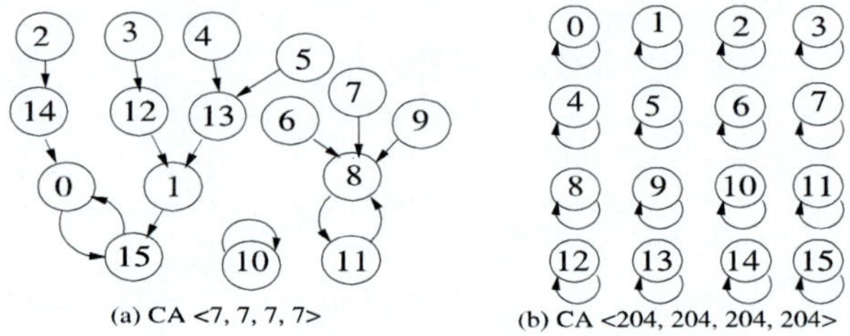


**Fig. 2** State transitions of SLCA

**Definition 4** *(Attractor)*
A set of states of a CA can form cycle ($10 \rightarrow 10$ and $8 \rightarrow 11 \rightarrow 8$ of Fig. 2a) and is called *attractor*. An attractor forms a basin with the states that lead to the attractor (($8 \rightarrow 11 \rightarrow 8$)-basin of Fig. 2a contains 5 states including the attractor states 8 and 11). The cycle $10 \rightarrow 10$ is a single length cycle attractor and called fixed point.

**Definition 5** *(Depth of CA)*
The depth of a CA is the length of the longest path traversed from any state to an attractor during its state transitions. The depth of the CA shown in Fig. 2a is 3 ($5 \rightarrow 13 \rightarrow 1 \rightarrow 15$).

The special class of irreversible CA with fixed point attractors is of our current interest.

**Definition 6** *(SLCA)*
The CA having at least one fixed point attractor is referred to as Single Length Cycle Attractor CA (SLCA). An SLCA may or may not have additional multi-length cycle attractor(s).

The CA of Fig. 2a has a fixed point attractor (state 10) and two multi-length cycle attractors ($8 \rightarrow 11 \rightarrow 8$ and $0 \rightarrow 15 \rightarrow 0$) and it is an SLCA. The state transitions of a special class of SLCA with $2^n$ ($n$ = length of CA) fixed point attractors is shown in Fig. 2b.

**Definition 7** *(SACA)*
The SLCA producing connected graph during its state transitions -that is, having only one fixed point attractor and no multi-length cycle, is called single length cycle single attractor CA (SACA).

An $n$-cell uniform CA with rule 16 forms a connected graph with only one fixed point attractor for any $n$. Therefore, the uniform CA with rule 16 is an SACA (Fig. 3a) and the rule 16 is an SACA rule.

**Definition 8** *(TACA)*
In its state transition, if an SLCA forms two connected components each with one fixed point attractor, then it is a single length cycle 2-attractor CA (TACA).

Uniform CA configured with rule 38 forms exactly two connected components, each with one fixed point attractor (as shown in Fig. 3b for 4-cell CA). Thus, it is a TACA and the rule 38 is a TACA rule.

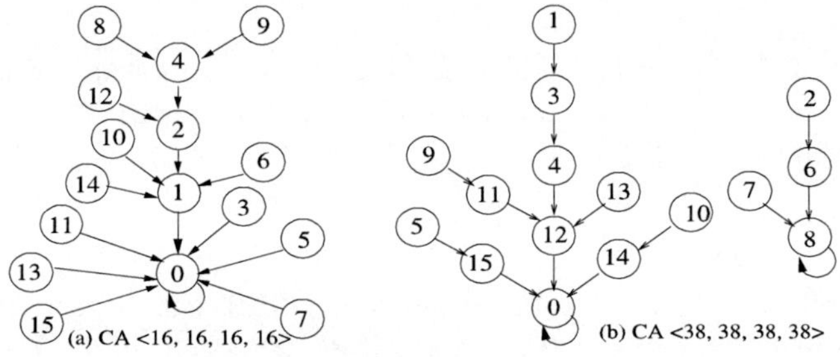


**Fig. 3** State transitions of the SACA and TACA

## 3 Next State RMT Transition Diagram

The graph based tool referred to as Next State RMT Transition Diagram (NSRTD) is used in this work to get more insight into the state transition behavior of SLCA. The NSRTD provides a generic methodology for determining the presence of fixed point attractors and the multi-length cycle attractors in a CA and, therefore, enables identification of rules that form only fixed point attractors. Some of the most relevant terminologies are described bellow.

**Definition 9** *(NCR)*
If $T_i^t$ is the RMT of $i$th-cell rule based on which the cell $i$ changes its state at $t$th time step, then the next cell RMTs (NCRs) of $T_i^t$ are the RMTs $T_{i+1}^t = (2 \times T_i^t \bmod 8)$, $((2 \times T_i^t + 1) \bmod 8)$ of $(i + 1)$th-cell rule based on which the $(i + 1)$th cell can change its state at $t$th time step.

For example, if the RMT of $i$th-cell rule $R_i$ is T(2), then its NCRs are T(4) (=2 × T(2) mod 8) and T(5) (=(2 × T(2) + 1) mod 8). All such NCRs of the RMTs are shown in Table 2.

**Definition 10** *(NSR)*
If $T_i^t$ (respectively $T_i^{t+1}$) be the RMT of $i$th-cell rule $R_i$ based on which cell $i$ changes its state at time $t$ (respectively at time $t+1$), then $T_i^{t+1}$ is the Next State RMT (NSR) of $T_i^t$ for cell $i$.

**Table 2** Relationship between $T_i$ and $T_{i+1}$ (NCR)

| RMT $T_i$ ($i$th cell rule) | NCR of RMT $T_i$ (RMTs of $(i+1)$th cell rule) |
|---|---|
| T(0)/T(4) | T(0), T(1) |
| T(1)/T(5) | T(2), T(3) |
| T(2)/T(6) | T(4), T(5) |
| T(3)/T(7) | T(6), T(7) |

**Definition 11** *(NSRS)*
The sequence of NSRs based on which the $i$th-cell changes its state during the state transitions of the CA is called Next State RMT Sequence (NSRS). That is, an NSRS for $i$th-cell (denoted as $NSRS_i$) is $T_i^0$, $T_i^1$, ..., $T_i^t$, $T_i^{t+1}$, ..., where $T_i^{t+1}$ is the NSR of $T_i^t$.

**Definition 12** *(NSRS Graph)*
For the $i$th-cell rule $R_i$ of a CA, the all possible NSRSs can be represented by a directed graph G(V,E), called NSRS graph (NSRS-G), where all the $T_i^t$s $\in$ V and $e_i(T_i^t, T_i^{t+1}) \in$ E iff $T_i^{t+1}$ is the NSR of $T_i^t$.

**Definition 13** *(NSRTD)*
The NCR, NSR, NSRS and NSRS-G of a CA enable construction of the set of directed graphs, called NSRTD, as 2-dimensional arrangement of nodes. Each node is an RMT and each graph in NSRTD represents an attractor of the CA. The edge ($T_i^t$, $T_i^{t+1}$) $\in$ E, such that $T_i^{t+1}$ is the NSR of $T_i^t$ and also the edge ($T_i^t$, $T_{i+1}^t$) $\in$ E, such that $T_{i+1}^t$ is the NCR of $T_i^t$. Further, the nodes $T_i^t$ ($\forall$ $t$) form an NSRS for $i$th-cell.

**Definition 14** *(Compatibility class of NSRSs)*
Two NSRSs defined for the $i$th and $(i+1)$th CA cell ($NSRS_i$ and $NSRS_{i+1}$) are called compatible if $j$th NSR of the two are related in the sense that $j$th NSR of $NSRS_{i+1}$ is the NCR (Table 2) of $j$th NSR of $NSRS_i$.

The following subsections explain the NSR, NSRS, NSRS-G, compatibility class of NSRSs and the construction of NSRTD.

## 3.1 Identification of NSRs

If xyz is the RMT of rule $R_i$ based on which the $i$th-cell changes its state, and $l$ and $r$ be the PS of cell ($i$-2) and cell ($i+2$) respectively, then the NSR of the cell can be $d_l d d_r$, where the RMT xyz of $R_i$ is $d$ and $d_l$ = RMT lxy of the $(i-1)$th-cell rule $R_{i-1}$ and $d_r$ = RMT yzr of the $(i+1)$th-cell rule $R_{i+1}$. For uniform CA with rule R, the $d_l$, d, and $d_r$ are the RMTs lxy, xyz and yzr of R. In a null-boundary $n$-cell ($cell_1$, $cell_2$, ... $cell_i$, ..., $cell_n$) CA, the following conditions are satisfied.

1. **For 1st cell** ($i$ = **1**): $d_l = 0$, x=0, r=0/1.
2. **For 2nd cell** ($i$= **2**): l=0, r=0/1.
3. **For Intermediate cell** ($i$ = **3**, ... ($n$- **2**)): l=r=0/1.
4. **For 2nd last cell** ($i$ = $n$- **1**): l=0/1, r=0.
5. **For Last cell** ($i$ = $n$): l=0/1, z=0, $d_r$ =0.

The derivation of NSRs for an $n$-cell null-boundary uniform CA with rule 7 is shown in Fig. 4 and Table 3.

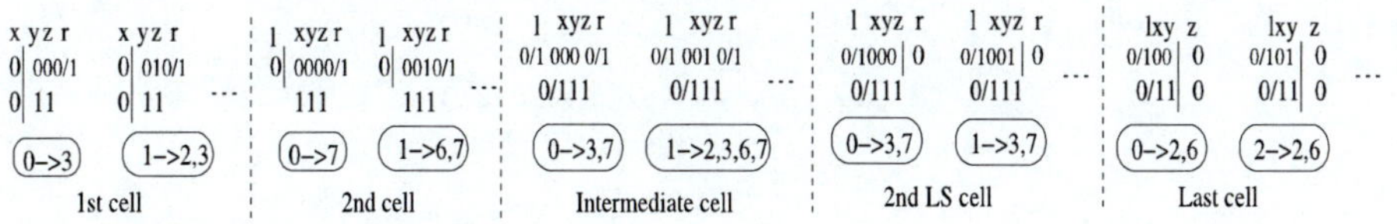


**Fig. 4** Identification of NSRs for uniform CA with rule 7

**Table 3** NSRs of the cells of an uniform CA with rule 7

| 1st cell | | 2nd cell | | Intermediate cells | | 2nd last cell | | Last cell | |
|---|---|---|---|---|---|---|---|---|---|
| RMT | NSR | RMT | NSR | RMT | NSR | RMT | NSR | RMT | NSR |
| (1) | (2) | (3) | (4) | (5) | (6) | (7) | (8) | (9 | (10) |
| T(0) | T(3) | T(0) | T(7) | T(0) | T(3), T(7) | T(0) | T(3), T(7) | T(0) | T(2), T(6) |
| T(1) | T(2), T(3) | T(1) | T(6), T(7) | T(1) | T(2), T(3), T(6), T(7) | T(1) | T(3), T(7) | T(2) | T(2), T(6) |
| T(2) | T(2) | T(2) | T(6) | T(2) | T(2), T(6) | T(2) | T(2), T(6) | T(4) | T(0), T(4) |
| T(3) | T(0) | T(3) | T(4) | T(3) | T(0), T(4) | T(3) | T(0), T(4) | T(6) | T(0) |
| | | T(4) | T(5) | T(4) | T(1), T(5) | T(4) | T(1), T(5) | | |
| | | T(5) | T(4), T(5) | T(5) | T(0), T(1), T(4), T(5) | T(5) | T(1), T(5) | | |
| | | T(6) | T(0) | T(6) | T(0) | T(6) | T(0) | | |
| | | T(7) | T(0) | T(7) | T(0) | T(7) | T(0) | | |

## 3.2 *Construction of NSRTD*

The NSRTD of a uniform CA with rule 7 is shown in Fig. 5. It has 3 graphs namely $G_1$, $G_2$ and $G_3$, where $G_1$ corresponds to a fixed point attractor and $G_2$ and $G_3$ correspond to multi-length cycle attractors.

Given a rule (R), length of CA ($n$), and the NCRs of the 8 RMTs (Table 2), the NSRs for different cells (1st, 2nd, intermediate, 2nd last and last cell) are identified and the NSRS-Gs are formed. The cycles (LPs) in NSRS-Gs are then found. From the set of cycles $LP_1$, $LP_2$, $LP_I$, $LP_{(n-1)}$, $LP_n$ for the 1st, 2nd, intermediate, 2nd last and last cell respectively, the compatibility classes of cycle (Def. 14) are computed. Then the sequence of RMTS $P_u = P_{1u} \rightarrow P_{2u} \rightarrow P_{3u} \rightarrow P_{4u} \rightarrow \ldots P_{(n-2)u} \rightarrow P_{(n-1)u} \rightarrow P_{nu}$ representing a graph in the NSRTD is extracted. The number of such n-length paths ($P_u$) corresponds to number of attractors of the CA. If all the NSRSs $P_{iu}$, for i

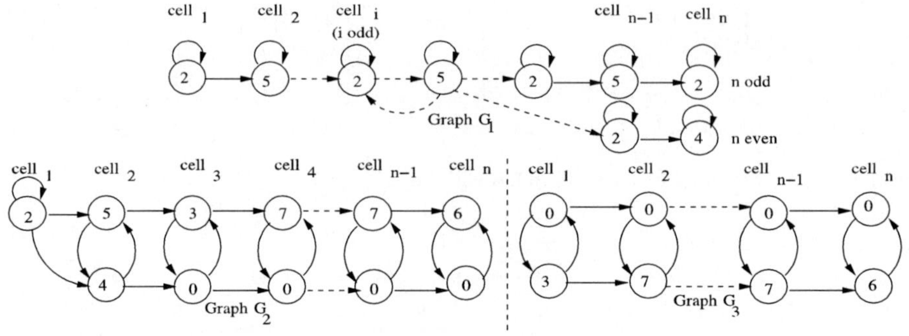


**Fig. 5** NSRTD for uniform CA with rule 7

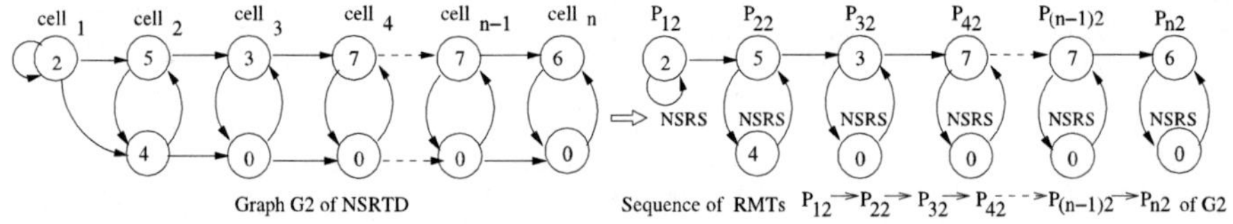


**Fig. 6** Sequence of RMTs for NSRTD graph

= 1 to n, in a $P_u$ is of single length, then $P_u$ correspond to a fixed point; otherwise it is a multi-length cycle attractor (see Fig. 6).

## 4 Proposed Methodology

This section reports the synthesis of CA with fixed points. Here, we consider the synthesis of Single Length Cycle Two Attractor CA (TACA). However, the proposed methodology can also be extended for the synthesis of CA with any number of fixed points. TACA having only two-point states is of immense importance in various applications, specifically in classifying faulty and non-faulty states. Devising a solution to such a problem necessitates the synthesis of the required TACA. Synthesis of such TACA, using all the 256 rules, is of exponential complexity. To reduce the search space, in this work, we consider hybridization of only 15 uniform

**Table 4** TACA rules

| Group | Rule for TACA |
|---|---|
| 3 | 38, 52 |
| 4 | 46, 106, 116, 120, 166, 180, 235, 249 |
| 5 | 174, 239, 244, 253, 254 |

TACA rules as identified in [4] and reported in Table 4 to configure a non-uniform TACA. However, the worst-case time complexity to synthesize non-uniform/hybrid TACA of n-length is still $O(15^n)$. In the following, we report the synthesis of scalable TACA structure involving all the 15 TACA rules in 3-neighborhood.

We next present Algorithms (Algorithm 1 and Algorithm 2) for the synthesis of a scalable TACA structure involving the 15 uniform TACA rules in 3-neighborhood null-boundary.

---

**Algorithm 1** CountCyclesInCA

---

**Input** Rule R, NCRs for RMTs T(0), T(1), T(2),..., T(7) forming the sets $NCR_0$, $NCR_1$, ... $NCR_7$.
**Output** Cycles in NRSR-Gs.

1: Start;
2: Find NSRs for 1st, 2nd, intermediate, 2nd last and last cell for rule R.
3: Construct NSRS-Gs for 1st cell (NSRS-$G_1$), 2nd cell (NSRS-$G_2$), intermediate cells (NSRS-$G_I$), 2nd last cell (NSRS-$G_{(n-1)}$) and last cell (NSRS-$G_n$);
NSRS-G = {NSRS-$G_1$, NSRS-$G_2$, NSRS-$G_I$, NSRS-$G_{(n-1)}$, NSRS-$G_n$ }.
4: Find cycles (Simple cycles) in the set NSRS-G
/*These include self-loops as well as multi-length cycles*/
$LP_1 = LP_{11}, LP_{12}$, .... for 1st cell; /*cycles in NSRS-$G_1$*/
$LP_2 = LP_{21}, LP_{22}$, .... for 2nd cell; /*cycles in NSRS-$G_2$*/
$LP_I = LP_{I1}, LP_{I2}$, .... for intermediate cells; /*cycles in NSRS-$G_I$*/
$LP_{(n-1)} = LP_{(n-1)1}, LP_{(n-1)2}$, .... for 2nd last cell; /*cycles in NSRS-$G_{(n-1)}$*/
$LP_n = LP_{n1}, LP_{n2}$, .... for last cell; /*cycles in NSRS-$G_n$*/
LP = {$LP_1$, $LP_2$, $LP_I$, $LP_{(n-1)}$, $LP_n$.};
5: Stop.

---

Algorithm 1 accepts rule (R), and NCRs of the 8 RMTs (Table 2) as input. It finds NSRs of the 1st, 2nd, intermediate, 2nd last and last cells and then computes the respective NSRS-Gs from which NSRSs ($LP_{ij}$s) are identified. The set LP contains all NSRSs from the 1st, 2nd, intermediate, 2nd last and last cells, which are used in Algorithm 2.

Algorithm 2 accepts the length of the CA (n), NCRs of the 8 RMTs (Table 4) and the cycles in the NSRS-Gs, which may include single-length as well as a multi-length cycle (s). In step 1, Algorithm 2 chooses a candidate rule for $R_1$ randomly from Table 4. Next, in step 2, it checks the compatibility classes between $LP_1$ and $LP_2$: $C_{12}$ and chooses a rule $R_2$ such that $C_{12}$ is having exactly two pairs of self-loop NSRSs and no pair of multi-length NSRSs. In the same way, Algorithm 2 chooses the rule $R_3$ in step 12. If the length of the CA is greater than 5, then it will continue to select $R_I$s by consulting the compatibility class $C_{I(I+1)}$ for length 3 to $(n-3)$. It chooses the next cell rules where compatibility classes between $LP_I$ and $LP_{I+1}$: $C_{I(I+1)}$ have exactly two self-loop and no multi-length NSRSs in steps 18 and 19. Next, in step 27, it checks the compatibility class between the 2nd last cell rule and $(n-3)$th cell rule and selects those rules that have exactly two self-loop and no multi-length NSRSs. At last, it chooses the last rule by checking the compatibility class $C_{(n-1)n}$ having exactly two self-loop and no multi-length NSRSs in step 34.

**Algorithm 2** ConstructHybridTACA

**Input:** Length of CA ($n$), NCRs, LPs.
/* LPs are computed by Algorithm 2 */
**Output:** The CA rule vector $R =< R_1, R_2, ...R_n >$, where R is a TACA

1: Choose Rule $R_1$ from Table 4 randomly;
2: Compute compatibility classes $C_{12}$ between $LP_1$ ($R_1$) and $LP_2$ ($R_p$) for p=1 to 15 as $C_{12} = \{(LP_{1i}, LP_{2j})$, such that $LP_{2j}$ is the NCR of $LP_{1i}$, for all i, j};
3: Candidate ($R_2$)= {$R_p$ such that $C_{12}$ have exactly two pairs of self-loop NSRSs and no pair of multi-length NSRSs};
4: **if** Candidate ($R_2$)$\neq$ Null **then**
5: Choose Rule $R_2$ from the Candidate ($R_2$);
6: **else**
7: exit
8: **end if**
9: Compute compatibility classes $C_{2I}$ between $LP_2$ ($R_2$) and $LP_3$ ($R_p$) for p=1 to 15 as $C_{2I} = \{(LP_{2k}, LP_{3l})$, such that $LP_{3l}$ is the NCR of $LP_{2k}$ for all k, l};
10: Candidate ($R_3$)= {$R_p$ such that $C_{2I}$ have exactly two pairs of self-loop NSRSs and no pair of multi-length NSRSs};
11: **if** Candidate ($R_3$)$\neq$ Null **then**
12: Choose Rule $R_3$ from the Candidate ($R_3$);
13: **else**
14: exit
15: **end if**
16: **if** $n > 5$ **then**
17: **for** $i = 3$ to $(n - 3)$
18: Compute compatibility classes $C_{I(I+1)}$ between $LP_I$ ($R_I$) and $LP_{I+1}$ ($R_p$) for p=1 to 15 as $C_{I(I+1)} = \{(LP_{Im}, LP_{(I+1)o})$, such that $LP_{(I+1)o}$ is the NCR of $LP_{Im}$ for all m, o};
19: Candidate ($R_{I+1}$)= {$R_p$ such that $C_{I(I+1)}$ have exactly two pairs of self-loop NSRSs and no pair of multi-length NSRSs};
20: **if** Candidate ($R_{I+1}$)$\neq$ Null **then**
21: Choose Rule $R_{I+1}$ from the Candidate ($R_{I+1}$);
22: **else**
23: exit
24: **end if**
25: **end for**
26: **end if**
27: Compute compatibility classes $C_{I(n-1)}$ between $LP_I$ ($R_I$) and $LP_{(n-1)}$ ($R_p$) for p=1 to 15 as $C_{I(n-1)} = \{(LP_{Iq}, LP_{(n-1)r})$, such that $LP_{(n-1)r}$ is the NCR of $LP_{Iq}$ for all q, r};
28: Candidate ($R_{n-1}$)= {$R_p$ such that $C_{I(n-1)}$ have exactly two pairs of self-loop NSRSs and no pair of multi-length NSRSs};
29: **if** Candidate ($R_{n-1}$)$\neq$ Null **then**
30: Choose Rule $R_{n-1}$ from the Candidate ($R_{n-1}$);
31: **else**
32: exit
33: **end if**
34: Compute compatibility classes $C_{(n-1)n}$ between $LP_{(n-1)}$ ($R_{(n-1)}$) and $LP_n$: $C_{(n-1)n} = \{(LP_{(n-1)s}, LP_{nt})$, such that $LP_{nt}$ is the NCR of $LP_{(n-1)s}$ for all s, t};
35: Candidate ($R_n$)= {$R_p$ such that $C_{(n-1)n}$ have exactly two pairs of self-loop NSRSs and no pair of multi-length NSRSs};
36: **if** Candidate ($R_n$)$\neq$ Null **then**
37: Choose Rule $R_n$ from the Candidate ($R_n$);
38: **else**
39: exit
40: **end if**
41: Output: $< R_1, R_2, ..., R_n >$

**Example 1** Let us consider synthesis of a 5-length hybrid TACA. So, following step 1 of Algorithm 2, the first cell rule is chosen randomly from Table 4. Suppose, rule 180 is chosen. Next, in step 2, it checks the compatibility class between $LP_1$ and $LP_2$ ($C_{12}$) and chooses rule 52 as $C_{12} = \{(R_0, R_0), (R_0, R_1)\}$ has exactly two self-loops. Similarly, in step 5, the next cell rule is chosen by checking the compatibility

**Fig. 7** NSRS-Gs for the hybrid CA with rule vector <180 52, 116, 244, 52>

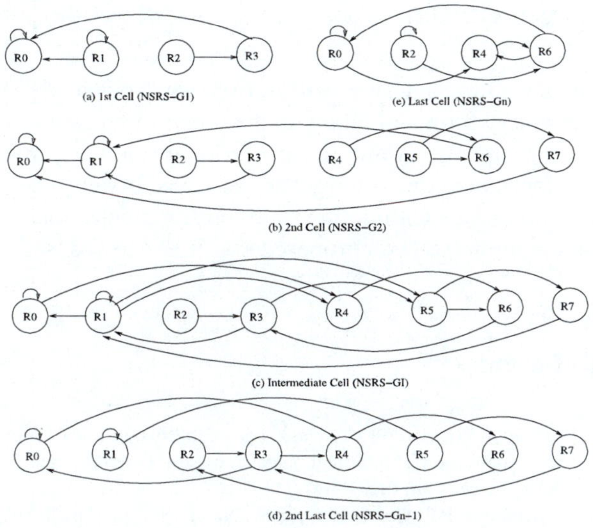


**Fig. 8** NSRTD for the hybrid CA with rule vector <180, 52, 116, 244, 52>

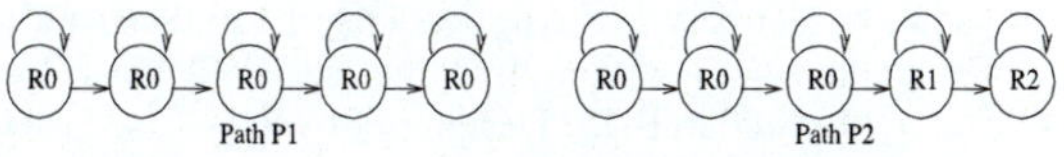


class between $LP_2$ and $LP_3$ ($C_{23}$). Compatibility classes between rule 52 and 116 have exactly two self-loops ($C_{23} = \{(R_0, R_0), (R_0, R_1)\}$). So, the rule 116 can be chosen as $3^{rd}$ cell rule. Following similar strategy, rule 244 is chosen as 4th cell rule ($C_{34} = \{(R_0, R_0), (R_0, R_1)\}$). At step 15, it chooses the last cell rule 52 as the compatibility classes between rule 244 and 52 has exactly two self-loop members, i.e. $C_{45} = \{(R_0, R_0), (R_1, R_2)\}$ (Fig. 7). The NSRTD of the hybrid CA <180, 52, 116, 244, 52> is shown in Fig. 8 wherein the existence of only two paths with self-loop NSRSs, $P_1 = R_0 \rightarrow R_0 \rightarrow R_0 \rightarrow R_0 \rightarrow R_0$ and $P_2 = R_0 \rightarrow R_0 \rightarrow R_0 \rightarrow R_1 \rightarrow R_2$, can easily be observed confirming the 5-cell CA <180, 52, 116, 244, 52> to be a TACA.

**Analysis**: Step 1 of Algorithm 2 chooses a rule randomly out of the 15 rules in Table 4, and it requires constant time. Although, for choosing rule $R_2$ has 15 possibilities, but in a practical scenario, it is observed that the number of candidates is only a few. The candidate rules for $R_2$ are chosen in such a way that the compatibility class $C_{12}$ has exactly two self-loop NSRSs and no multi-length NSRSs. Similarly, we can choose the rest of the cell rules, and the candidate rules for each cell are only a few ($\ll$15). Let's assume the maximum size of candidate rule set for any cell (cell 1 to cell n) is P. So, the total time complexity for selecting rules of an n-cell CA will be upper bounded by $P^n$, which is actually much less than $15^n$.

## 5 Conclusion

Cellular Automata have evolved from von Neumann's early days through Wolfram's elementary form and finally to the present trends in research employing this attractive yet straightforward model. During its journey, various forms of CAs have been developed targeting different applications. In this paper, we have discussed NSRTD to get more insight into the state transition behaviour of CA with fixed points, and an algorithm has been proposed to synthesize TACA.

## References


1. Bacquey, N.: Complexity classes on spatially periodic cellular automata. In: 31st International Symposium on Theoretical Aspects of Computer Science (STACS 2014). Schloss Dagstuhl-Leibniz-Zentrum fuer Informatik (2014)
2. Chaudhuri, P.P., Chowdhury, D.R., Nandi, S., Chattopadhyay, S.: Additive Cellular Automata: Theory and Applications, vol. 43. Wiley (1997)
3. Dalui, M.: Theory and Applications of Cellular Automata for CMPs Cache System Protocol Design and Verification. Ph.D. thesis, IIEST, West Bengal, India (2014)
4. Dalui, M., Sikdar, B.K.: Design of directory based cache coherence protocol verification logic in CMPs around TACA. In: 2013 International Conference on High Performance Computing & Simulation (HPCS), pp. 318–325. IEEE (2013)
5. Das, S., Mukherjee, S., Naskar, N., Sikdar, B.K.: Characterization of single cycle ca and its application in pattern classification. Electron. Notes Theor. Comput. Sci. **252**, 181–203 (2009)
6. Das, S., Naskar, N.N., Mukherjee, S., Dalui, M., Sikdar, B.K.: Characterization of CA rules for SACA targeting detection of faulty nodes in WSN. In: International Conference on Cellular Automata, pp. 300–311. Springer (2010)
7. Maiti, N.S., Ghosh, S., Munshi, S., Pal Chaudhuri, P.: Linear time algorithm for identifying the invertibility of null-boundary three neighborhood cellular automata. Complex Syst. **19**(1), 89 (2010)
8. Maitii, N.S., Ghosh, S., Sikdar, B.K., Chaudhuri, P.P.: Rule vector graph (RVG) to design linear time algorithm for identifying the invertibility of periodic-boundary three neighborhood cellular automata. J. Cell. Autom. **7**(4) (2012)
9. McIntosh, H.V.: Linear cellular automata via de Bruijn diagrams. Citeseer (1989)
10. Mukherjee, M., Ganguly, N., Chaudhuri, P.P.: Cellular automata based authentication (CAA). In: International Conference on Cellular Automata, pp. 259–269. Springer (2002)
11. Wolfram, S.: Cellular Automata and Complexity: Collected Papers. CRC Press (2018)

# Author Index





S. Das and G. J. Martinez (eds.), *Proceedings of First Asian Symposium on Cellular Automata Technology*, Advances in Intelligent Systems and Computing 1425,
https://doi.org/10.1007/978-981-19-0542-1